T0176127

CATEGORY THEORY IN CONTEXT

Emily Riehl

DOVER PUBLICATIONS, Garden City, New York

Chapter 6 is adapted with permission from Chapter 1 of
Categorical Homotopy Theory by Emily Riehl, Cambridge University Press.
Copyright © 2014 by Emily Riehl

Bibliographical Note
Category Theory in Context is a new work,
first published by Dover Publications in 2016.

International Standard Book Number
ISBN-13: 978-0-486-80903-8
ISBN-10: 0-486-80903-X

Manufactured in the United States of America
80903011
www.doverpublications.com

To Peter Johnstone, whose beautiful Part III lectures provided my first acquaintance with category theory and form the skeleton of this book.

and

To Martin Hyland, who guided my initial explorations of this subject's frontiers and inspired my aspirations to think categorically.

The aim of theory really is, to a great extent, that of systematically organizing past experience in such a way that the next generation, our students and their students and so on, will be able to absorb the essential aspects in as painless a way as possible, and this is the only way in which you can go on cumulatively building up any kind of scientific activity without eventually coming to a dead end.

M.F. Atiyah, "How research is carried out"
[Ati74]

Contents

PREFACE

Atiyah described mathematics as the "science of analogy." In this vein, the purview of category theory is *mathematical analogy*. Category theory provides a cross-disciplinary language for mathematics designed to delineate general phenomena, which enables the transfer of ideas from one area of study to another. The category-theoretic perspective can function as a simplifying[1] abstraction, isolating propositions that hold for formal reasons from those whose proofs require techniques particular to a given mathematical discipline.[2]

A subtle shift in perspective enables mathematical content to be described in language that is relatively indifferent to the variety of objects being considered. Rather than characterize the objects directly, the categorical approach emphasizes the transformations between objects of the same general type. A fundamental lemma in category theory implies that any mathematical object can be characterized by its *universal property*—loosely by a representation of the morphisms to or from other objects of a similar form. For example, tensor products, "free" constructions, and localizations are characterized by universal properties in appropriate *categories*, or mathematical contexts. A universal property typically expresses one of the mathematical roles played by the object in question. For instance, one universal property associated to the unit interval identifies self-homeomorphisms of this space with re-parameterizations of paths. Another highlights the operation of gluing two intervals end to end to obtain a new interval, the construction used to define composition of paths.

Certain classes of universal properties define blueprints which specify how a new object may be built out of a collection of existing ones. A great variety of mathematical constructions fit into this paradigm: products, kernels, completions, free products, "gluing" constructions, and quotients are all special cases of the general category-theoretic notion of *limits* or *colimits*, a characterization that makes it easy to define transformations to or from the objects so-defined. The input data for these constructions are *commutative diagrams*, which are themselves a vehicle for mathematical definitions, e.g., of rings or algebras, representations of a group, or chain complexes.

Important technical differences between particular varieties of mathematical objects can be described by the distinctive properties of their categories: that rings have all limits and colimits while fields have few, that a continuous bijection defines an isomorphism of compact Hausdorff spaces but not of generic topological spaces. Constructions that convert mathematical objects of one type into objects of another type often define transformations between categories, called *functors*. Many of the basic objects of study in modern algebraic topology and algebraic geometry involve functors and would be impossible to define without category-theoretic language.

---

[1]In his mathematical notebooks, Hilbert formulated a "24th problem" (inspired by his work on syzygies) to develop a criterion of simplicity for evaluating competing proofs of the same result [**TW02**].

[2]For example, the standard properties of induced representations (Frobenius reciprocity, transitivity of induction, even the explicit formula) are true of any construction defined as a *left Kan extension*; character tables, however, are non-formal.

Category theory also contributes new proof techniques, such as *diagram chasing* or arguments by duality; Steenrod called these methods "abstract nonsense."[3] The aim of this text is to introduce the language, philosophy, and basic theorems of category theory. A complementary objective is to put this theory into practice: studying functoriality in algebraic topology, naturality in group theory, and universal properties in algebra.

Practitioners often assert that the hard part of category theory is to state the correct definitions. Once these are established and the categorical style of argument is sufficiently internalized, proving the theorems tends to be relatively easy.[4] Indeed, the proofs of several propositions appearing in this text are left as exercises, with confidence that the reader will eventually find it more efficient to supply their own arguments than to read the author's.[5] The relative simplicity of the proofs of major theorems occasionally leads detractors to assert that there are no theorems in category theory. This is not at all the case! Counterexamples abound in the text that follows. A short list of further significant theorems, beyond the scope of a first course but not too far to be out of the reach of comprehension, appears as an epilogue.

Sample corollaries

It is difficult to preview the main theorems in category theory before developing fluency in the language needed to state them. (A reader possessing such fluency might wish to glance ahead to §E.1.) Instead, here are a few corollaries, results in other areas of mathematics that follow trivially as special cases of general categorical results that are proven in this text.

As an application of the theory of equivalence between categories:

COROLLARY 1.5.13. *In a path-connected space, any choice of basepoint yields an isomorphic fundamental group.*

A fundamental lemma in category theory has the following two results as corollaries:

COROLLARY 2.2.9. *Every row operation on matrices with n rows is defined by left multiplication by some $n \times n$ matrix, namely the matrix obtained by performing the row operation on the identity matrix.*

COROLLARY 2.2.10. *Any group is isomorphic to a subgroup of a permutation group.*

A special case of a general result involving the interchange of limits and colimits is:

COROLLARY 3.8.4. *For any pair of sets X and Y and any function $f : X \times Y \to \mathbb{R}$*

$$\sup_{x \in X} \inf_{y \in Y} f(x, y) \leq \inf_{y \in Y} \sup_{x \in X} f(x, y)$$

---

[3]Lang's *Algebra* [**Lan02**, p. 759] supports the general consensus that this was not intended as an epithet:

> In the forties and fifties (mostly in the works of Cartan, Eilenberg, MacLane, and Steenrod, see [**CE56**]), it was realized that there was a systematic way of developing certain relations of linear algebra, depending only on fairly general constructions which were mostly arrow-theoretic, and were affectionately called **abstract nonsense** by Steenrod.

[4]A famous exercise in Lang's *Algebra* asks the reader to "Take any book on homological algebra, and prove all the theorems without looking at the proofs given in that book" [**Lan84**, p. 175]. Homological algebra is the subject whose development induced Eilenberg and Mac Lane to introduce the general notions of category, functor, and natural transformation.

[5]In the first iteration of the course that inspired the writing of these lecture notes, the proofs of several major theorems were also initially left to the exercises, with a type-written version appearing only after the problem set was due.

whenever these infima and suprema exist.

The following five results illustrate a few of the many corollaries of a common theorem, which describes one consequence of a type of "duality" enjoyed by certain pairs of mathematical constructions:

COROLLARY 4.5.4. *For any function $f: A \to B$, the inverse image function $f^{-1}: PB \to PA$ between the power sets of A and B preserves both unions and intersections, while the direct image function $f_*: PA \to PB$ only preserves unions.*

COROLLARY 4.5.5. *For any vector spaces U, V, W,*

$$U \otimes (V \oplus W) \cong (U \otimes V) \oplus (U \otimes W).$$

COROLLARY 4.5.6. *For any cardinals α, β, γ, cardinal arithmetic satisfies the laws:*

$$\alpha \times (\beta + \gamma) = (\alpha \times \beta) + (\alpha \times \gamma) \qquad (\beta \times \gamma)^{\alpha} = \beta^{\alpha} \times \gamma^{\alpha} \qquad \alpha^{\beta+\gamma} = \alpha^{\beta} \times \alpha^{\gamma}.$$

COROLLARY 4.5.7. *The free group on the set $X \sqcup Y$ is the free product of the free groups on the sets X and Y.*

COROLLARY 4.5.8. *For any R–S bimodule M, the tensor product $M \otimes_S -$ is right exact.*

Finally, a general theorem that recognizes categories whose objects bear some sort of "algebraic" structure has a number of consequences, including:

COROLLARY 5.6.2. *Any bijective continuous function between compact Hausdorff spaces is a homeomorphism.*

This is not to say that category theory necessarily provides a more efficient proof of these results. In many cases, the proof that general consensus designates the "most elegant" reflects the categorical argument. The point is that the category-theoretic perspective allows for an efficient packaging of general arguments that can be used over and over again and eliminates contextual details that can safely be ignored. For instance, our proof that the tensor product commutes with the direct sum of vector spaces will not make use of any bases, but appeals instead to the universal properties of the tensor product and direct sum constructions.

A tour of basic categorical notions

> ... the science of mathematics exemplifies the interdependence of its parts.
>
> ---
>
> Saunders Mac Lane, "Topology and logic as a source of algebra" [ML76]

A *category* is a context for the study of a particular class of mathematical objects. Importantly, a category is not simply a type signature, it has both "nouns" and "verbs," containing specified collections of objects and transformations, called *morphisms*,[6] between them. Groups, modules, topological spaces, measure spaces, ordinals, and so forth form categories, but these classifications are not the main point. Rather, the action of packaging each variety of objects into a category shifts one's perspective from the particularities of each mathematical sub-discipline to potential commonalities between them. A basic observation along these lines is that there is a single categorical definition of *isomorphism*

---

[6]The term "morphism" is derived from *homomorphism*, the name given in algebra to a structure-preserving function. Synonyms include "arrow" (because of the notation "→") and "map" (adopting the standard mathematical colloquialism).

that specializes to define isomorphisms of groups, homeomorphisms of spaces, order isomorphisms of posets, and even isomorphisms between categories (see Definition 1.1.9).

Mathematics is full of constructions that translate mathematical objects of one kind into objects of another kind. A construction that converts the objects in one category into objects in another category is *functorial* if it can be extended to a mapping on morphisms in such a way that composites and identity morphisms are preserved. Such constructions define morphisms between categories, called *functors*. Functoriality is often a key property: for instance, the chain rule from multivariable calculus expresses the functoriality of the derivative (see Example 1.3.2(x)). In contrast with earlier numerical invariants in topology, functorial invariants (the fundamental group, homology) tend both to be more easily computable and also provide more precise information. While the Euler characteristic can distinguish between the closed unit disk and its boundary circle, an easy proof by contradiction involving the functoriality of their fundamental groups proves that any continuous endomorphism of the disk must have a fixed point (see Theorem 1.3.3).

On occasion, functoriality is achieved by *categorifying* an existing mathematical construction. "Categorification" refers to the process of turning sets into categories by adding morphisms, whose introduction typically demands a re-interpretation of the elements of the sets as related mathematical objects. A celebrated knot invariant called the Jones polynomial must vanish for any knot diagram that presents the unknot, but its *categorification*, a functor[7] called *Khovanov homology*, detects the unknot in the sense that any knot diagram whose Khovanov homology vanishes must represent the unknot. Khovanov homology converts an oriented link diagram into a chain complex whose graded Euler characteristic is the Jones polynomial.

A functor may describe an *equivalence of categories*, in which case the objects in one category can be translated into and reconstructed from the objects of another. For instance, there is an equivalence between the category of finite-dimensional vector spaces and linear maps and a category whose objects are natural numbers and whose morphisms are matrices (see Corollary 1.5.11). This process of conversion from college linear algebra to high school linear algebra defines an equivalence of categories; eigenvalues and eigenvectors can be developed for matrices or for linear transformations, it makes no difference.

Treating categories as mathematical objects in and of themselves, a basic observation is that the process of formally "turning around all the arrows" in a category produces another category. In particular, any theorem proven for all categories also applies to these *opposite categories*; the re-interpretation of the result in the opposite of an opposite category yields the statement of the *dual theorem*. Categorical constructions also admit duals: for instance, in Zermelo–Fraenkel set theory, a function $f: X \to Y$ is defined via its *graph*, a subset of $X \times Y$ isomorphic to X. The dual presentation represents a function via its *cograph*, a Y-indexed partition of $X \sqcup Y$. Categorically-proven properties of the graph representation will dualize to describe properties of the cograph representation.

Categories and functors were introduced by Eilenberg and Mac Lane with the goal of giving precise meaning to the colloquial usage of "natural" to describe families of isomorphisms. For example, for any triple of $\Bbbk$-vector spaces U, V, W, there is an isomorphism

$$(0.0.1) \qquad \mathrm{Vect}_{\Bbbk}(U \otimes_{\Bbbk} V, W) \cong \mathrm{Vect}_{\Bbbk}(U, \mathrm{Hom}(V, W))$$

between the set of linear maps $U \otimes_{\Bbbk} V \to W$ and the set of linear maps from U to the vector space $\mathrm{Hom}(V, W)$ of linear maps from V to W. This isomorphism is natural in all three

---

[7]Morally, one could argue that functoriality is the main innovation in this construction, but making this functoriality precise is somewhat subtle [**CMW09**].

variables, meaning it defines an isomorphism not simply between these sets of maps but between appropriate set-valued functors of U, V, and W. Chapter 1 introduces the basic language of category theory, defining categories, functors, natural transformations, and introducing the principle of duality, equivalences of categories, and the method of proof by diagram chasing.

In fact, the isomorphism (0.0.1) *defines* the vector space $U \otimes_{\Bbbk} V$ by declaring that linear maps $U \otimes_{\Bbbk} V \to W$ correspond to linear maps $U \to \mathrm{Hom}(V, W)$, i.e., to bilinear maps $U \times V \to W$. This definition is sufficiently robust that important properties of the tensor product—for instance its symmetry and associativity—can be proven without reference to any particular construction (see Proposition 2.3.9 and Exercise 2.3.ii). The advantages of this approach compound as the mathematical objects so-described become more complicated.

In Chapter 2, we study such definitions abstractly. A characterization of the morphisms either to or from a fixed object describes its *universal property*; the cases of "to" or "from" are dual. By the Yoneda lemma—which, despite its innocuous statement, is arguably the most important result in category theory—every object is characterized by either of its universal properties. For example, the Sierpinski space is characterized as a topological space by the property that continuous functions $X \to S$ correspond naturally to open subsets of X. The complete graph on n vertices is characterized by the property that graph homomorphisms $G \to K_n$ correspond to n-colorings of the vertices of the graph G with the property that adjacent vertices are assigned distinct colors. The polynomial ring $\mathbb{Z}[x_1, \ldots, x_n]$ is characterized as a commutative unital ring by the property that ring homomorphisms $\mathbb{Z}[x_1, \ldots, x_n] \to R$ correspond to n-tuples of elements $(r_1, \ldots, r_n) \in R$. Modern algebraic geometry begins from the observation that a commutative ring can be identified with the functor that it represents.

The idea of probing a fixed object using morphisms abutting to it from other objects in the category gives rise to a notion of "generalized elements" (see Remark 3.4.15). The elements of a set A are in bijection with functions $* \to A$ with domain a singleton set; a *generalized element* of A is a morphism $X \to A$ with generic domain. In the category of directed graphs, a parallel pair of graph homomorphisms $\phi, \psi \colon A \rightrightarrows B$ can be distinguished by considering generalized elements of A whose domain is the free-living vertex or the free-living directed edge.[8] A related idea leads to the representation of a topological space via its singular complex.

The Yoneda lemma implies that a general mathematical object can be *represented* as a functor valued in the category of sets. A related classical antecedent is a result that comforted those who were troubled by the abstract definition of a group: namely that any group is isomorphic to a subgroup of a permutation group (see Corollary 2.2.10). A deep consequence of these functorial representations is that proofs that general categorically-described constructions are isomorphic reduce to the construction of a bijection between their set-theoretical analogs (for instance, see the proof of Theorem 3.4.12).

Chapter 3 studies a special case of definitions by universal properties, which come in two dual forms, referred to as *limits* and *colimits*. For example, aggregating the data of the cyclic p-groups $\mathbb{Z}/p^n$ and homomorphisms between them, one can build more complicated abelian groups. Limit constructions build new objects in a category by "imposing equations" on existing ones. For instance, the diagram of quotient homomorphisms

$$\cdots \twoheadrightarrow \mathbb{Z}/p^n \twoheadrightarrow \cdots \twoheadrightarrow \mathbb{Z}/p^3 \twoheadrightarrow \mathbb{Z}/p^2 \twoheadrightarrow \mathbb{Z}/p$$

---

[8]The incidence relation in the graph A can be recovered by also considering the homomorphisms between these graphs.

has a limit, namely the group $\mathbb{Z}_p$ of *p-adic integers*: its elements can be understood as tuples of elements $(a_n \in \mathbb{Z}/p^n)_{n \in \omega}$ that are compatible modulo congruence. There is a categorical explanation for the fact that $\mathbb{Z}_p$ is a commutative ring and not merely an abelian group: each of these quotient maps is a ring homomorphism, and so this diagram and also its limit lifts to the category of rings.[9]

By contrast, colimit constructions build new objects by "gluing together" existing ones. The colimit of the sequence of inclusions

$$\mathbb{Z}/p \hookrightarrow \mathbb{Z}/p^2 \hookrightarrow \mathbb{Z}/p^3 \hookrightarrow \cdots \hookrightarrow \mathbb{Z}/p^n \hookrightarrow \cdots$$

is the *Prüfer p-group* $\mathbb{Z}[\frac{1}{p}]/\mathbb{Z}$, an abelian group which can be presented via generators and relations as

(0.0.2) $\mathbb{Z}[\frac{1}{p}]/\mathbb{Z} := \left\langle g_1, g_2, \ldots \,\middle|\, pg_1 = 0, pg_2 = g_1, pg_3 = g_2, \ldots \right\rangle.$

The inclusion maps are not ring homomorphisms (failing to preserve the multiplicative identity) and indeed it turns out that the Prüfer *p*-group does not admit any non-trivial multiplicative structure.

Limits and colimits are accompanied by universal properties that generalize familiar universal properties in analysis. A poset $(A, \leq)$ may be regarded as a category whose objects are the elements $a \in A$ and in which a morphism $a \to a'$ is present if and only if $a \leq a'$. The *supremum* of a collection of elements $\{a_i\}_{i \in I}$, an example of a colimit in the category $(A, \leq)$, has a universal property: namely to prove that

$$\sup_{i \in I} a_i \leq a$$

is equivalent to proving that $a_i \leq a$ for all $i \in I$. The universal property of a generic colimit is a generalization of this, where the collection of morphisms $(a_i \to a)_{i \in I}$ is regarded as data, called a *cone* under the diagram, rather than simply a family of conditions. Limits have a dual universal property that specializes to the universal property of the infimum of a collection of elements in a poset.

Chapter 4 studies a generalization of the notion of equivalence of categories, in which a pair of categories are connected by a pair of opposite-pointing translation functors called an *adjunction*. An adjunction expresses a kind of "duality" between a pair of functors, first recognized in the case of the construction of the tensor product and hom functors for abelian groups (see Example 4.3.11). Any adjunction restricts to define an equivalence between certain subcategories, but categories connected by adjunctions need not be equivalent. For instance, there is an adjunction connecting the poset of subsets of $\mathbb{C}^n$ and the poset of subsets of the ring $\mathbb{C}[x_1, \ldots, x_n]$ that restricts to define an equivalence between Zariski closed subsets and radical ideals (see Example 4.3.2). Another adjunction encodes a duality between the constructions of the suspension and of the loop space of a based topological space (see Example 4.3.14).

When a "forgetful" functor admits an adjoint, that adjoint defines a "free" (or, less commonly, the dual "cofree") construction. Such functors define universal solutions to optimization problems, e.g., of adjoining a multiplicative unit to a non-unital ring. The existence of free groups or free rings have implications for the constructions of limits in these categories (namely, Theorem 4.5.2); the dual properties for colimits do not hold because there are no "cofree" groups or rings in general. A category-theoretic re-interpretation of the construction of the Stone–Čech compactification of a topological space defines a

---

[9]The lifting of the limit is considerably more subtle than the lifting of the diagram. Results of this nature motivate Chapter 5.

left adjoint to any limit-preserving functor between any pair of categories with similar set-theoretic properties (see Theorem 4.6.10 and Example 4.6.12).

Many familiar varieties of "algebraic" objects—such as groups, rings, modules, pointed sets, or sets acted on by a group—admit a "free–forgetful" adjunction with the category of sets. A special property of these adjoint functors explains many of the common features of the categories of algebras that are presented in this manner. Chapter 5 introduces the categorical approach to universal algebra, which distinguishes the categories of rings, compact Hausdorff spaces, and lattices from the set-theoretically similar categories of fields, generic topological spaces, and posets. The former categories, but not the latter, are *categories of algebras* over the category of sets.

The notion of *algebra* is given a precise meaning in relation to a *monad*, an endofunctor that provides a syntactic encoding of algebraic structure that may be borne by objects in the category on which it acts. Monads are also used to construct categories whose morphisms are partially-defined or non-deterministic functions, such as Markov kernels (see Example 5.2.10), and are separately of interest in computer science. A key result in categorical universal algebra is a vast generalization of the notion of a presentation of a group via generators and relations, such as in (0.0.2), which demonstrates that an algebra of any variety can be presented canonically as a *coequalizer*[10] of a pair of maps from a free algebra on the "relations" to a free algebra on the "generators."

The concluding Chapter 6 introduces a general formalism that can be used to redefine all of the basic categorical notions introduced in the first part of the text. Special cases of *Kan extensions* define representable functors, limits, colimits, adjoint functors, and monads, and their study leads to a generalization of, as well as a dualization of, the Yoneda lemma. In the most important cases, a Kan extension can be computed by a particular formula, which specializes to give the construction of a representation for a group induced from a representation for a subgroup (see Example 6.2.8), to provide a new way to think about the collection of ultrafilters on a set (see Example 6.5.12), and to define an equivalence of categories connecting sheaves on a space with étale spaces over that space (see Exercise 6.5.iii).

A brief detour introduces *derived functors*, which are certain special Kan extensions that are of great importance in homological algebra and algebraic topology. A recent categorical discovery reveals that a common mechanism for constructing "point-set level" derived functors yields total derived functors with superior universal properties (see Propositions 6.4.12 and 6.4.13). A final motivation for the study of Kan extensions reaches beyond the scope of this book. The calculus of Kan extensions facilitates the extension of basic category theory to *enriched*, *internal*, *fibered*, or *higher-dimensional* contexts, which provide natural homes for more sophisticated varieties of mathematical objects whose transformations have some sort of higher-dimensional structure.

Note to the reader

The text that follows is littered with examples drawn from a broad range of mathematical areas. The examples are included for color or historical context but are never essential for understanding the abstract category theory. In principle, one could study category theory immediately after learning some basic set theory and logic, as no other prerequisites are strictly required, but without some level of mathematical maturity it would be difficult to see what the point of it all is. We hope that the majority of examples are comprehensible in outline, even if the details are unfamiliar, but if this is not the case, it is not worth stressing

---

[10]A coequalizer is a generalization of a cokernel to contexts that may lack a "zero" homomorphism.

over. Inevitably, given the diversity of mathematical tastes and experiences, the examples presented here will seldom be optimized for any particular individual, and indeed, each reader is encouraged to search for their own contexts in which to explore categorical ideas.

Notational conventions

An arrow symbol "$\to$," either in a display or in text, is only ever used to denote a morphism in an appropriate category. In particular, the objects surrounding it necessarily lie in a common category. Double arrows "$\Rightarrow$" are reserved for natural transformations, the notation used to suggest the intuition that these are some variety of "2-dimensional" morphisms. The symbol "$\mapsto$," read as "maps to," appears occasionally when defining a function between sets by specifying its action on particular elements. The symbol "$\rightsquigarrow$" is used in a less technical sense to mean something along the lines of "yields" or "leads to" or "can be used to construct." If the presence of certain morphisms implies the existence of another morphism, the latter is often depicted with a dashed arrow "$\dashrightarrow$" to suggest the correct order of inference.[11]

We use "$\rightrightarrows$" as an abbreviation for a **parallel pair** of morphisms, i.e., for a pair of morphisms with common source and target, and "$\rightleftarrows$" as an abbreviation for an **opposing pair** of morphisms with sources and targets swapped.

Italics are used occasionally for emphasis and to highlight technical terms. Boldface signals that a technical term is being defined by its surrounding text.

The symbol "$=$" is reserved for genuine equality (with "$:=$" used for definitional equality), with "$\cong$" used instead for isomorphism in the appropriate ambient category, by far the more common occurrence.

Acknowledgments

Many of the theorems appearing here are standard fare for a first course on category theory, but the examples are not. Rather than rely solely on my own generative capacity, I consulted a great many people while preparing this text and am grateful for their generosity in sharing ideas.

To begin, I would like to thank the following people who responded to a call for examples on the n-Category Café and the categories mailing list: John Baez, Martin Brandenburg, Ronnie Brown, Tyler Bryson, Tim Campion, Yemon Choi, Adrian Clough, Samuel Dean, Josh Drum, David Ellerman, Tom Ellis, Richard Garner, Sameer Gupta, Gejza Jenča, Mark Johnson, Anders Kock, Tom LaGatta, Paul Levy, Fred Linton, Aaron Mazel-Gee, Jesse McKeown, Kimmo Rosenthal, Mike Shulman, Peter Smith, Arnaud Spiwack, Ross Street, John Terilla, Todd Trimble, Mozibur Ullah, Enrico Vitale, David White, Graham White, and Qiaochu Yuan.

In particular, Anders Kock suggested a more general formulation of "the chain rule expresses the functoriality of the derivative" than appears in Example 1.3.2(x). The expression of the fundamental theorem of Galois theory as an isomorphism of categories that appears as Example 1.3.15 is a favorite exercise of Peter May's. Charles Blair suggested Exercise 1.3.iii and a number of expository improvements to the first chapter. I learned about the unnatural isomorphism of Proposition 1.4.4 from Mitya Boyarchenko. Peter Haine suggested Example 1.4.6, expressing the Riesz representation theorem as a natural isomorphism of Banach space-valued functors; Example 3.6.2, constructing the path-components functor, and Example 3.8.6. He also contributed Exercises 1.2.v, 1.5.v, 1.6.iv, 3.1.xii, and 4.5.v and served as my LaTeX consultant.

---

[11]Readers who dislike this convention can simply connect the dots.

John Baez reminded me that the groupoid of finite sets is a categorification of the natural numbers, providing a suitable framework in which to prove certain basic equations in elementary arithmetic; see Example 1.4.9 and Corollary 4.5.6. Juan Climent Vidal suggested using the axiom of regularity to define the non-trivial part of the equivalence of categories presented in Example 1.5.6 and contributed the equivalence of plane geometries that appears as Exercise 1.5.viii. Samuel Dean suggested Corollary 1.5.13, Fred Linton suggested Corollary 2.2.9, and Ronnie Brown suggested Example 3.5.8. Ralf Meyer suggested Example 2.1.5(vi), Example 5.2.6(ii), Corollary 4.5.8, and a number of exercises including 2.1.ii and 2.4.v, which he used when teaching a similar course. He also pointed me toward a simpler proof of Proposition 6.4.12.

Mozibur Ullah suggested Exercise 3.5.vii. I learned of the non-natural objectwise isomorphism appearing in Example 3.6.5 from Martin Brandenburg who acquired it from Tom Leinster. Martin also suggested the description of the real exponential function as a Kan extension appearing in Example 6.2.7. Andrew Putman pointed out that Lang's *Algebra* constructs the free group on a set using the construction of the General Adjoint Functor Theorem, recorded as Example 4.6.6. Paul Levy suggested using affine spaces to motivate the category of algebras over a monad, as discussed in Section 5.2; a similar example was suggested by Enrico Vitale. Dominic Verity suggested something like Exercise 5.5.vii. Vladimir Sotirov pointed out that the appropriate size hypotheses were missing from the original statement of Theorem 6.3.7 and directed me toward a more elegant proof of Lemma 4.6.5.

Marina Lehner, while writing her undergraduate senior thesis under my direction, showed me that an entirely satisfactory account of Kan extensions can be given without the calculus of ends and coends. I have enjoyed, and this book has been enriched by, several years of impromptu categorical conversations with Omar Antolín Camarena. I am extremely appreciative of the careful readings undertaken by Tobias Barthel, Martin Brandenburg, Benjamin Diamond, Darij Grinberg, Peter Haine, Ralf Meyer, Peter Smith, and Juan Climent Vidal, who each sent detailed lists of corrections and suggestions. I would also like to thank John Grafton and Janet Kopito, the acquisitions and in-house editors at Dover Publications, and David Gargaro for his meticulous copyediting.

I am grateful for the perspicacious comments and questions from those who attended the first iteration of this course at Harvard—Paul Bamberg, Nathan Gupta, Peter Haine, Andrew Liu, Wyatt Mackey, Nat Mayer, Selorm Ohene, Jacob Seidman, Aaron Slipper, Alma Steingart, Nithin Tumma, Jeffrey Yan, Liang Zhang, and Michael Fountaine, who served as the course assistant—and the second iteration at Johns Hopkins—Benjamin Diamond, Nathaniel Filardo, Alex Grounds, Hanveen Koh, Alex Rozenshteyn, Xiyuan Wang, Shengpei Yan, and Zhaoning Yang. I would also like to thank the Departments of Mathematics at both institutions for giving me opportunities to teach this course and my colleagues there, who have created two extremely pleasant working environments. While these notes were being revised, I received financial support from the National Science Foundation Division of Mathematical Sciences DMS-1509016.

My enthusiasm for mathematical writing and patience for editing were inherited from Peter May, my PhD supervisor. I appreciate the indulgence of my collaborators, my friends, and my family while this manuscript was in its final stages of preparation. This book is dedicated to Peter Johnstone and Martin Hyland, whose tutelage I was fortunate to come under in a pivotal year spent at Cambridge supported by the Winston Churchill Foundation of the United States.

CATEGORY
THEORY
IN
CONTEXT

CHAPTER 1

Categories, Functors, Natural Transformations

Frequently in modern mathematics there occur
phenomena of "naturality".

Samuel Eilenberg and Saunders Mac Lane,
"Natural isomorphisms in group theory"
[EM42b]

A **group extension** of an abelian group H by an abelian group G consists of a group E together with an inclusion of $G \hookrightarrow E$ as a normal subgroup and a surjective homomorphism $E \twoheadrightarrow H$ that displays H as the quotient group E/G. This data is typically displayed in a diagram of group homomorphisms:

$$0 \to G \to E \to H \to 0.^1$$

A pair of group extensions E and E' of G and H are considered to be equivalent whenever there is an isomorphism $E \cong E'$ that *commutes with* the inclusions of G and quotient maps to H, in a sense that is made precise in §1.6. The set of equivalence classes of *abelian* group extensions E of H by G defines an abelian group $\text{Ext}(H, G)$.

In 1941, Saunders Mac Lane gave a lecture at the University of Michigan in which he computed for a prime p that $\text{Ext}(\mathbb{Z}[\frac{1}{p}]/\mathbb{Z}, \mathbb{Z}) \cong \mathbb{Z}_p$, the group of p-adic integers, where $\mathbb{Z}[\frac{1}{p}]/\mathbb{Z}$ is the Prüfer p-group. When he explained this result to Samuel Eilenberg, who had missed the lecture, Eilenberg recognized the calculation as the homology of the 3-sphere complement of the p-adic solenoid, a space formed as the infinite intersection of a sequence of solid tori, each wound around p times inside the preceding torus. In teasing apart this connection, the pair of them discovered what is now known as the **universal coefficient theorem** in algebraic topology, which relates the *homology* H_* and *cohomology groups* H^* associated to a space X via a group extension [**ML05**]:

$$(1.0.1) \qquad 0 \to \text{Ext}(H_{n-1}(X), G) \to H^n(X, G) \to \text{Hom}(H_n(X), G) \to 0.$$

To obtain a more general form of the universal coefficient theorem, Eilenberg and Mac Lane needed to show that certain isomorphisms of abelian groups expressed by this group extension extend to spaces constructed via direct or inverse limits. And indeed this is the case, precisely because the homomorphisms in the diagram (1.0.1) are *natural* with respect to continuous maps between topological spaces.

The adjective "natural" had been used colloquially by mathematicians to mean "defined without arbitrary choices." For instance, to define an isomorphism between a finite-dimensional vector space V and its **dual**, the vector space of linear maps from V to the

---

[1]The zeros appearing on the ends provide no additional data. Instead, the first zero implicitly asserts that the map $G \to E$ is an inclusion and the second that the map $E \to H$ is a surjection. More precisely, the displayed sequence of group homomorphisms is **exact**, meaning that the kernel of each homomorphism equals the image of the preceding homomorphism.

ground field $\Bbbk$, requires a choice of basis. However, there is an isomorphism between V and its double dual that requires no choice of basis; the latter, but not the former, is *natural*.

To give a rigorous proof that their particular family of group isomorphisms extended to inverse and direct limits, Eilenberg and Mac Lane sought to give a mathematically precise definition of the informal concept of "naturality." To that end, they introduced the notion of a *natural transformation*, a parallel collection of homomorphisms between abelian groups in this instance. To characterize the source and target of a natural transformation, they introduced the notion of a *functor*.[2] And to define the source and target of a functor in the greatest generality, they introduced the concept of a *category*. This work, described in "The general theory of natural equivalences" [**EM45**], published in 1945, marked the birth of category theory.

While categories and functors were first conceived as auxiliary notions, needed to give a precise meaning to the concept of naturality, they have grown into interesting and important concepts in their own right. Categories suggest a particular perspective to be used in the study of mathematical objects that pays greater attention to the maps between them. Functors, which translate mathematical objects of one type into objects of another, have a more immediate utility. For instance, the Brouwer fixed point theorem translates a seemingly intractable problem in topology to a trivial one ($0 \neq 1$) in algebra. It is to these topics that we now turn.

Categories are introduced in §1.1 in two guises: firstly as universes categorizing mathematical objects and secondly as mathematical objects in their own right. The first perspective is used, for instance, to define a general notion of *isomorphism* that can be specialized to mathematical objects of every conceivable variety. The second perspective leads to the observation that the axioms defining a category are self-dual.[3] Thus, as explored in §1.2, for any proof of a theorem about all categories from these axioms, there is a dual proof of the dual theorem obtained by a syntactic process that is interpreted as "turning around all the arrows."

Functors and natural transformations are introduced in §1.3 and §1.4 with examples intended to shed light on the linguistic and practical utility of these concepts. The category-theoretic notions of *isomorphism*, *monomorphism*, and *epimorphism* are invariant under certain classes of functors, including in particular the *equivalences of categories*, introduced in §1.5. At a high level, an equivalence of categories provides a precise expression of the intuition that mathematical objects of one type are "the same as" objects of another variety: an equivalence between the category of matrices and the category of finite-dimensional vector spaces equates high school and college linear algebra.

In addition to providing a new language to describe emerging mathematical phenomena, category theory also introduced a new proof technique: that of the diagram chase. The introduction to the influential book [**ES52**] presents *commutative diagrams* as one of the "new techniques of proof" appropriate for their axiomatic treatment of homology theory. The technique of diagram chasing is introduced in §1.6 and applied in §1.7 to construct new natural transformations as *horizontal* or *vertical composites* of given ones.

---

[2]A brief account of functors and natural isomorphisms in group theory appeared in a 1942 paper [**EM42b**].

[3]As is the case for the duality in projective plane geometry, this duality can be formulated precisely as a feature of the first-order theories that axiomatize these structures.

1.1. Abstract and concrete categories

> It frames a possible template for any
> mathematical theory: the theory should have
> *nouns* and *verbs*, i.e., objects, and morphisms,
> and there should be an explicit notion of
> composition related to the morphisms; the theory
> should, in brief, be packaged by a category.
>
> Barry Mazur, "When is one thing equal to some
> other thing?" [**Maz08**]

DEFINITION 1.1.1. A **category** consists of

- a collection of **objects** $X, Y, Z, \ldots$
- a collection of **morphisms** $f, g, h, \ldots$

so that:

- Each morphism has specified **domain** and **codomain** objects; the notation $f: X \to Y$ signifies that f is a morphism with domain X and codomain Y.
- Each object has a designated **identity morphism** $1_X: X \to X$.
- For any pair of morphisms f, g with the codomain of f equal to the domain of g, there exists a specified **composite morphism**[4] gf whose domain is equal to the domain of f and whose codomain is equal to the codomain of g, i.e.,:

$$f: X \to Y, \quad g: Y \to Z \quad \rightsquigarrow \quad gf: X \to Z.$$

This data is subject to the following two axioms:

- For any $f: X \to Y$, the composites $1_Y f$ and $f1_X$ are both equal to f.
- For any composable triple of morphisms f, g, h, the composites $h(gf)$ and $(hg)f$ are equal and henceforth denoted by hgf.

$$f: X \to Y, \quad g: Y \to Z, \quad h: Z \to W \quad \rightsquigarrow \quad hgf: X \to W.$$

That is, the composition law is associative and unital with the identity morphisms serving as two-sided identities.

REMARK 1.1.2. The objects of a category are in bijective correspondence with the identity morphisms, which are uniquely determined by the property that they serve as two-sided identities for composition. Thus, one can define a category to be a collection of morphisms with a partially-defined composition operation that has certain special morphisms, which are used to recognize composable pairs and which serve as two-sided identities; see [**Ehr65**, §I.1] or [**FS90**, §1.1]. But in practice it is not so hard to specify both the objects and the morphisms and this is what we shall do.

It is traditional to name a category after its objects; typically, the preferred choice of accompanying structure-preserving morphisms is clear. However, this practice is somewhat contrary to the basic philosophy of category theory: that mathematical objects should always be considered in tandem with the morphisms between them. By Remark 1.1.2, the algebra of morphisms determines the category, so of the two, the objects and morphisms, the morphisms take primacy.

EXAMPLE 1.1.3. Many familiar varieties of mathematical objects assemble into a category.

---

[4]The composite may be written less concisely as $g \cdot f$ when this adds typographical clarity.

(i) Set has sets as its objects and functions, with specified domain and codomain,[5] as its morphisms.

(ii) Top has topological spaces as its objects and continuous functions as its morphisms.

(iii) Set$_*$ and Top$_*$ have sets or spaces with a specified basepoint[6] as objects and basepoint-preserving (continuous) functions as morphisms.

(iv) Group has groups as objects and group homomorphisms as morphisms. This example lent the general term "morphisms" to the data of an abstract category. The categories Ring of associative and unital rings and ring homomorphisms and Field of fields and field homomorphisms are defined similarly.

(v) For a fixed unital but not necessarily commutative ring R, Mod$_R$ is the category of left R-modules and R-module homomorphisms. This category is denoted by Vect$_{\Bbbk}$ when the ring happens to be a field $\Bbbk$ and abbreviated as Ab in the case of Mod$_{\mathbb{Z}}$, as a $\mathbb{Z}$-module is precisely an abelian group.

(vi) Graph has graphs as objects and graph morphisms (functions carrying vertices to vertices and edges to edges, preserving incidence relations) as morphisms. In the variant DirGraph, objects are directed graphs, whose edges are now depicted as arrows, and morphisms are directed graph morphisms, which must preserve sources and targets.

(vii) Man has smooth (i.e., infinitely differentiable) manifolds as objects and smooth maps as morphisms.

(viii) Meas has measurable spaces as objects and measurable functions as morphisms.

(ix) Poset has partially-ordered sets as objects and order-preserving functions as morphisms.

(x) Ch$_R$ has chain complexes of R-modules as objects and chain homomorphisms as morphisms.[7]

(xi) For any *signature* σ, specifying constant, function, and relation symbols, and for any collection of well formed sentences $\mathbb{T}$ in the first-order language associated to σ, there is a category Model$_{\mathbb{T}}$ whose objects are σ-structures that *model* $\mathbb{T}$, i.e., sets equipped with appropriate constants, relations, and functions satisfying the axioms $\mathbb{T}$. Morphisms are functions that preserve the specified constants, relations, and functions, in the usual sense.[8] Special cases include (iv), (v), (vi), (ix), and (x).

The preceding are all examples of *concrete categories*, those whose objects have underlying sets and whose morphisms are functions between these underlying sets, typically the "structure-preserving" morphisms. A more precise definition of a concrete category is given in 1.6.17. However, "abstract" categories are also prevalent:

EXAMPLE 1.1.4.

(i) For a unital ring R, Mat$_R$ is the category whose objects are positive integers and in which the set of morphisms from n to m is the set of $m \times n$ matrices with values in

---

[5][**EM45**, p. 239] emphasizes that the data of a function should include specified sets of inputs and potential outputs, a perspective that was somewhat radical at the time.

[6]A **basepoint** is simply a chosen distinguished point in the set or space.

[7]A **chain complex** $C_\bullet$ is a collection $(C_n)_{n\in\mathbb{Z}}$ of R-modules equipped with R-module homomorphisms $d\colon C_n \to C_{n-1}$, called **boundary homomorphisms**, with the property that $d^2 = 0$, i.e., the composite of any two boundary maps is the zero homomorphism. A map of chain complexes $f\colon C_\bullet \to C'_\bullet$ is comprised of a collection of homomorphisms $f_n\colon C_n \to C'_n$ so that $df_n = f_{n-1}d$ for all $n \in \mathbb{Z}$.

[8]Model theory pays greater attention to other types of morphisms, for instance the *elementary embeddings*, which are (automatically injective) functions that preserve and reflect satisfaction of first-order formulae.

R. Composition is by matrix multiplication

$$n \xrightarrow{A} m, \quad m \xrightarrow{B} k \quad \rightsquigarrow \quad n \xrightarrow{B \cdot A} k$$

with identity matrices serving as the identity morphisms.

(ii) A group G (or, more generally, a monoid[9]) defines a category $\mathsf{B}G$ with a single object. The group elements are its morphisms, each group element representing a distinct endomorphism of the single object, with composition given by multiplication. The identity element $e \in G$ acts as the identity morphism for the unique object in this category.

$$\mathsf{B}S_3 =$$

(iii) A poset $(\mathsf{P}, \leq)$ (or, more generally, a preorder[10]) may be regarded as a category. The elements of P are the objects of the category and there exists a unique morphism $x \to y$ if and only if $x \leq y$. Transitivity of the relation "$\leq$" implies that the required composite morphisms exist. Reflexivity implies that identity morphisms exist.

(iv) In particular, any ordinal $\alpha = \{\beta \mid \beta < \alpha\}$ defines a category whose objects are the smaller ordinals. For example, $\mathbb{0}$ is the category with no objects and no morphisms. $\mathbb{1}$ is the category with a single object and only its identity morphism. $\mathbb{2}$ is the category with two objects and a single non-identity morphism, conventionally depicted as $0 \to 1$. ω is the category *freely generated by the graph*

$$0 \to 1 \to 2 \to 3 \to \cdots$$

in the sense that every non-identity morphism can be uniquely factored as a composite of morphisms in the displayed graph; a precise definition of the notion of free generation is given in Example 4.1.13.

(v) A set may be regarded as a category in which the elements of the set define the objects and the only morphisms are the required identities. A category is **discrete** if every morphism is an identity.

(vi) Htpy, like Top, has spaces as its objects but morphisms are homotopy classes of continuous maps. Htpy$_*$ has based spaces as its objects and basepoint-preserving homotopy classes of based continuous maps as its morphisms.

(vii) Measure has measure spaces as objects. One reasonable choice for the morphisms is to take equivalence classes of measurable functions, where a parallel pair of functions are equivalent if their domain of difference is contained within a set of measure zero.

Thus, the philosophy of category theory is extended. The categories listed in Example 1.1.3 suggest that mathematical objects ought to be considered together with the appropriate notion of morphism between them. The categories listed in Example 1.1.4 illustrate that

---

[9]A **monoid** is a set M equipped with an associative binary multiplication operation $M \times M \to M$ and an identity element $e \in M$ serving as a two-sided identity. In other words, a monoid is precisely a one-object category.

[10]A **preorder** is a set with a binary relation $\leq$ that is reflexive and transitive. In other words, a preorder is precisely a category in which there are no parallel pairs of distinct morphisms between any fixed pair of objects. A **poset** is a preorder that is additionally antisymmetric: $x \leq y$ and $y \leq x$ implies that $x = y$.

these morphisms are not always functions.[11] The morphisms in a category are also called **arrows** or **maps**, particularly in the contexts of Examples 1.1.4 and 1.1.3, respectively.

REMARK 1.1.5. Russell's paradox implies that there is no set whose elements are "all sets." This is the reason why we have used the vague word "collection" in Definition 1.1.1. Indeed, in each of the examples listed in 1.1.3, the collection of objects is not a set. Eilenberg and Mac Lane address this potential area of concern as follows:

> ... the whole concept of a category is essentially an auxiliary one; our basic concepts are essentially those of a *functor* and of a natural transformation .... The idea of a category is required only by the precept that every function should have a definite class as domain and a definite class as range, for the categories are provided as the domains and ranges of functors. Thus one could drop the category concept altogether and adopt an even more intuitive standpoint, in which a functor such as "Hom" is not defined over the category of "all" groups, but for each particular pair of groups which may be given. [**EM45**]

The set-theoretical issues that confront us while defining the notion of a category will compound as we develop category theory further. For that reason, common practice among category theorists is to work in an extension of the usual Zermelo–Fraenkel axioms of set theory, with new axioms allowing one to distinguish between "small" and "large" sets, or between sets and classes. The search for the most useful set-theoretical foundations for category theory is a fascinating topic that unfortunately would require too long of a digression to explore.[12] Instead, we sweep these foundational issues under the rug, not because these issues are not serious or interesting, but because they distract from the task at hand.[13]

For the reasons just discussed, it is important to introduce adjectives that explicitly address the size of a category.

DEFINITION 1.1.6. A category is **small** if it has only a set's worth of arrows.

By Remark 1.1.2, a small category has only a set's worth of objects. If C is a small category, then there are functions

$$\text{mor}\, C \underset{\text{cod}}{\overset{\text{dom}}{\underset{\longleftarrow \text{id} \longrightarrow}{\rightrightarrows}}} \text{ob}\, C$$

that send a morphism to its domain and its codomain and an object to its identity.

---

[11]Reid's *Undergraduate algebraic geometry* emphasizes that the morphisms are not always functions, writing "Students who disapprove are recommended to give up at once and take a reading course in category theory instead" [**Rei88**, p. 4].

[12]The preprint [**Shu08**] gives an excellent overview, though it is perhaps better read after Chapters 1–4.

[13]If pressed, let us assume that there exists a countable sequence of **inaccessible cardinals**, meaning uncountable cardinals that are **regular** and **strong limit**. A cardinal κ is **regular** if every union of fewer than κ sets each of cardinality less than κ has cardinality less than κ, and **strong limit** if $\lambda < \kappa$ implies that $2^\lambda < \kappa$. Inaccessibility means that sets of size less than κ are closed under power sets and κ-small unions. If κ is inaccessible, then the κ-stage of the *von Neumann hierarchy*, the set V_κ of sets of rank less than κ, is a model of Zermelo–Fraenkel set theory with choice (ZFC); the set V_κ is a *Grothendieck universe*. The assumption that there exists a countable sequence of inaccessible cardinals means that we can "do set theory" inside the universe V_κ, and then enlarge the universe if necessary as often as needed.

If ZFC is consistent, these axioms cannot prove the existence of an inaccessible cardinal or the consistency of the assumption that one exists (by Gödel's second incompleteness theorem). Nonetheless, from the perspective of the hierarchy of large cardinal axioms, the existence of inaccessibles is a relatively mild hypothesis.

None of the categories in Example 1.1.3 are small—each has too many objects—but "locally" they resemble small categories in a sense made precise by the following notion:

DEFINITION 1.1.7. A category is **locally small** if between any pair of objects there is only a set's worth of morphisms.

It is traditional to write

(1.1.8) $\mathsf{C}(X, Y)$ or $\mathrm{Hom}(X, Y)$

for the set of morphisms from X to Y in a locally small category C.[14] The set of arrows between a pair of fixed objets in a locally small category is typically called a **hom-set**, whether or not it is a set of "homomorphisms" of any particular kind. Because the notation (1.1.8) is so convenient, it is also adopted for the collection of morphisms between a fixed pair of objects in a category that is not necessarily locally small.

A category provides a context in which to answer the question "When is one thing the same as another thing?" Almost universally in mathematics, one regards two objects of the same category to be "the same" when they are isomorphic, in a precise categorical sense that we now introduce.

DEFINITION 1.1.9. An **isomorphism** in a category is a morphism $f\colon X \to Y$ for which there exists a morphism $g\colon Y \to X$ so that $gf = 1_X$ and $fg = 1_Y$. The objects X and Y are **isomorphic** whenever there exists an isomorphism between X and Y, in which case one writes $X \cong Y$.

An **endomorphism**, i.e., a morphism whose domain equals its codomain, that is an isomorphism is called an **automorphism**.

EXAMPLE 1.1.10.

  (i) The isomorphisms in Set are precisely the **bijections**.
 (ii) The isomorphisms in Group, Ring, Field, or Mod_R are the bijective homomorphisms.
(iii) The isomorphisms in the category Top are the **homeomorphisms**, i.e., the continuous functions with continuous inverse, which is a stronger property than merely being a bijective continuous function.
 (iv) The isomorphisms in the category Htpy are the **homotopy equivalences**.
  (v) In a poset $(\mathsf{P}, \leq)$, the axiom of antisymmetry asserts that $x \leq y$ and $y \leq x$ imply that $x = y$. That is, the only isomorphisms in the category $(\mathsf{P}, \leq)$ are identities.

Examples 1.1.10(ii) and (iii) suggest the following general question: In a concrete category, when are the isomorphisms precisely those maps in the category that induce bijections between the underlying sets? We will see an answer in Lemma 5.6.1.

DEFINITION 1.1.11. A **groupoid** is a category in which every morphism is an isomorphism.

EXAMPLE 1.1.12.

  (i) A **group** is a groupoid with one object.[15]
 (ii) For any space X, its **fundamental groupoid** $\Pi_1(X)$ is a category whose objects are the points of X and whose morphisms are endpoint-preserving homotopy classes of paths.

---

[14]Mac Lane credits Emmy Noether for emphasizing the importance of homomorphisms in abstract algebra, particularly the homomorphism onto a quotient group, which plays an integral role in the statement of her first isomorphism theorem. His recollection is that the arrow notation first appeared around 1940, perhaps due to Hurewicz [**ML88**]. The notation $\mathrm{Hom}(X, Y)$ was first used in [**EM42a**] for the set of homomorphisms between a pair of abelian groups.

[15]This is not simply an example; it is a definition.

A **subcategory** D of a category C is defined by restricting to a subcollection of objects and subcollection of morphisms subject to the requirements that the subcategory D contains the domain and codomain of any morphism in D, the identity morphism of any object in D, and the composite of any composable pair of morphisms in D. For example, there is a subcategory CRing ⊂ Ring of commutative unital rings. Both of these form subcategories of the category Rng of not-necessarily unital rings and homomorphisms that need not preserve the multiplicative unit.[16]

LEMMA 1.1.13. *Any category C contains a **maximal groupoid**, the subcategory containing all of the objects and only those morphisms that are isomorphisms.*

PROOF. Exercise 1.1.ii.   □

For instance, Fin_{iso}, the category of finite sets and bijections, is the maximal subgroupoid of the category Fin of finite sets and all functions. Example 1.4.9 will explain how this groupoid can be regarded as a categorification of the natural numbers, providing a vantage point from which to prove the laws of elementary arithmetic.

Exercises.

EXERCISE 1.1.i.

(i) Show that a morphism can have at most one inverse isomorphism.
(ii) Consider a morphism $f\colon x \to y$. Show that if there exists a pair of morphisms $g, h\colon y \rightrightarrows x$ so that $gf = 1_x$ and $fh = 1_y$, then $g = h$ and f is an isomorphism.

EXERCISE 1.1.ii. Let C be a category. Show that the collection of isomorphisms in C defines a subcategory, the **maximal groupoid** inside C.

EXERCISE 1.1.iii. For any category C and any object $c \in$ C, show that:

(i) There is a category c/C whose objects are morphisms $f\colon c \to x$ with domain c and in which a morphism from $f\colon c \to x$ to $g\colon c \to y$ is a map $h\colon x \to y$ between the codomains so that the triangle

commutes, i.e., so that $g = hf$.

(ii) There is a category C/c whose objects are morphisms $f\colon x \to c$ with codomain c and in which a morphism from $f\colon x \to c$ to $g\colon y \to c$ is a map $h\colon x \to y$ between the domains so that the triangle

commutes, i.e., so that $f = gh$.

The categories c/C and C/c are called **slice categories** of C **under** and **over** c, respectively.

---

[16]To justify our default notion of ring, see Poonen's "Why all rings should have a 1" [**Poo14**]. The relationship between unital and non-unital rings is explored in greater depth in §4.6.

1.2. Duality

> The dual of any axiom for a category is also an
> axiom . . . A simple metamathematical argument
> thus proves the *duality principle*. If any statement
> about a category is deducible from the axioms for
> a category, the dual statement is likely deducible.

Saunders Mac Lane, "Duality for groups" [**ML50**]

Upon first acquaintance, the primary role played by the notion of a category might appear to be taxonomic: vector spaces and linear maps define one category, manifolds and smooth functions define another. But a category, as defined in 1.1.1, is also a mathematical object in its own right, and as with any mathematical definition, this one is worthy of further consideration. Applying a mathematician's gaze to the definition of a category, the following observation quickly materializes. If we visualize the morphisms in a category as arrows pointing from their domain object to their codomain object, we might imagine simultaneously reversing the directions of every arrow. This leads to the following notion.

DEFINITION 1.2.1. Let C be any category. The **opposite category** C^{op} has

- the same objects as in C, and
- a morphism f^{op} in C^{op} for each a morphism f in C so that the domain of f^{op} is defined to be the codomain of f and the codomain of f^{op} is defined to be the domain of f: i.e.,

$$f^{\mathrm{op}} \colon X \to Y \quad \in C^{\mathrm{op}} \qquad \leftrightsquigarrow \qquad f \colon Y \to X \quad \in C.$$

That is, C^{op} has the same objects and morphisms as C, except that "each morphism is pointing in the opposite direction." The remaining structure of the category C^{op} is given as follows:

- For each object X, the arrow 1_X^{op} serves as its identity in C^{op}.
- To define composition, observe that a pair of morphisms $f^{\mathrm{op}}, g^{\mathrm{op}}$ in C^{op} is composable precisely when the pair g, f is composable in C, i.e., precisely when the codomain of g equals the domain of f. We then define $g^{\mathrm{op}} \cdot f^{\mathrm{op}}$ to be $(f \cdot g)^{\mathrm{op}}$: i.e.,

$$f^{\mathrm{op}} \colon X \to Y, \ g^{\mathrm{op}} \colon Y \to Z \quad \in C^{\mathrm{op}} \quad \rightsquigarrow \quad g^{\mathrm{op}} f^{\mathrm{op}} \colon X \to Z \quad \in C^{\mathrm{op}}$$
$$\updownarrow \qquad\qquad\qquad\qquad\qquad\qquad\qquad \updownarrow$$
$$g \colon Z \to Y, \ f \colon Y \to X \quad \in C \quad \rightsquigarrow \quad fg \colon Z \to X \quad \in C$$

The data described in Definition 1.2.1 defines a category C^{op}—i.e., the composition law is associative and unital—if and only if C defines a category. In summary, the process of "turning around the arrows" or "exchanging domains and codomains" exhibits a syntactical self-duality satisfied by the axioms for a category. Note that the category C^{op} contains precisely the same information as the category C. Questions about the one can be answered by examining the other.

EXAMPLE 1.2.2.

(i) $\mathsf{Mat}_R^{\mathrm{op}}$ is the category whose objects are non-zero natural numbers and in which a morphism from m to n is an $m \times n$ matrix with values in R. The upshot is that a reader who would have preferred the opposite handedness conventions when defining Mat_R would have lost nothing by adopting them.

(ii) When a preorder $(\mathsf{P}, \leq)$ is regarded as a category, its opposite category is the category that has a morphism $x \to y$ if and only if $y \leq x$. For example, ω^{op} is the category freely generated by the graph

$$\cdots \to 3 \to 2 \to 1 \to 0.$$

(iii) If G is a group, regarded as a one-object groupoid, the category $(BG)^{\text{op}} \cong B(G^{\text{op}})$ is again a one-object groupoid, and hence a group. The group G^{op} is called the **opposite group** and is used to define right actions as a special case of left actions; see Example 1.3.9.

This syntactical duality has a very important consequence for the development of category theory. Any theorem containing a universal quantification of the form "for all categories C" also necessarily applies to the opposites of these categories. Interpreting the result in the dual context leads to a **dual theorem**, proven by the dual of the original proof, in which the direction of each arrow appearing in the argument is reversed. The result is a two-for-one deal: any proof in category theory simultaneously proves two theorems, the original statement and its dual.[17] For example, the reader may have found Exercise 1.1.iii redundant, precisely because the statements (i) and (ii) are dual; see Exercise 1.2.i.

To illustrate the principle of duality in category theory, let us consider the following result, which provides an important characterization of the isomorphisms in a category.

LEMMA 1.2.3. *The following are equivalent:*

(i) $f: x \to y$ is an isomorphism in C.

(ii) For all objects $c \in$ C, post-composition with f defines a bijection

$$f_*: \mathsf{C}(c, x) \to \mathsf{C}(c, y).$$

(iii) For all objects $c \in$ C, pre-composition with f defines a bijection

$$f^*: \mathsf{C}(y, c) \to \mathsf{C}(x, c).$$

REMARK 1.2.4. In language introduced in Chapter 2, Lemma 1.2.3 asserts that isomorphisms in a locally small category are defined *representably* in terms of isomorphisms in the category of sets. That is, a morphism $f: x \to y$ in an arbitrary locally small category C is an isomorphism if and only if the post-composition function $f_*: \mathsf{C}(c, x) \to \mathsf{C}(c, y)$ between hom-sets defines an isomorphism in Set for each object $c \in$ C.

In set theoretical foundations that permit the definition of functions between large sets, the proof given here applies also to non-locally small categories. In our exposition, the set theoretical hypotheses of smallness and local smallness will only appear when there are essential subtleties concerning the sizes of the categories in question. This is not one of those occasions.

PROOF OF LEMMA 1.2.3. We will prove the equivalence (i) ⟺ (ii) and conclude the equivalence (i) ⟺ (iii) by duality.

Assuming (i), namely that $f: x \to y$ is an isomorphism with inverse $g: y \to x$, then, as an immediate application of the associativity and identity laws for composition in a category, post-composition with g defines an inverse function

$$g_*: \mathsf{C}(c, y) \to \mathsf{C}(c, x)$$

to f_* in the sense that the composites

$$g_* f_*: \mathsf{C}(c, x) \to \mathsf{C}(c, x) \quad \text{and} \quad f_* g_*: \mathsf{C}(c, y) \to \mathsf{C}(c, y)$$

[17]More generally, the proof of a statement of the form "for all categories $C_1, C_2, \ldots, C_n$" leads to 2^n dual theorems. In practice, however, not all of the dual statements will differ meaningfully from the original; see e.g., §4.3.

are both the identity function: for any $h: c \to x$ and $k: c \to y$, $g_* f_*(h) = gfh = h$ and $f_* g_*(k) = fgk = k$.

Conversely, assuming (ii), there must be an element $g \in C(y, x)$ whose image under $f_*: C(y, x) \to C(y, y)$ is 1_y. By construction, $1_y = fg$. But now, by associativity of composition, the elements $gf, 1_x \in C(x, x)$ have the common image f under the function $f_*: C(x, x) \to C(x, y)$, whence $gf = 1_x$. Thus, f and g are inverse isomorphisms.

We have just proven the equivalence (i) $\Leftrightarrow$ (ii) for all categories and in particular for the category C^{op}: i.e., a morphism $f^{op}: y \to x$ in C^{op} is an isomorphism if and only if

(1.2.5) $f_*^{op}: C^{op}(c, y) \to C^{op}(c, x)$ is an isomorphism for all $c \in C^{op}$.

Interpreting the data of C^{op} in its opposite category C, the statement (1.2.5) expresses the same mathematical content as

(1.2.6) $f^*: C(y, c) \to C(x, c)$ is an isomorphism for all $c \in C$.

That is: $C^{op}(c, x) = C(x, c)$, post-composition with f^{op} in C^{op} translates to pre-composition with f in the opposite category C. The notion of isomorphism, as defined in 1.1.9, is self-dual: $f^{op}: y \to x$ is an isomorphism in C^{op} if and only if $f: x \to y$ is an isomorphism in C. So the equivalence (i) $\Leftrightarrow$ (ii) in C^{op} expresses the equivalence (i) $\Leftrightarrow$ (iii) in C.[18] $\square$

Concise expositions of the duality principle in category theory may be found in [Awo10, §3.1] and [HS97, §II.3]. As we become more comfortable with arguing by duality, dual proofs and eventually also dual statements will seldom be described in this much detail.

Categorical definitions also have duals; for instance:

DEFINITION 1.2.7. A morphism $f: x \to y$ in a category is

  (i) a **monomorphism** if for any parallel morphisms $h, k: w \rightrightarrows x$, $fh = fk$ implies that $h = k$; or

  (ii) an **epimorphism** if for any parallel morphisms $h, k: y \rightrightarrows z$, $hf = kf$ implies that $h = k$.

Note that a monomorphism or epimorphism in C is, respectively, an epimorphism or monomorphism in C^{op}. In adjectival form, a monomorphism is **monic** and an epimorphism is **epic**. In common shorthand, a monomorphism is a **mono** and an epimorphism is an **epi**. For graphical emphasis, monos are often decorated with a tail "$\rightarrowtail$" while epis may be decorated at their head "$\twoheadrightarrow$."

The following dual statements re-express Definition 1.2.7:

  (i) $f: x \to y$ is a monomorphism in C if and only if for all objects $c \in C$, post-composition with f defines an injection $f_*: C(c, x) \to C(c, y)$.

  (ii) $f: x \to y$ is an epimorphism in C if and only if for all objects $c \in C$, pre-composition with f defines an injection $f^*: C(y, c) \to C(x, c)$.

EXAMPLE 1.2.8. Suppose $f: X \to Y$ is a monomorphism in the category of sets. Then, in particular, given any two maps $x, x': 1 \rightrightarrows X$, whose domain is the singleton set, if $fx = fx'$ then $x = x'$. Thus, monomorphisms are injective functions. Conversely, any injective function can easily be seen to be a monomorphism.

Similarly, a function $f: X \to Y$ is an epimorphism in the category of sets if and only if it is surjective. Given functions $h, k: Y \rightrightarrows Z$, the equation $hf = kf$ says exactly that h is equal to k on the image of f. This only implies that $h = k$ in the case where the image is all of Y.

---

[18] A similar translation, as just demonstrated between the statements (1.2.5) and (1.2.6), transforms the proof of (i) $\Leftrightarrow$ (ii) into a proof of (i) $\Leftrightarrow$ (iii).

Thus, monomorphisms and epimorphisms should be regarded as categorical analogs of the notions of injective and surjective functions. In practice, if C is a category in which objects have "underlying sets," then any morphism that induces an injective or surjective function between these defines a monomorphism or epimorphism; see Exercise 1.6.iii for a precise discussion. However, even in such categories, the notions of monomorphism and epimorphism can be more general, as demonstrated in Exercise 1.6.v.

EXAMPLE 1.2.9. Suppose that $x \xrightarrow{s} y \xrightarrow{r} x$ are morphisms so that $rs = 1_x$. The map s is a **section** or **right inverse** to r, while the map r defines a **retraction** or **left inverse** to s. The maps s and r express the object x as a **retract** of the object y.

In this case, s is always a monomorphism and, dually, r is always an epimorphism. To acknowledge the presence of these one-sided inverses, s is said to be a **split monomorphism** and r is said to be a **split epimorphism**.[19]

EXAMPLE 1.2.10. By the previous example, an isomorphism is necessarily both monic and epic, but the converse need not hold in general. For example, the inclusion $\mathbb{Z} \hookrightarrow \mathbb{Q}$ is both monic and epic in the category Ring, but this map is not an isomorphism: there are no ring homomorphisms from $\mathbb{Q}$ to $\mathbb{Z}$.

Since the notions of monomorphism and epimorphism are dual, their abstract categorical properties are also dual, such as exhibited by the following lemma.

LEMMA 1.2.11.

(i) If $f: x \rightarrowtail y$ and $g: y \rightarrowtail z$ are monomorphisms, then so is $gf: x \rightarrowtail z$.
(ii) If $f: x \rightarrow y$ and $g: y \rightarrow z$ are morphisms so that gf is monic, then f is monic.

Dually:

(i') If $f: x \twoheadrightarrow y$ and $g: y \twoheadrightarrow z$ are epimorphisms, then so is $gf: x \twoheadrightarrow z$.
(ii') If $f: x \rightarrow y$ and $g: y \rightarrow z$ are morphisms so that gf is epic, then g is epic.

PROOF. Exercise 1.2.iii.   □

Exercises.

EXERCISE 1.2.i. Show that $C/c \cong (c/(C^{op}))^{op}$. Defining C/c to be $(c/(C^{op}))^{op}$, deduce Exercise 1.1.iii(ii) from Exercise 1.1.iii(i).

EXERCISE 1.2.ii.
  (i) Show that a morphism $f: x \rightarrow y$ is a split epimorphism in a category C if and only if for all $c \in C$, post-composition $f_*: C(c, x) \rightarrow C(c, y)$ defines a surjective function.
  (ii) Argue by duality that f is a split monomorphism if and only if for all $c \in C$, pre-composition $f^*: C(y, c) \rightarrow C(x, c)$ is a surjective function.

EXERCISE 1.2.iii. Prove Lemma 1.2.11 by proving either (i) or (i') and either (ii) or (ii'), then arguing by duality. Conclude that the monomorphisms in any category define a subcategory of that category and dually that the epimorphisms also define a subcategory.

EXERCISE 1.2.iv. What are the monomorphisms in the category of fields?

EXERCISE 1.2.v. Show that the inclusion $\mathbb{Z} \hookrightarrow \mathbb{Q}$ is both a monomorphism and an epimorphism in the category Ring of rings. Conclude that a map that is both monic and epic need not be an isomorphism.

EXERCISE 1.2.vi. Prove that a morphism that is both a monomorphism and a split epimorphism is necessarily an isomorphism. Argue by duality that a split monomorphism that is an epimorphism is also an isomorphism.

---

[19]The axiom of choice asserts that every epimorphism in the category of sets is a split epimorphism.

EXERCISE 1.2.vii. Regarding a poset $(P, \leq)$ as a category, define the supremum of a subcollection of objects $A \in P$ in such a way that the dual statement defines the infimum. Prove that the supremum of a subset of objects is unique, whenever it exists, in such a way that the dual proof demonstrates the uniqueness of the infimum.

1.3. Functoriality

> ... every sufficiently good analogy is yearning to become a functor.

> John Baez, "Quantum Quandaries: A Category-Theoretic Perspective" [**Bae06**]

A key tenet in category theory, motivating the very definition of a category, is that any mathematical object should be considered together with its accompanying notion of structure-preserving morphism. In "General theory of natural equivalences" [**EM45**], Eilenberg and Mac Lane argue further:

> ... whenever new abstract objects are constructed in a specified way out of given ones, it is advisable to regard the construction of the corresponding induced mappings on these new objects as an integral part of their definition.

Categories are themselves mathematical objects, if of a somewhat unfamiliar sort, which leads to a question: What is a morphism between categories?

DEFINITION 1.3.1. A **functor** $F: C \to D$, between categories C and D, consists of the following data:

- An object $Fc \in D$, for each object $c \in C$.
- A morphism $Ff: Fc \to Fc' \in D$, for each morphism $f: c \to c' \in C$, so that the domain and codomain of Ff are, respectively, equal to F applied to the domain or codomain of f.

The assignments are required to satisfy the following two **functoriality axioms**:

- For any composable pair f, g in C, $Fg \cdot Ff = F(g \cdot f)$.
- For each object c in C, $F(1_c) = 1_{Fc}$.

Put concisely, a functor consists of a mapping on objects and a mapping on morphisms that preserves all of the structure of a category, namely domains and codomains, composition, and identities.[20]

EXAMPLE 1.3.2.

(i) There is an endofunctor[21] $P: \mathsf{Set} \to \mathsf{Set}$ that sends a set A to its power set $PA = \{A' \subset A\}$ and a function $f: A \to B$ to the direct-image function $f_*: PA \to PB$ that sends $A' \subset A$ to $f(A') \subset B$.

(ii) Each of the categories listed in Example 1.1.3 has a **forgetful functor**, a general term that is used for any functor that forgets structure, whose codomain is the category of sets. For example, $U: \mathsf{Group} \to \mathsf{Set}$ sends a group to its underlying set and a group homomorphism to its underlying function. The functor $U: \mathsf{Top} \to \mathsf{Set}$ sends a space to its set of points. There are two natural forgetful functors $V, E: \mathsf{Graph} \rightrightarrows \mathsf{Set}$ that send a graph to its vertex or edge sets, respectively; if desired, these can be combined

---

[20]While a functor should be regarded as a mapping from the data of one category to the data of another, parentheses are used as seldom as possible unless demanded for notational clarity.

[21]An **endofunctor** is a functor whose domain is equal to its codomain.

to define a single functor $V \sqcup E$: Graph $\rightarrow$ Set that carries a graph to the disjoint union of its vertex and edge sets. These mappings are functorial because in each instance a morphism in the domain category has an underlying function.

(iii) There are intermediate forgetful functors $\mathsf{Mod}_R \rightarrow$ Ab and Ring $\rightarrow$ Ab that forget some but not all of the algebraic structure. The inclusion functors Ab $\hookrightarrow$ Group and Field $\hookrightarrow$ Ring may also be regarded as "forgetful." Note that the latter two, but neither of the former, are injective on objects: a group is either abelian or not, but an abelian group might admit the structure of a ring in multiple ways.

(iv) Similarly, there are forgetful functors Group $\rightarrow$ Set$_*$ and Ring $\rightarrow$ Set$_*$ that take the basepoint to be the identity and zero elements, respectively. These assignments are functorial because group and ring homomorphisms necessarily preserve these elements.

(v) There are functors Top $\rightarrow$ Htpy and Top$_*$ $\rightarrow$ Htpy$_*$ that act as the identity on objects and send a (based) continuous function to its homotopy class.

(vi) The *fundamental group* defines a functor π_1: Top$_*$ $\rightarrow$ Group; a continuous function $f: (X, x) \rightarrow (Y, y)$ of based spaces induces a group homomorphism $f_*: \pi_1(X, x) \rightarrow \pi_1(Y, y)$ and this assignment is functorial, satisfying the two functoriality axioms described above. A precise expression of the statement that "the fundamental group is a homotopy invariant" is that this functor factors through the functor Top$_*$ $\rightarrow$ Htpy$_*$ to define a functor π_1: Htpy$_*$ $\rightarrow$ Group.

(vii) A related functor Π_1: Top $\rightarrow$ Groupoid assigns an unbased topological space its fundamental groupoid, the category defined in Example 1.1.12(ii). A continuous function $f: X \rightarrow Y$ induces a functor $f_*: \Pi_1(X) \rightarrow \Pi_1(Y)$ that carries a point $x \in X$ to the point $f(x) \in Y$. This mapping extends to morphisms in $\Pi_1(X)$ because continuous functions preserve paths and path homotopy classes.

(viii) For each $n \in \mathbb{Z}$, there are functors Z_n, B_n, H_n: Ch$_R$ $\rightarrow$ Mod$_R$. The functor Z_n computes the n-**cycles** defined by $Z_n C_\bullet = \ker(d: C_n \rightarrow C_{n-1})$. The functor B_n computes the n-**boundary** defined by $B_n C_\bullet = \mathrm{im}(d: C_{n+1} \rightarrow C_n)$. The functor H_n computes the nth **homology** $H_n C_\bullet := Z_n C_\bullet / B_n C_\bullet$. We leave it to the reader to verify that each of these three constructions is functorial. Considering all degrees simultaneously, the cycle, boundary, and homology functors assemble into functors Z_*, B_*, H_*: Ch$_R$ $\rightarrow$ GrMod$_R$ from the category of chain complexes to the category of graded R-modules. The singular homology of a topological space is defined by precomposing H_* with a suitable functor Top $\rightarrow$ Ch$_R$.

(ix) There is a functor F: Set $\rightarrow$ Group that sends a set X to the **free group** on X. This is the group whose elements are finite "words" whose letters are elements $x \in X$ or their formal inverses x^{-1}, modulo an equivalence relation that equates the words xx^{-1} and $x^{-1}x$ with the empty word. Multiplication is by concatenation, with the empty word serving as the identity. This is one instance of a large family of "free" functors studied in Chapter 4.

(x) The chain rule expresses the functoriality of the derivative. Let Euclid$_*$ denote the category whose objects are pointed finite-dimensional Euclidean spaces $(\mathbb{R}^n, a)$—or, better, open subsets thereof—and whose morphisms are pointed differentiable functions. The **total derivative** of $f: \mathbb{R}^n \rightarrow \mathbb{R}^m$, evaluated at the designated basepoint $a \in \mathbb{R}^n$, gives rise to a matrix called the **Jacobian matrix** defining the directional derivatives of f at the point a. If f is given by component functions $f_1, \ldots, f_m: \mathbb{R}^n \rightarrow \mathbb{R}$, the (i, j)-entry of this matrix is $\frac{\partial}{\partial x_j} f_i(a)$. This defines the action on morphisms of a functor D: Euclid$_*$ $\rightarrow$ Mat$_\mathbb{R}$; on objects, D assigns a pointed Euclidean space its dimension. Given $g\colon \mathbb{R}^m \to \mathbb{R}^k$ carrying the designated basepoint $f(a) \in \mathbb{R}^m$ to $gf(a) \in \mathbb{R}^k$, functoriality of D asserts that the product of the Jacobian of f at a with the Jacobian of g at $f(a)$ equals the Jacobian of gf at a. This is the chain rule from multivariable calculus.[22]

(xi) Any commutative monoid M can be used to define a functor $M^-\colon \mathsf{Fin}_* \to \mathsf{Set}$. Writing $n_+ \in \mathsf{Fin}_*$ for the set with n non-basepoint elements, define M^{n_+} to be M^n, the n-fold cartesian product of the set M with itself. By convention, M^{0_+} is a singleton set. For any based map $f\colon m_+ \to n_+$, define the ith component of the corresponding function $M^f\colon M^m \to M^n$ by projecting from M^m to the coordinates indexed by elements in the fiber $f^{-1}(i)$ and then multiplying these using the commutative monoid structure; if the fiber is empty, the function M^f inserts the unit element in the ith coordinate. Note each of the sets M^n itself has a basepoint, the n-tuple of unit elements, and each of the maps in the image of the functor are based. It follows that the functor M^- lifts along the forgetful functor $U\colon \mathsf{Set}_* \to \mathsf{Set}$.

There is a special property satisfied by this construction that allows one to extract the commutative monoid M from the functor $\mathsf{Fin}_* \to \mathsf{Set}$. This observation was used by Segal to introduce a suitable notion of "commutative monoid" into algebraic topology [**Seg74**].

More examples of functors will appear shortly, but first we illustrate the utility of knowing that the assignment of a mathematical object of one type to mathematical objects of another type is *functorial*. Applying the functoriality of the fundamental group construction $\pi_1\colon \mathsf{Top}_* \to \mathsf{Group}$, one can prove:

THEOREM 1.3.3 (Brouwer Fixed Point Theorem). *Any continuous endomorphism of a 2-dimensional disk D^2 has a fixed point.*

PROOF. Assuming $f\colon D^2 \to D^2$ is such that $f(x) \neq x$ for all $x \in D^2$, there is a continuous function $r\colon D^2 \to S^1$ that carries a point $x \in D^2$ to the intersection of the ray from $f(x)$ to x with the boundary circle S^1. Note that the function r fixes the points on the boundary circle $S^1 \subset D^2$. Thus, r defines a retraction of the inclusion $i\colon S^1 \hookrightarrow D^2$, which is to say, the composite $S^1 \xrightarrow{i} D^2 \xrightarrow{r} S^1$ is the identity.

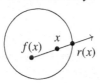

Pick any basepoint on the boundary circle S^1 and apply the functor π_1 to obtain a composable pair of group homomorphisms:

$$\pi_1(S^1) \xrightarrow{\pi_1(i)} \pi_1(D^2) \xrightarrow{\pi_1(r)} \pi_1(S^1).$$

By the functoriality axioms, we must have

$$\pi_1(r) \cdot \pi_1(i) = \pi_1(ri) = \pi_1(1_{S^1}) = 1_{\pi_1(S^1)}.$$

However, a computation involving covering spaces reveals that $\pi_1(S^1) = \mathbb{Z}$, while $\pi_1(D^2) = 0$, the trivial group. The composite endomorphism $\pi_1(r) \cdot \pi_1(i)$ of $\mathbb{Z}$ must be zero, since it factors through the trivial group. Thus, it cannot equal the identity homomorphism, which

---

[22]Taking a more sophisticated perspective, the derivative defines the action on morphisms of a functor from the category Man_* to the category of real vector spaces that sends a pointed manifold to its tangent space.

carries the generator $1 \in \mathbb{Z}$ to itself $(0 \neq 1)$. This contradiction proves that the retraction r cannot exist, and so f must have a fixed point.[23]   □

Functoriality also plays a key role in the emerging area of topological data analysis.

EXAMPLE 1.3.4 (in search of a clustering functor). A *clustering algorithm* is a function that converts a finite metric space into a partition of its points into sets of "clusters." An impossibility theorem of Kleinberg proves that there are no clustering algorithms that satisfy three reasonable axioms [**Kle03**].[24] A key insight of Carlsson and Mémoli is that these axioms can be encoded as morphisms in a category of finite metric spaces in such a way that what is desired is not a clustering function but a clustering functor into a suitable category [**CM13**]. Ghrist's *Elementary Applied Topology* [**Ghr14**, p. 216] describes this move as follows:

> What is the good of this? Category theory is criticized as an esoteric language: formal and fruitless for conversation. *This is not so.*[25] The virtue of reformulating (the negative) Theorem [of Kleinberg] functorially is a clearer path to a positive statement. If the goal is to have a theory of clustering; if clustering is, properly, a nontrivial functor; if no nontrivial functors between the proposed categories exist; then, naturally, the solution is to alter the domain or codomain categories and classify the ensuing functors. One such modification is to consider a category of persistent clusters.

One pair of categories considered in [**CM13**] are the categories FinMetric, of finite metric spaces and distance non-increasing functions, and Cluster, of clusters and refinements. An object in Cluster is a partitioned set. Given a function $f \colon X \to Y$, the preimages of a partition of the set Y define a partition of X. A morphism in Cluster is a function $f \colon X \to Y$ of underlying sets so that the given partition on X refines the partition on X defined by the preimages of the given partition on Y.

Carlsson and Mémoli observe that the only scale-invariant functors FinMetric $\to$ Cluster either assign each metric space the discrete partition (into singletons) or the indiscrete partition (into a single cluster); both cases fail to satisfy Kleinberg's surjectivity condition. This suggests that clusters should be replaced by a notion of "persistent" clusters. A *persistent cluster* on X is a functor from the poset $([0, \infty), \leq)$ to the poset of clusters on X, where $\phi \leq \psi$ if and only if the partition ϕ refines the partition ψ. The idea is that when the parameter $r \in [0, \infty)$ is small, the partition on X might be very fine, but the clusters are allowed to coalesce as one "zooms out," i.e., as r increases.

There is a category PCluster whose objects are persistent clusters and whose morphisms are functions of underlying sets $f \colon X \to Y$ that define morphisms in Cluster for each $r \in [0, \infty)$. Carlsson and Mémoli prove that there is a unique functor FinMet $\to$ PCluster, which takes the metric space with two points of distance r to the persistent cluster with one cluster for $t \geq r$ and two clusters for $0 \leq t < r$ and satisfies two other reasonable conditions; see [**CM13**] for the details.

---

[23]The same argument, with the nth homotopy group functor $\pi_n \colon \mathsf{Top}_* \to \mathsf{Group}$ in place of π_1, proves that any continuous endomorphism of an n-dimensional disk has a fixed point.

[24]Namely, there are no clustering algorithms that are invariant under rigid scaling, consistent under alterations to the distance function that "sharpen" the point clusters, and have the property that some distance function realizes each possible partition.

[25]Emphasis his.

The functors defined in 1.3.1 are called **covariant** so as to distinguish them from another variety of functor that we now introduce.

DEFINITION 1.3.5. A **contravariant functor** F from C to D is a functor $F: \mathsf{C}^{\mathrm{op}} \to \mathsf{D}$.[26] Explicitly, this consists of the following data:

- An object $Fc \in \mathsf{D}$, for each object $c \in \mathsf{C}$.
- A morphism $Ff: Fc' \to Fc \in \mathsf{D}$, for each morphism $f: c \to c' \in \mathsf{C}$, so that the domain and codomain of Ff are, respectively, equal to F applied to the codomain or domain of f.

The assignments are required to satisfy the following two **functoriality axioms**:

- For any composable pair f, g in C, $Ff \cdot Fg = F(g \cdot f)$.
- For each object c in C, $F(1_c) = 1_{Fc}$.

NOTATION 1.3.6. To avoid unnatural arrow-theoretic representations, a morphism in the domain of a functor $F: \mathsf{C}^{\mathrm{op}} \to \mathsf{D}$ will always be depicted as an arrow $f: c \to c'$ in C, pointing from its domain in C to its codomain in C. Similarly, its image will always be depicted as an arrow $Ff: Fc' \to Fc$ in D, pointing from its domain to its codomain. Note that these conventions require that the domain and codomain objects switch their relative places, from left to right, but in examples, for instance in the case where C and D are concrete categories, these positions are the familiar ones. Graphically, the mapping on morphisms given by a contravariant functor is depicted as follows:

$$
\mathsf{C}^{\mathrm{op}} \xrightarrow{\quad F \quad} \mathsf{D}
$$

$$
\begin{array}{ccc}
c & \mapsto & Fc \\
{\scriptstyle f}\big\downarrow & \mapsto & \big\uparrow{\scriptstyle Ff} \\
c' & \mapsto & Fc'
\end{array}
$$

In accordance with this convention, if $f: c \to c'$ and $g: c' \to c''$ are morphisms in C, their composite will always be written as $gf: c \to c''$. The image of this morphism under the contravariant functor $F: \mathsf{C}^{\mathrm{op}} \to \mathsf{D}$ is $F(gf): Fc'' \to Fc$, the composite $Ff \cdot Fg$ of $Fg: Fc'' \to Fc'$ and $Ff: Fc' \to Fc$.

In summary, even in the presence of opposite categories, we always make an effort to draw arrows pointing in the "correct way" and depict composition in the usual order.[27]

EXAMPLE 1.3.7.

(i) The contravariant power set functor $P: \mathsf{Set}^{\mathrm{op}} \to \mathsf{Set}$ sends a set A to its power set PA and a function $f: A \to B$ to the inverse-image function $f^{-1}: PB \to PA$ that sends $B' \subset B$ to $f^{-1}(B') \subset A$.

(ii) There is a functor $(-)^*: \mathsf{Vect}_{\Bbbk}^{\mathrm{op}} \to \mathsf{Vect}_{\Bbbk}$ that carries a vector space to its **dual vector space** $V^* = \mathrm{Hom}(V, \Bbbk)$. A vector in V^* is a **linear functional** on V, i.e., a linear map

---

[26] In this text, a contravariant functor F from C to D will always be written as $F: \mathsf{C}^{\mathrm{op}} \to \mathsf{D}$. Some mathematicians omit the "op" and let the context or surrounding verbiage convey the variance. We think this is bad practice, as the co- or contravariance is an essential part of the data of a functor, which is not necessarily determined by its assignation on objects. More to the point, we find that this notational convention helps mitigate the consequences of temporary distraction. Seeing $F: \mathsf{C}^{\mathrm{op}} \to \mathsf{D}$ written on a chalkboard immediately conveys that F is a contravariant functor from C to D, even to the most spaced-out observer. A similar principle will motivate other notational conventions introduced in Definition 3.1.15 and Notation 4.1.5.

[27] Of course, technically there is no meaning to the phrase "opposite category": every category is the opposite of some other category (its opposite category). But in practice, there is no question which of Set and Set^{op} is the "opposite category," and sufficiently many of the other cases can be deduced from this one.

$V \to \Bbbk$. This functor is contravariant, with a linear map $\phi \colon V \to W$ sent to the linear map $\phi^* \colon W^* \to V^*$ that pre-composes a linear functional $W \xrightarrow{\omega} \Bbbk$ with ϕ to obtain a linear functional $V \xrightarrow{\phi} W \xrightarrow{\omega} \Bbbk$.

(iii) The functor $O \colon \mathsf{Top}^{\mathrm{op}} \to \mathsf{Poset}$ that carries a space X to its poset $O(X)$ of open subsets is contravariant on the category of spaces: a continuous map $f \colon X \to Y$ gives rise to a function $f^{-1} \colon O(Y) \to O(X)$ that carries an open subset $U \subset Y$ to its preimage $f^{-1}(U)$, which is open in X; this is the definition of **continuity**. A similar functor $C \colon \mathsf{Top}^{\mathrm{op}} \to \mathsf{Poset}$ carries a space to its poset of closed subsets.

(iv) There is a contravariant functor $\mathrm{Spec} \colon \mathsf{CRing}^{\mathrm{op}} \to \mathsf{Top}$ that sends a commutative ring R to its set $\mathrm{Spec}(R)$ of prime ideals given the **Zariski topology**. The closed subsets in the Zariski topology are those subsets $V_I \subset \mathrm{Spec}(R)$ of prime ideals containing a fixed ideal $I \subset R$. This construction is contravariantly functorial: for any ring homomorphism $\phi \colon R \to S$ and prime ideal $\mathrm{p} \subset S$, the inverse image $\phi^{-1}(\mathrm{p}) \subset R$ defines a prime ideal of R, and the inverse image function $\phi^{-1} \colon \mathrm{Spec}(S) \to \mathrm{Spec}(R)$ is continuous with respect to the Zariski topology.

(v) For a generic small category C, a functor $\mathsf{C}^{\mathrm{op}} \to \mathsf{Set}$ is called a (set-valued) **presheaf** on C. A typical example is the functor $O(X)^{\mathrm{op}} \to \mathsf{Set}$ whose domain is the poset $O(X)$ of open subsets of a topological space X and whose value at $U \subset X$ is the set of continuous real-valued functions on U. The action on morphisms is by restriction. This presheaf is a *sheaf*, if it satisfies an axiom that is introduced in Definition 3.3.4.

(vi) Presheaves on the category Δ, of finite non-empty ordinals and order-preserving maps, are called **simplicial sets**. Δ is also called the **simplex category**. The ordinal $n + 1 = \{0, 1, \ldots, n\}$ may be thought of as a direct version of the topological n-simplex and, with this interpretation in mind, is typically denoted by "$[n]$" by algebraic topologists.

The following result, which appears immediately after functors are first defined in **[EM42b]**, is arguably the first lemma in category theory.

LEMMA 1.3.8. *Functors preserve isomorphisms.*

PROOF. Consider a functor $F \colon \mathsf{C} \to \mathsf{D}$ and an isomorphism $f \colon x \to y$ in C with inverse $g \colon y \to x$. Applying the two functoriality axioms:

$$F(g)F(f) = F(gf) = F(1_x) = 1_{Fx}.$$

Thus, $Fg \colon Fy \to Fx$ is a left inverse to $Ff \colon Fx \to Fy$. Exchanging the roles of f and g (or arguing by duality) shows that Fg is also a right inverse. $\square$

EXAMPLE 1.3.9. Let G be a group, regarded as a one-object category BG. A functor $X \colon BG \to \mathsf{C}$ specifies an object $X \in \mathsf{C}$ (the unique object in its image) together with an endomorphism $g_* \colon X \to X$ for each $g \in G$. This assignment must satisfy two conditions:

(i) $h_* g_* = (hg)_*$ for all $g, h \in G$.

(ii) $e_* = 1_X$, where $e \in G$ is the identity element.

In summary, the functor $BG \to \mathsf{C}$ defines an **action** of the group G on the object $X \in \mathsf{C}$. When $\mathsf{C} = \mathsf{Set}$, the object X endowed with such an action is called a G-**set**. When $\mathsf{C} = \mathsf{Vect}_\Bbbk$, the object X is called a G-**representation**. When $\mathsf{C} = \mathsf{Top}$, the object X is called a G-**space**. Note the utility of this categorical language for defining several analogous concepts simultaneously.

The action specified by a functor $BG \to \mathsf{C}$ is sometimes called a **left action**. A **right action** is a functor $BG^{\mathrm{op}} \to \mathsf{C}$. As before, each $g \in G$ determines an endomorphism

$g^*: X \to X$ in C and the identity element must act trivially. But now, for a pair of elements $g, h \in G$ these actions must satisfy the composition rule $(hg)^* = g^*h^*$.

Because the elements $g \in G$ are isomorphisms when regarded as morphisms in the 1-object category BG that represents the group, their images under any such functor must also be isomorphisms in the target category. In particular, in the case of a G-representation $V: BG \to \text{Vect}_k$, the linear map $g_*: V \to V$ must be an *automorphism* of the vector space V. The point is that the functoriality axioms (i) and (ii) imply automatically that each g_* is an automorphism and that $(g^{-1})_* = (g_*)^{-1}$; the proof is a special case of Lemma 1.3.8.

In summary:

COROLLARY 1.3.10. *When a group G acts functorially on an object X in a category* C, *its elements g must act by automorphisms $g_*: X \to X$ and, moreover, $(g_*)^{-1} = (g^{-1})_*$.*

A functor may or may not preserve monomorphisms or epimorphisms, but an argument similar to the proof of Lemma 1.3.8 shows that a functor necessarily preserves split monomorphisms and split epimorphisms. The retraction or section defines an "equational witness" for the mono or the epi.

DEFINITION 1.3.11. If C is locally small, then for any object $c \in$ C we may define a pair of covariant and contravariant **functors represented by** c:

$$C \xrightarrow{C(c,-)} \text{Set} \qquad C^{\text{op}} \xrightarrow{C(-,c)} \text{Set}$$

$$
\begin{array}{ccc}
x & \mapsto & C(c, x) \\
f \downarrow & \mapsto & \downarrow f_* \\
y & \mapsto & C(c, y)
\end{array}
\qquad
\begin{array}{ccc}
x & \mapsto & C(x, c) \\
f \downarrow & \mapsto & \uparrow f^* \\
y & \mapsto & C(y, c)
\end{array}
$$

The notation suggests the action on objects: the functor $C(c, -)$ carries $x \in$ C to the set $C(c, x)$ of arrows from c to x in C. Dually, the functor $C(-, c)$ carries $x \in$ C to the set $C(x, c)$.

The functor $C(c, -)$ carries a morphism $f: x \to y$ to the post-composition function $f_*: C(c, x) \to C(c, y)$ introduced in Lemma 1.2.3(ii). Dually, the functor $C(-, c)$ carries f to the pre-composition function $f^*: C(y, c) \to C(x, c)$ introduced in 1.2.3(iii). Note that post-composition defines a *covariant* action on hom-sets, while pre-composition defines a *contravariant* action. There are no choices involved here; post-composition is always a covariant operation, while pre-composition is always a contravariant one. This is just the natural order of things.

We leave it to the reader to verify that the assignments just described satisfy the two functoriality axioms. Note that Lemma 1.3.8 specializes in the case of represented functors to give a proof of the implications (i) $\Rightarrow$ (ii) and (i) $\Rightarrow$ (iii) of Lemma 1.2.3. These functors will play a starring role in Chapter 2, where a number of examples in disguise are discussed.

The data of the covariant and contravariant functors introduced in Definition 1.3.11 may be encoded in a single **bifunctor**, which is the name for a functor of two variables. Its domain is given by the product of a pair of categories.

DEFINITION 1.3.12. For any categories C and D, there is a category C $\times$ D, their **product**, whose

- objects are ordered pairs (c, d), where c is an object of C and d is an object of D,
- morphisms are ordered pairs $(f, g): (c, d) \to (c', d')$, where $f: c \to c' \in$ C and $g: d \to d' \in$ D, and

- in which composition and identities are defined componentwise.

DEFINITION 1.3.13. If C is locally small, then there is a **two-sided represented functor**

$$C(-,-) \colon C^{op} \times C \to Set$$

defined in the evident manner. A pair of objects (x, y) is mapped to the hom-set $C(x, y)$. A pair of morphisms $f \colon w \to x$ and $h \colon y \to z$ is sent to the function

$$C(x, y) \xrightarrow{(f^*, h_*)} C(w, z)$$

$$g \quad \mapsto \quad hgf$$

that takes an arrow $g \colon x \to y$ and then pre-composes with f and post-composes with h to obtain $hgf \colon w \to z$.

At the beginning of this section, it was suggested that functors define morphisms between categories. Indeed, categories and functors assemble into a category. Here the size issues are even more significant than we have encountered thus far. To put a lid on things, define Cat to be the category whose objects are small categories and whose morphisms are functors between them. This category is locally small but not small: it contains Set, Poset, Monoid, Group, and Groupoid as proper subcategories (see Exercises 1.3.i and 1.3.ii). However, none of these categories are *objects* of Cat.

The non-small categories of Example 1.1.3 are objects of CAT, some category of "large" categories and functors between them. Russell's paradox suggests that CAT should not be so large as to contain itself, so we require the objects in CAT to be locally small categories; the category CAT defined in this way is not locally small, and so is thus excluded. There is an inclusion functor Cat ↪ CAT but no obvious functor pointing in the other direction.

The category of categories gives rise to a notion of an **isomorphism of categories**, defined by interpreting Definition 1.1.9 in Cat or in CAT. Namely, an isomorphism of categories is given by a pair of inverse functors $F \colon C \to D$ and $G \colon D \to C$ so that the composites GF and FG, respectively, equal the identity functors on C and on D. An isomorphism induces a bijection between the objects of C and objects of D and likewise for the morphisms.

EXAMPLE 1.3.14. For instance:

(i) The functor $(-)^{op} \colon CAT \to CAT$ defines a non-trivial automorphism of the category of categories. Note that a functor $F \colon C \to D$ also defines a functor $F \colon C^{op} \to D^{op}$.

(ii) For any group G, the categories BG and BG^{op} are isomorphic via the functor $(-)^{-1}$ that sends each morphism $g \in G$ to its inverse. Any right action can be converted into a left action by precomposing with this isomorphism, which has the effect of "inserting inverses in the formula" defining the endomorphism associated to a particular group element.

(iii) Similarly, any groupoid is isomorphic to its opposite category via the functor that acts as the identity on objects and sends a morphism to its unique inverse morphism.

(iv) Any ring R has an opposite ring R^{op} with the same underlying abelian group but with the product of elements r and s in R^{op} defined to be the product $s \cdot r$ of the elements s and r in R. A left R-module is the same thing as a right R^{op}-module, which is to say there is a covariant isomorphism of categories $Mod_R \cong {}_{R^{op}}Mod$ between the category of left R-modules and the category of right R^{op}-modules.

(v) For any space X, there is a contravariant isomorphism of poset categories $O(X) \cong C(X)^{op}$ that associates an open subset of X to its closed complement.

(vi) The category Mat_R is isomorphic to its opposite via an identity-on-objects functor that carries a matrix to its transpose.

Contrary to the impression created by Examples 1.3.14 (ii), (iii), and (vi), a category is not typically isomorphic to its opposite category.

EXAMPLE 1.3.15. Let E/F be a finite **Galois extension**: this means that F is a finite-index subfield of E and that the size of the group $\mathrm{Aut}(E/F)$ of automorphisms of E fixing every element of F is at least (in fact, equal to) the index $[E : F]$. In this case, $G := \mathrm{Aut}(E/F)$ is called the **Galois group** of the Galois extension E/F.

Consider the **orbit category** O_G associated to the group G. Its objects are subgroups $H \subset G$, which we identify with the left G-set G/H of left cosets of H. Morphisms $G/H \to G/K$ are G-equivariant maps, i.e., functions that commute with the left G-action. By an elementary exercise left to the reader, every morphism $G/H \to G/K$ has the form $gH \mapsto g\gamma K$, where $\gamma \in G$ is an element so that $\gamma^{-1} H \gamma \subset K$.

Let Field_F^E denote the subcategory of F/Field whose objects are intermediate fields $F \subset K \subset E$. A morphism $K \to L$ is a field homomorphism that fixes the elements of F pointwise. Note that the group of automorphisms of the object $E \in \mathsf{Field}_F^E$ is the Galois group $G = \mathrm{Aut}(E/F)$.

We define a functor $\Phi \colon O_G^{\mathrm{op}} \to \mathsf{Field}_F^E$ that sends $H \subset G$ to the subfield of E of elements that are fixed by H under the action of the Galois group. If $G/H \to G/K$ is induced by γ, then the field homomorphism $x \mapsto \gamma x$ sends an element $x \in E$ that is fixed by K to an element $\gamma x \in E$ that is fixed by H. This defines the action of the functor Φ on morphisms. The **fundamental theorem of Galois theory** asserts that Φ defines a bijection on objects but in fact more is true: Φ defines an isomorphism of categories $O_G^{\mathrm{op}} \cong \mathsf{Field}_F^E$.

These examples aside, the notion of isomorphism of categories is somewhat unnatural. To illustrate, consider the category Set^∂ of sets and partially-defined functions. A **partial function** $f \colon X \to Y$ is a function from a (possibly-empty) subset $X' \subset X$ to Y; the subset X' is the **domain of definition** of the partial function f. The composite of two partial functions $f \colon X \to Y$ and $g \colon Y \to Z$ is the partial function whose domain of definition is the intersection of the domain of definition of f with the preimage of the domain of definition of g.

There is a functor $(-)_+ \colon \mathsf{Set}^\partial \to \mathsf{Set}_*$, whose codomain is the category of pointed sets, that sends a set X to the pointed set X_+, which is defined to be the disjoint union of X with a freely-added basepoint. By the axiom of regularity, we might define $X_+ := X \cup \{X\}$.[28] A partial function $f \colon X \to Y$ gives rise to a pointed function $f_+ \colon X_+ \to Y_+$ that sends every point outside of the domain of definition of f to the formally added basepoint of Y_+. The inverse functor $U \colon \mathsf{Set}_* \to \mathsf{Set}^\partial$ discards the basepoint and sends a based function $f \colon (X, x) \to (Y, y)$ to the partial function $X\backslash\{x\} \to Y\backslash\{y\}$ with the maximal possible domain of definition.

By construction, we see that the composite $U(-)_+$ is the identity endofunctor of the category Set^∂. By contrast, the other composite $(U-)_+ \colon \mathsf{Set}_* \to \mathsf{Set}_*$ sends a pointed set (X, x) to $(X\backslash\{x\} \cup \{X\backslash\{x\}\}, X\backslash\{x\})$. These sets are isomorphic but they are not identical. Nor is another set-theoretical construction of the "freely added basepoint" likely to define a genuine inverse to the functor $U \colon \mathsf{Set}_* \to \mathsf{Set}^\partial$. It is too restrictive to ask for the categories Set^∂ and Set_* to be isomorphic.

---

[28]In the axioms of Zermelo-Fraenkel set theory, elements of sets (like everything else in its mathematical universe) are themselves sets. The axiom of regularity prohibits a set from being an element of itself. As $X \notin X$, we are free to add the element X as a disjoint basepoint.

Indeed, there is a better way to decide whether two categories may safely be regarded as "the same." To define it, we must relax the identities $GF = 1_C$ and $FG = 1_D$ between functors $F: C \to D$ and $G: D \to C$ that define an isomorphism of categories. This is possible because the collections Hom(C, C) and Hom(D, D) are not mere (possibly large) sets: they have higher-dimensional structure. For any pair of categories C and D, the collection Hom(C, D) of functors is itself a category. To explain this, we introduce what in French is called a *morphisme de foncteurs*, the notion that launched the entire subject of category theory: a *natural transformation*.

Exercises.

EXERCISE 1.3.i. What is a functor between groups, regarded as one-object categories?

EXERCISE 1.3.ii. What is a functor between preorders, regarded as categories?

EXERCISE 1.3.iii. Find an example to show that the objects and morphisms in the image of a functor $F: C \to D$ do not necessarily define a subcategory of D.

EXERCISE 1.3.iv. Verify that the constructions introduced in Definition 1.3.11 are functorial.

EXERCISE 1.3.v. What is the difference between a functor $C^{op} \to D$ and a functor $C \to D^{op}$? What is the difference between a functor $C \to D$ and a functor $C^{op} \to D^{op}$?

EXERCISE 1.3.vi. Given functors $F: D \to C$ and $G: E \to C$, show that there is a category, called the **comma category** $F \downarrow G$, which has

- as objects, triples $(d \in D, e \in E, f: Fd \to Ge \in C)$, and
- as morphisms $(d, e, f) \to (d', e', f')$, a pair of morphisms $(h: d \to d', k: e \to e')$ so that the square

$$\begin{array}{ccc} Fd & \xrightarrow{f} & Ge \\ {\scriptstyle Fh}\downarrow & & \downarrow{\scriptstyle Gk} \\ Fd' & \xrightarrow[f']{} & Ge' \end{array}$$

commutes in C, i.e., so that $f' \cdot Fh = Gk \cdot f$.

Define a pair of projection functors dom: $F \downarrow G \to D$ and cod: $F \downarrow G \to E$.

EXERCISE 1.3.vii. Define functors to construct the slice categories c/C and C/c of Exercise 1.1.iii as special cases of comma categories constructed in Exercise 1.3.vi. What are the projection functors?

EXERCISE 1.3.viii. Lemma 1.3.8 shows that functors preserve isomorphisms. Find an example to demonstrate that functors need not **reflect isomorphisms**: that is, find a functor $F: C \to D$ and a morphism f in C so that Ff is an isomorphism in D but f is not an isomorphism in C.

EXERCISE 1.3.ix. For any group G, we may define other groups:

- the **center** $Z(G) = \{h \in G \mid hg = gh \, \forall g \in G\}$, a subgroup of G,
- the **commutator subgroup** $C(G)$, the subgroup of G generated by elements $ghg^{-1}h^{-1}$ for any $g, h \in G$, and
- the **automorphism group** Aut(G), the group of isomorphisms $\phi: G \to G$ in Group.

Trivially, all three constructions define a functor from the discrete category of groups (with only identity morphisms) to Group. Are these constructions functorial in

- the isomorphisms of groups? That is, do they extend to functors Group$_{iso}$ → Group?

- the epimorphisms of groups[29]? That is, do they extend to functors $\mathsf{Group}_{epi} \to \mathsf{Group}$?
- all homomorphisms of groups? That is, do they extend to functors $\mathsf{Group} \to \mathsf{Group}$?

EXERCISE 1.3.x. Show that the construction of the set of conjugacy classes of elements of a group is functorial, defining a functor Conj: $\mathsf{Group} \to \mathsf{Set}$. Conclude that any pair of groups whose sets of conjugacy classes of elements have differing cardinalities cannot be isomorphic.

1.4. Naturality

> It is not too misleading, at least historically, to say that categories are what one must define in order to define functors, and that functors are what one must define in order to define natural transformations.

Peter Freyd, *Abelian categories* [**Fre03**]

Any finite-dimensional $\Bbbk$-vector space V is isomorphic to its **linear dual**, the vector space $V^* = \mathrm{Hom}(V, \Bbbk)$ of linear maps $V \to \Bbbk$, because these vector spaces have the same dimension. This can be proven through the construction of an explicit **dual basis**: choose a basis $e_1, \ldots, e_n$ for V and then define $e_1^*, \ldots, e_n^* \in V^*$ by

$$e_i^*(e_j) = \begin{cases} 1 & i = j \\ 0 & i \neq j. \end{cases}$$

The collection $e_1^*, \ldots, e_n^*$ defines a basis for V^* and the map $e_i \mapsto e_i^*$ extends by linearity to define an isomorphism $V \cong V^*$.

Now consider a related construction of the **double dual** $V^{**} = \mathrm{Hom}(\mathrm{Hom}(V, \Bbbk), \Bbbk)$ of V. If V is finite dimensional, then the isomorphism $V \cong V^*$ is carried by the dual vector space functor $(-)^*: \mathsf{Vect}_{\Bbbk}^{op} \to \mathsf{Vect}_{\Bbbk}$ to an isomorphism $V^* \cong V^{**}$. The composite isomorphism $V \cong V^{**}$ sends the basis $e_1, \ldots, e_n$ to the dual dual basis $e_1^{**}, \ldots, e_n^{**}$.

As it turns out, this isomorphism has a simpler description. For any $v \in V$, the "evaluation function"

$$f \mapsto f(v): V^* \xrightarrow{\mathrm{ev}_v} \Bbbk$$

defines a linear functional on V^*. It turns out the assignment $v \mapsto \mathrm{ev}_v$ defines a linear isomorphism $V \cong V^{**}$, this time requiring no "unnatural" choice of basis.[30]

What distinguishes the isomorphism between a finite-dimensional vector space and its double dual from the isomorphism between a finite-dimensional vector space and its single dual is that the former assembles into the components of a *natural transformation* in the sense that we now introduce.

DEFINITION 1.4.1. Given categories C and D and functors $F, G: \mathsf{C} \rightrightarrows \mathsf{D}$, a **natural transformation** $\alpha: F \Rightarrow G$ consists of:

- an arrow $\alpha_c: Fc \to Gc$ in D for each object $c \in \mathsf{C}$, the collection of which define the **components** of the natural transformation,

---

[29]A non-trivial theorem demonstrates that a homomorphism $\phi: G \to H$ is an epimorphism in Group if and only if its underlying function is surjective; see [**Lin70**].

[30]In fact, $e_i^{**}(e_j^*) = e_j^*(e_i) = \mathrm{ev}_{e_i}(e_j^*)$, and so the two isomorphisms $V \cong V^{**}$ are the same—it is only our description that has improved.

so that, for any morphism $f\colon c \to c'$ in C, the following square of morphisms in D

(1.4.2)
$$
\begin{array}{ccc}
Fc & \xrightarrow{\ \alpha_c\ } & Gc \\
{\scriptstyle Ff}\downarrow & & \downarrow{\scriptstyle Gf} \\
Fc' & \xrightarrow[\ \alpha_{c'}\ ]{} & Gc'
\end{array}
$$

commutes, i.e., has a common composite $Fc \to Gc'$ in D.

A **natural isomorphism** is a natural transformation $\alpha\colon F \Rightarrow G$ in which every component α_c is an isomorphism. In this case, the natural isomorphism may be depicted as $\alpha\colon F \cong G$.

In practice, it is usually most elegant to define a natural transformation by saying that "the arrows X are natural," which means that the collection of arrows defines the components of a natural transformation, leaving implicit the correct choices of domain and codomain functors, and source and target categories. Here X should be a collection of morphisms in a clearly identifiable (target) category, whose domains and codomains are defined using a common "variable" (an object of the source category). If this variable is c, one might say "the arrows X are natural in c" to emphasize the domain object whose component is being described. However, the totality of the data of the source and target categories, the parallel pair of functors, and the components should always be considered part of the natural transformation. The naturality condition (1.4.2) cannot be stated precisely with any less: it refers to every object and every morphism in the domain category and is described using the images in the codomain category under the action of both functors. The "boundary data" needed to define a natural transformation α is often displayed in a globular diagram:

The globular depiction of a natural transformation makes the notions of composable natural transformations that are introduced in §1.7 particularly intuitive.

EXAMPLE 1.4.3.

(i) For vector spaces of any dimension, the map $\mathrm{ev}\colon V \to V^{**}$ that sends $v \in V$ to the linear function $\mathrm{ev}_v\colon V^* \to \Bbbk$ defines the components of a natural transformation from the identity endofunctor on $\mathsf{Vect}_\Bbbk$ to the double dual functor. To check that the naturality square

$$
\begin{array}{ccc}
V & \xrightarrow{\ \mathrm{ev}\ } & V^{**} \\
{\scriptstyle \phi}\downarrow & & \downarrow{\scriptstyle \phi^{**}} \\
W & \xrightarrow[\ \mathrm{ev}\ ]{} & W^{**}
\end{array}
$$

commutes for any linear map $\phi\colon V \to W$, it suffices to consider the image of a generic vector $v \in V$. By definition, $\mathrm{ev}_{\phi v}\colon W^* \to \Bbbk$ carries a functional $f\colon W \to \Bbbk$ to $f(\phi v)$. Recalling the definition of the action of the dual functor of Example 1.3.7(ii) on morphisms, we see that $\phi^{**}(\mathrm{ev}_v)\colon W^* \to \Bbbk$ carries a functional $f\colon W \to \Bbbk$ to $f\phi(v)$, which amounts to the same thing.

(ii) By contrast, the identity functor and the single dual functor on finite-dimensional vector spaces are not naturally isomorphic. One technical obstruction is somewhat beside the point: the identity functor is covariant while the dual functor is con-travariant.[31] More significant is the essential failure of naturality. The isomorphisms $V \cong V^*$ that can be defined whenever V is finite dimensional require a choice of basis, which is preserved by essentially no linear maps, indeed by no non-identity linear endomorphism.[32]

(iii) There is a natural transformation $\eta \colon 1_{\mathsf{Set}} \Rightarrow P$ from the identity to the covariant power set functor whose components $\eta_A \colon A \to PA$ are the functions that carry $a \in A$ to the singleton subset $\{a\} \in PA$.

(iv) For G a group, Example 1.3.9 shows that a functor $X \colon \mathsf{B}G \to \mathsf{C}$ corresponds to an object $X \in \mathsf{C}$ equipped with a left action of G, which suggests a question: What is a natural transformation between a pair $X, Y \colon \mathsf{B}G \rightrightarrows \mathsf{C}$ of such functors? Because the category $\mathsf{B}G$ has only one object, the data of $\alpha \colon X \Rightarrow Y$ consists of a single morphism $\alpha \colon X \to Y$ in C that is G-**equivariant**, meaning that for each $g \in G$, the diagram

$$
\begin{array}{ccc}
X & \xrightarrow{\ \alpha\ } & Y \\
{\scriptstyle g_*}\big\downarrow & & \big\downarrow{\scriptstyle g_*} \\
X & \xrightarrow[\ \alpha\ ]{} & Y
\end{array}
$$

commutes.

(v) The open and closed subset functors described in Example 1.3.7(iii) are naturally isomorphic when they are regarded as functors $O, C \colon \mathsf{Top}^{\mathrm{op}} \rightrightarrows \mathsf{Set}$ valued in the category of sets. The components $O(X) \cong C(X)$ of the natural isomorphism are defined by taking an open subset of X to its complement, which is closed. Naturality asserts that the process of forming complements commutes with the operation of taking preimages.

(vi) The construction of the opposite group described in Example 1.2.2(iii) defines a (covariant!) endofunctor $(-)^{\mathrm{op}} \colon \mathsf{Group} \to \mathsf{Group}$ of the category of groups; a homomorphism $\phi \colon G \to H$ induces a homomorphism $\phi^{\mathrm{op}} \colon G^{\mathrm{op}} \to H^{\mathrm{op}}$ defined by $\phi^{\mathrm{op}}(g) = \phi(g)$. This functor is naturally isomorphic to the identity. Define $\eta_G \colon G \to G^{\mathrm{op}}$ to be the homomorphism that sends $g \in G$ to its inverse $g^{-1} \in G^{\mathrm{op}}$; this mapping does not define an automorphism of G, because it fails to commute with the group multiplication, but it does define a homomorphism $G \to G^{\mathrm{op}}$. Now given any homomorphism $\phi \colon G \to H$, the diagram

$$
\begin{array}{ccc}
G & \xrightarrow{\ \eta_G\ } & G^{\mathrm{op}} \\
{\scriptstyle \phi}\big\downarrow & & \big\downarrow{\scriptstyle \phi^{\mathrm{op}}} \\
H & \xrightarrow[\ \eta_H\ ]{} & H^{\mathrm{op}}
\end{array}
$$

commutes because $\phi^{\mathrm{op}}(g^{-1}) = \phi(g^{-1}) = \phi(g)^{-1}$.

(vii) Define an endofunctor of $\mathsf{Vect}_{\Bbbk}$ by $V \mapsto V \otimes V$. There is a natural transformation from the identity functor to this endofunctor whose components are the zero maps, but this is the only such natural transformation: there is no basis-independent way to define a linear map $V \to V \otimes V$. The same result is true for the category of Hilbert

---

[31]A more flexible notion of *extranatural transformation* can accommodate functors with conflicting variance [**ML98a**, IX.4]; see Exercise 1.4.vi.

[32]A proof that there exists no extranatural isomorphism between the identity and dual functors on the categories of finite-dimensional vector spaces is given in [**EM45**, p. 234].

spaces and linear operators between them, in which context it is related to the "no cloning theorem" in quantum physics.[33]

Another familiar isomorphism that is not natural arises in the classification of finitely generated abelian groups, objects of a category $\mathsf{Ab}_{\mathrm{fg}}$. Let TA denote the **torsion subgroup** of an abelian group A, the subgroup of elements with finite order. In classifying finitely generated groups, one proves that every finitely generated abelian group A is isomorphic to the direct sum $TA \oplus (A/TA)$, the summand A/TA being the **torsion-free** part of A. However, these isomorphisms are not natural, as we now demonstrate.

PROPOSITION 1.4.4. *The isomorphisms $A \cong TA \oplus (A/TA)$ are not natural[34] in $A \in \mathsf{Ab}_{\mathrm{fg}}$.*

PROOF. Suppose the isomorphisms $A \cong TA \oplus (A/TA)$ were natural in A. Then the composite

$$(1.4.5) \qquad\qquad A \twoheadrightarrow A/TA \rightarrowtail TA \oplus (A/TA) \cong A$$

of the canonical quotient map, the inclusion into the direct sum, and the hypothesized natural isomorphism would define a natural endomorphism of the identity functor on $\mathsf{Ab}_{\mathrm{fg}}$. We shall see that this is impossible.

To derive the contradiction, we first show that every natural endomorphism of the identity functor on $\mathsf{Ab}_{\mathrm{fg}}$ is multiplication by some $n \in \mathbb{Z}$. Clearly the component of $\alpha\colon 1_{\mathsf{Ab}_{\mathrm{fg}}} \Rightarrow 1_{\mathsf{Ab}_{\mathrm{fg}}}$ at $\mathbb{Z}$ has this description for some n. Now observe that homomorphisms $\mathbb{Z} \xrightarrow{a} A$ correspond bijectively to elements $a \in A$, choosing a to be the image of $1 \in \mathbb{Z}$. Thus, commutativity of

$$
\begin{array}{ccc}
\mathbb{Z} & \xrightarrow{\;\alpha_{\mathbb{Z}} = n \cdot -\;} & \mathbb{Z} \\
{\scriptstyle a}\big\downarrow & & \big\downarrow{\scriptstyle a} \\
A & \xrightarrow[\;\alpha_A\;]{} & A
\end{array}
$$

forces us to define $\alpha_A(a) = n \cdot a$.

In the case where α is the natural transformation defined by (1.4.5), by examining the component at $A = \mathbb{Z}$, we can see that $n \neq 0$. Finally, consider $A = \mathbb{Z}/2n\mathbb{Z}$. This group is torsion, so any map, such as $\alpha_{\mathbb{Z}/2n\mathbb{Z}}$, which factors through the quotient by its torsion subgroup is zero. But $n \neq 0 \in \mathbb{Z}/2n\mathbb{Z}$, a contradiction. □

EXAMPLE 1.4.6. The Riesz representation theorem can be expressed as a natural isomorphism of functors from the category cHaus of compact Hausdorff spaces and continuous maps to the category Ban of real Banach spaces and continuous linear maps. Let $\Sigma\colon \mathsf{cHaus} \to \mathsf{Ban}$ be the functor that carries a compact Hausdorff space X to the Banach space $\Sigma(X)$ of signed Baire measures on X and sends a continuous map $f\colon X \to Y$ to the map $\mu \mapsto \mu \circ f^{-1}\colon \Sigma(X) \to \Sigma(Y)$. Let $C^*\colon \mathsf{cHaus} \to \mathsf{Ban}$ be the functor that carries X to the linear dual $C(X)^*$ of the Banach space $C(X)$ of continuous real-valued functions on X.

Now for each $\mu \in \Sigma(X)$, there is a linear functional $\phi_\mu\colon C(X) \to \mathbb{R}$ defined by

$$\phi_\mu(g) := \int_X g \, d\mu, \qquad g \in C(X).$$

---

[33]The states in a quantum mechanical system are modeled by vectors in a Hilbert space and the observables are operators on that space. See [**Bae06**] for more.

[34]Any finitely generated abelian group A has a short exact sequence $0 \to TA \to A \to A/TA \to 0$. Proposition 1.4.4 asserts that there is no natural splitting.

For each $\mu \in \Sigma(X)$, $f: X \to Y$, and $h \in C(Y)$ the equation

$$\int_X hf \, d\mu = \int_Y h \, d(\mu \circ f^{-1})$$

says that the assignment $\mu \mapsto \phi_\mu$ defines the components of a natural transformation $\eta: \Sigma \Rightarrow C^*$. The **Riesz representation theorem** asserts that this natural transformation is a natural isomorphism; see [**Har83**].

EXAMPLE 1.4.7. Consider morphisms $f: w \to x$ and $h: y \to z$ in a locally small category C. Post-composition by h and pre-composition by f define functions between hom-sets

(1.4.8)
$$
\begin{array}{ccc}
C(x, y) & \xrightarrow{\;h\cdot-\;} & C(x, z) \\
{\scriptstyle -\cdot f}\big\downarrow & & \big\downarrow{\scriptstyle -\cdot f} \\
C(w, y) & \xrightarrow[\;h\cdot-\;]{} & C(w, z)
\end{array}
$$

In Definition 1.3.13 and elsewhere, $h \cdot -$ was denoted by h_* and $- \cdot f$ was denoted by f^*, but we find this less-concise notation to be move evocative here. Associativity of composition implies that this diagram commutes: for any $g: x \to y$, the common image is $hgf: w \to z$.

Interpreting the vertical arrows as the images of f under the actions of the functors $C(-, y)$ and $C(-, z)$, the square (1.4.8) demonstrates that there is a natural transformation

$$h_*: C(-, y) \Rightarrow C(-, z)$$

whose components are defined by post-composition with $h: y \to z$. Flipping perspectives and interpreting the horizontal arrows as the images of h under the actions of the functors $C(x, -)$ and $C(w, -)$, the square (1.4.8) demonstrates that there is a natural transformation

$$f^*: C(x, -) \Rightarrow C(w, -)$$

whose components are defined by pre-composition with $f: w \to x$.

A final example describes the natural isomorphisms that supply proofs of the fundamental laws of elementary arithmetic.

EXAMPLE 1.4.9 (a categorification of the natural numbers). For sets A and B, let $A \times B$ denote their cartesian product, let $A + B$ denote their disjoint union, and let A^B denote the set of functions from B to A. These constructions are related by natural isomorphisms

(1.4.10)
$$A \times (B + C) \cong (A \times B) + (A \times C) \qquad\qquad (A \times B)^C \cong A^C \times B^C$$
$$A^{B+C} \cong A^B \times A^C \qquad\qquad\qquad\qquad (A^B)^C \cong A^{B \times C}$$

In the first instance, the isomorphism defines the components of a natural transformation between a pair of functors $\mathsf{Set} \times \mathsf{Set} \times \mathsf{Set} \to \mathsf{Set}$. For the others, the variance in the variables appearing as "exponents" is contravariant. This is because the assignment $(B, A) \mapsto A^B$ defines a functor $\mathsf{Set}^{\mathrm{op}} \times \mathsf{Set} \to \mathsf{Set}$, namely the two-sided represented functor introduced in Definition 1.3.13.

The displayed natural isomorphisms restrict to the category $\mathsf{Fin}_{\mathrm{iso}}$ of finite sets and bijections, a category which serves as the domain for the cardinality functor $|-|: \mathsf{Fin}_{\mathrm{iso}} \to \mathbb{N}$, whose codomain is the discrete category of natural numbers.[35] Writing $a = |A|$, $b = |B|$,

---

[35]Mathematical invariants often take the form of a functor from a groupoid to a discrete category.

and $c = |C|$, the cardinality functor carries these natural isomorphisms to the equations

$$a \times (b + c) = (a \times b) + (a \times c) \qquad\qquad (a \times b)^c = a^c \times b^c$$
$$a^{b+c} = a^b \times a^c \qquad\qquad (a^b)^c = a^{(b \times c)}$$

through a process called **decategorification**. Reversing directions, $\mathsf{Fin}_{\mathrm{iso}}$ is a **categorification** of the natural numbers, which reveals that the familiar laws of arithmetic follow from more fundamental natural isomorphisms between various constructions on sets. A slick proof of each of the displayed natural isomorphisms (1.4.10) appears in Corollary 4.5.6.

Exercises.

EXERCISE 1.4.i. Suppose $\alpha\colon F \Rightarrow G$ is a natural isomorphism. Show that the inverses of the component morphisms define the components of a natural isomorphism $\alpha^{-1}\colon G \Rightarrow F$.

EXERCISE 1.4.ii. What is a natural transformation between a parallel pair of functors between groups, regarded as one-object categories?

EXERCISE 1.4.iii. What is a natural transformation between a parallel pair of functors between preorders, regarded as categories?

EXERCISE 1.4.iv. In the notation of Example 1.4.7, prove that distinct parallel morphisms $f, g\colon c \rightrightarrows d$ define distinct natural transformations

$$f_*, g_*\colon \mathsf{C}(-, c) \Rightarrow \mathsf{C}(-, d) \quad \text{and} \quad f^*, g^*\colon \mathsf{C}(d, -) \Rightarrow \mathsf{C}(c, -)$$

by post- and pre-composition.

EXERCISE 1.4.v. Recall the construction of the comma category for any pair of functors $F\colon \mathsf{D} \to \mathsf{C}$ and $G\colon \mathsf{E} \to \mathsf{C}$ described in Exercise 1.3.vi. From this data, construct a canonical natural transformation $\alpha\colon F\,\mathrm{dom} \Rightarrow G\,\mathrm{cod}$ between the functors that form the boundary of the square

$$
\begin{array}{ccc}
F\downarrow G & \xrightarrow{\ \mathrm{cod}\ } & \mathsf{E} \\
{\scriptstyle\mathrm{dom}}\big\downarrow & {\scriptstyle\alpha}\!\nearrow & \big\downarrow {\scriptstyle G} \\
\mathsf{D} & \xrightarrow[\ F\ ]{} & \mathsf{C}
\end{array}
$$

EXERCISE 1.4.vi. Given a pair of functors $F\colon \mathsf{A} \times \mathsf{B} \times \mathsf{B}^{\mathrm{op}} \to \mathsf{D}$ and $G\colon \mathsf{A} \times \mathsf{C} \times \mathsf{C}^{\mathrm{op}} \to \mathsf{D}$, a family of morphisms

$$\alpha_{a,b,c}\colon F(a, b, b) \to G(a, c, c)$$

in D defines the components of an **extranatural transformation** $\alpha\colon F \Rightarrow G$ if for any $f\colon a \to a'$, $g\colon b \to b'$, and $h\colon c \to c'$ the following diagrams commute in D:

$$
\begin{array}{ccc}
F(a,b,b) \xrightarrow{\alpha_{a,b,c}} G(a,c,c) & F(a,b,b') \xrightarrow{F(1_a,1_b,g)} F(a,b,b) & F(a,b,b) \xrightarrow{\alpha_{a,b,c'}} G(a,c',c') \\
{\scriptstyle F(f,1_b,1_b)}\big\downarrow \qquad \big\downarrow {\scriptstyle G(f,1_c,1_c)} & {\scriptstyle F(1_a,g,1_{b'})}\big\downarrow \qquad \big\downarrow {\scriptstyle \alpha_{a,b,c}} & \big\downarrow {\scriptstyle G(1_a,1_{c'},h)} \\
F(a',b,b) \xrightarrow[\alpha_{a',b,c}]{} G(a',c,c) & F(a,b',b') \xrightarrow[\alpha_{a,b',c}]{} G(a,c,c) & G(a,c,c) \xrightarrow[G(1_a,h,1_c)]{} G(a,c',c)
\end{array}
$$

The left-hand square asserts that the components $\alpha_{-,b,c}\colon F(-, b, b) \Rightarrow G(-, c, c)$ define a natural transformation in a for each $b \in \mathsf{B}$ and $c \in \mathsf{C}$. The remaining squares assert that the components $\alpha_{a,-,c}\colon F(a, -, -) \Rightarrow G(a, c, c)$ and $\alpha_{a,b,-}\colon F(a, b, b) \Rightarrow G(a, -, -)$ define transformations that are respectively extranatural in b and in c. Explain why the functors F and G must have a common target category for this definition to make sense.

1.5. Equivalence of categories

> ... la mathématique est l'art de donner le même
> nom à des choses différentes.
> ... mathematics is the art of giving the same
> name to different things.
>
> Henri Poincaré, "L'avenir des mathématiques"
> **[Poi08]**

There is an analogy between the notion of a natural transformation between a parallel pair of functors and the notion of a homotopy between a parallel pair of continuous functions with one important difference: natural transformations are not generally invertible.[36] As in Example 1.1.4(iv), let $\mathbb{1}$ denote the discrete category with a single object and let 2 denote the category with two objects $0, 1 \in 2$ and a single non-identity arrow $0 \to 1$. There are two evident functors $i_0, i_1 \colon \mathbb{1} \rightrightarrows 2$ whose subscripts designate the objects in their image.

LEMMA 1.5.1. *Fixing a parallel pair of functors* $F, G \colon \mathsf{C} \rightrightarrows \mathsf{D}$, *natural transformations* $\alpha \colon F \Rightarrow G$ *correspond bijectively to functors* $H \colon \mathsf{C} \times 2 \to \mathsf{D}$ *such that H restricts along i_0 and i_1 to the functors F and G, i.e., so that*

(1.5.2)

$$
\begin{array}{ccc}
\mathsf{C} \xrightarrow{\ i_0\ } & \mathsf{C} \times 2 & \xleftarrow{\ i_1\ } \mathsf{C} \\
& \downarrow{\scriptstyle H} & \\
F \searrow & \mathsf{D} & \swarrow G
\end{array}
$$

commutes.

Here i_0 denotes the functor defined on objects by $c \mapsto (c, 0)$; it may be regarded as the product of the identity functor on C with the functor i_0.

PROOF. Exercise 1.5.i.   □

For instance, if $\mathsf{C} = 2$, each functor $F, G \colon 2 \rightrightarrows \mathsf{D}$ picks out an arrow of D, which we also denote by F and G. The directed graph underlying the category 2×2 is

(1.5.3)

together with four identity endoarrows not depicted here; the diagonal serves as the common composite of the edges of the square. The functor H necessarily maps the top and bottom arrows of (1.5.3) to F and G, respectively. The vertical arrows define the components α_0 and α_1 of the natural transformation, and the diagonal arrow witnesses that the square formed by these four morphisms in D commutes.

If, in (1.5.2), the category 2 were replaced by the category $\mathbb{I}$ with two objects and a single arrow in each hom-set, necessarily an isomorphism, then "homotopies" with this interval correspond bijectively to natural *isomorphisms*. The category 2 defines the **walking arrow** or **free arrow**, while $\mathbb{I}$ defines the **walking isomorphism** or **free isomorphism**, in a sense that is explained in Examples 2.1.5(x) and (xi).

---

[36]A natural transformation is invertible if and only if each of its constituent arrows is an isomorphism, in which case the pointwise inverses assemble into a natural transformation by Exercise 1.4.i.

Natural isomorphisms are used to define the notion of *equivalence* of categories. An equivalence of categories is precisely a "homotopy equivalence" where the notion of homotopy is defined using the category $\mathbb{I}$.

DEFINITION 1.5.4. An **equivalence of categories** consists of functors $F\colon \mathsf{C} \leftrightarrows \mathsf{D}\colon G$ together with natural isomorphisms $\eta\colon 1_\mathsf{C} \cong GF$, $\epsilon\colon FG \cong 1_\mathsf{D}$.[37] Categories C and D are **equivalent**, written $\mathsf{C} \simeq \mathsf{D}$, if there exists an equivalence between them.

Unsurprisingly:

LEMMA 1.5.5. *The notion of equivalence of categories defines an equivalence relation. In particular, if $\mathsf{C} \simeq \mathsf{D}$ and $\mathsf{D} \simeq \mathsf{E}$, then $\mathsf{C} \simeq \mathsf{E}$.*

PROOF. Exercise 1.5.vi.   □

EXAMPLE 1.5.6. The functors $(-)_+\colon \mathsf{Set}^\partial \to \mathsf{Set}_*$ and $U\colon \mathsf{Set}_* \to \mathsf{Set}^\partial$ introduced in §1.3 define an equivalence of categories between the category of pointed sets and the category of sets and partial functions. The composite $U(-)_+$ is the identity on Set^∂, so one of the required natural isomorphisms is the identity. There is a natural isomorphism $\eta\colon 1_{\mathsf{Set}_*} \cong (U-)_+$ whose components

$$\eta_{(X,x)}\colon (X, x) \to (X\backslash\{x\} \cup \{X\backslash\{x\}\}, X\backslash\{x\})$$

are defined to be the based functions that act as the identity on $X\backslash\{x\}$.

Consider the categories $\mathsf{Mat}_\Bbbk$ and $\mathsf{Vect}_\Bbbk^{\mathrm{fd}}$ of $\Bbbk$-matrices and finite-dimensional non-zero $\Bbbk$-vector spaces together with an intermediate category $\mathsf{Vect}_\Bbbk^{\mathrm{basis}}$ whose objects are finite-dimensional vector spaces with chosen basis and whose morphisms are arbitrary (not necessarily basis-preserving) linear maps. These categories are related by the displayed sequence of functors:

$$\mathsf{Mat}_\Bbbk \underset{H}{\overset{\Bbbk^{(-)}}{\rightleftarrows}} \mathsf{Vect}_\Bbbk^{\mathrm{basis}} \underset{C}{\overset{U}{\rightleftarrows}} \mathsf{Vect}_\Bbbk^{\mathrm{fd}}$$

Here $U\colon \mathsf{Vect}_\Bbbk^{\mathrm{basis}} \to \mathsf{Vect}_\Bbbk^{\mathrm{fd}}$ is the forgetful functor. The functor $\Bbbk^{(-)}\colon \mathsf{Mat}_\Bbbk \to \mathsf{Vect}_\Bbbk^{\mathrm{basis}}$ sends n to the vector space $\Bbbk^n$, equipped with the standard basis. An $m{\times}n$-matrix, interpreted with respect to the standard bases on $\Bbbk^n$ and $\Bbbk^m$, defines a linear map $\Bbbk^n \to \Bbbk^m$ and this assignment is functorial. The functor H carries a vector space to its dimension and a linear map $\phi\colon V \to W$ to the matrix expressing the action of ϕ on the chosen basis of V using the chosen basis of W. The functor C is defined by choosing a basis for each vector space.

Our aim is to show that these functors display equivalences of categories

$$\mathsf{Mat}_\Bbbk \simeq \mathsf{Vect}_\Bbbk^{\mathrm{basis}} \simeq \mathsf{Vect}_\Bbbk^{\mathrm{fd}}.$$

The composite equivalence $\mathsf{Mat}_\Bbbk \simeq \mathsf{Vect}_\Bbbk^{\mathrm{fd}}$ expresses an equivalence between concrete and abstract presentations of linear algebra. A direct proof of these equivalences, by defining suitable natural isomorphisms, is not difficult, but we prefer to give an indirect proof via a useful general theorem characterizing those functors forming part of an equivalence of categories. Its statement requires a few definitions:

DEFINITION 1.5.7. A functor $F\colon \mathsf{C} \to \mathsf{D}$ is

- **full** if for each $x, y \in \mathsf{C}$, the map $\mathsf{C}(x, y) \to \mathsf{D}(Fx, Fy)$ is surjective;
- **faithful** if for each $x, y \in \mathsf{C}$, the map $\mathsf{C}(x, y) \to \mathsf{D}(Fx, Fy)$ is injective;

---

[37]The notion of equivalence of categories was introduced by Grothendieck in the form of what we would now call an *adjoint equivalence*; this definition appears in Proposition 4.4.5. This explains the directions we have adopted for the natural isomorphisms η and ϵ, which are otherwise immaterial (see Exercise 1.4.i).

- and **essentially surjective on objects** if for every object $d \in \mathsf{D}$ there is some $c \in \mathsf{C}$ such that d is isomorphic to Fc.

REMARK 1.5.8. Fullness and faithfulness are *local* conditions; a *global* condition, by contrast, applies "everywhere." A faithful functor need not be injective on morphisms; neither must a full functor be surjective on morphisms. A faithful functor that is injective on objects is called an **embedding** and identifies the domain category as a subcategory of the codomain; in this case, faithfulness implies that the functor is (globally) injective on arrows. A full and faithful functor, called **fully faithful** for short, that is injective-on-objects defines a **full embedding** of the domain category into the codomain category. The domain then defines a **full subcategory** of the codomain.

THEOREM 1.5.9 (characterizing equivalences of categories). *A functor defining an equivalence of categories is full, faithful, and essentially surjective on objects. Assuming the axiom of choice, any functor with these properties defines an equivalence of categories.*

The proof of Theorem 1.5.9 makes repeated use of the following elementary lemma.

LEMMA 1.5.10. *Any morphism $f\colon a \to b$ and fixed isomorphisms $a \cong a'$ and $b \cong b'$ determine a unique morphism $f'\colon a' \to b'$ so that any of—or, equivalently, all of—the following four diagrams commute:*

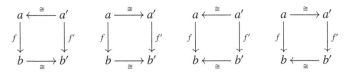

PROOF. The left-hand diagram defines f'. The commutativity of the remaining diagrams is left as Exercise 1.5.iii.   □

PROOF[38] OF THEOREM 1.5.9. First suppose that $F\colon \mathsf{C} \to \mathsf{D}$, $G\colon \mathsf{D} \to \mathsf{C}$, $\eta\colon 1_{\mathsf{C}} \cong GF$, and $\epsilon\colon FG \cong 1_{\mathsf{D}}$ define an equivalence of categories. For any $d \in \mathsf{D}$, the component of the natural isomorphism $\epsilon_d\colon FGd \cong d$ demonstrates that F is essentially surjective. Consider a parallel pair $f, g\colon c \rightrightarrows c'$ in C. If $Ff = Fg$, then both f and g define an arrow $c \to c'$ making the diagram

$$
\begin{array}{ccc}
c & \xrightarrow[\cong]{\eta_c} & GFc \\
{\scriptstyle f \text{ or } g}\downarrow & & \downarrow{\scriptstyle GFf=GFg} \\
c' & \xrightarrow[\cong]{\eta_{c'}} & GFc'
\end{array}
$$

that expresses the naturality of η commute. Lemma 1.5.10 implies that there is a unique arrow $c \to c'$ with this property, whence $f = g$. Thus, F is faithful, and by symmetry, so is G. Given $k\colon Fc \to Fc'$, by Lemma 1.5.10, Gk and the isomorphisms η_c and $\eta_{c'}$ define a unique $h\colon c \to c'$ for which both Gk and GFh make the diagram

$$
\begin{array}{ccc}
c & \xrightarrow[\cong]{\eta_c} & GFc \\
{\scriptstyle h}\downarrow & & \downarrow{\scriptstyle Gk \text{ or } GFh} \\
c' & \xrightarrow[\cong]{\eta_{c'}} & GFc'
\end{array}
$$

---

[38]The reader is strongly encouraged to stop reading here and attempt to prove this result on their own. Indeed, in the first iteration of the course that produced these lecture notes, this proof was left to the homework, without even the time-saving suggestion of Lemma 1.5.10.

commute. By Lemma 1.5.10 again, $GFh = Gk$, whence $Fh = k$ by faithfulness of G. Thus, F is full, faithful, and essentially surjective.

For the converse, suppose now that $F \colon \mathsf{C} \to \mathsf{D}$ is full, faithful, and essentially surjective on objects. Using essential surjectivity and the axiom of choice, choose, for each $d \in \mathsf{D}$, an object $Gd \in \mathsf{C}$ and an isomorphism $\epsilon_d \colon FGd \cong d$. For each $\ell \colon d \to d'$, Lemma 1.5.10 defines a unique morphism making the square

$$
\begin{array}{ccc}
FGd & \xrightarrow{\ \epsilon_d\ } & d \\
\big\downarrow & {\scriptstyle \cong} & \big\downarrow{\scriptstyle \ell} \\
FGd' & \xrightarrow[\ \epsilon_{d'}\ ]{\cong} & d'
\end{array}
$$

commute. Since F is fully faithful, there is a unique morphism $Gd \to Gd'$ with this image under F, which we define to be $G\ell$. This definition is arranged so that the chosen isomorphisms assemble into the components of a natural transformation $\epsilon \colon FG \Rightarrow 1_{\mathsf{D}}$. It remains to prove that the assignment of arrows $\ell \mapsto G\ell$ is functorial and to define the natural isomorphism $\eta \colon 1_{\mathsf{C}} \Rightarrow GF$.

Functoriality of G is another consequence of Lemma 1.5.10 and faithfulness of F. The morphisms $FG1_d$ and $F1_{Gd}$ both make

$$
\begin{array}{ccc}
FGd & \xrightarrow{\ \epsilon_d\ } & d \\
{\scriptstyle FG1_d \text{ or } F1_{Gd}}\big\downarrow & {\scriptstyle \cong} & \big\downarrow{\scriptstyle 1_d} \\
FGd & \xrightarrow[\ \epsilon_d\ ]{\cong} & d
\end{array}
$$

commute, whence $G1_d = 1_{Gd}$. Similarly, given $\ell' \colon d' \to d''$, both $F(G\ell' \cdot G\ell)$ and $FG(\ell'\ell)$ make

$$
\begin{array}{ccc}
FGd & \xrightarrow{\ \epsilon_d\ } & d \\
{\scriptstyle F(G\ell' \cdot G\ell) \text{ or } FG(\ell'\ell)}\big\downarrow & {\scriptstyle \cong} & \big\downarrow{\scriptstyle \ell'\ell} \\
FGd'' & \xrightarrow[\ \epsilon_{d''}\ ]{\cong} & d''
\end{array}
$$

commute, whence $G\ell' \cdot G\ell = G(\ell'\ell)$.

Finally, by full and faithfulness of F, we may define the isomorphisms $\eta_c \colon c \to GFc$ by specifying isomorphisms $F\eta_c \colon Fc \to FGFc$; see Exercise 1.5.iv. Define $F\eta_c$ to be ϵ_{Fc}^{-1}. For any $f \colon c \to c'$, the outer rectangle

$$
\begin{array}{ccccc}
Fc & \xrightarrow{\ F\eta_c\ } & FGFc & \xrightarrow{\ \epsilon_{Fc}\ } & Fc \\
{\scriptstyle Ff}\big\downarrow & & {\scriptstyle FGFf}\big\downarrow & & \big\downarrow{\scriptstyle Ff} \\
Fc' & \xrightarrow[\ F\eta_{c'}\ ]{} & FGFc' & \xrightarrow[\ \epsilon_{Fc'}\ ]{} & Fc'
\end{array}
$$

commutes, both composites being Ff. The right-hand square commutes by naturality of ϵ. Because $\epsilon_{Fc'}$ is an isomorphism, this implies that the left-hand square commutes; see Lemma 1.6.21. Faithfulness of F tells us that $\eta_{c'} \cdot f = GFf \cdot \eta_c$, i.e., that η is a natural transformation.   □

The proof of Theorem 1.5.9 is an example of a proof by "diagram chasing," a technique that is introduced more formally in §1.6. If some of the steps were hard to follow, we suggest having a second look after that section is absorbed.

COROLLARY 1.5.11 (an equivalence between abstract and concrete linear algebra). *For any field* $\Bbbk$, *the categories* $\mathsf{Mat}_\Bbbk$ *and* $\mathsf{Vect}_\Bbbk^{fd}$ *are equivalent.*

PROOF. Applying Theorem 1.5.9, it is easy to see that each of the functors

$$\mathsf{Mat}_\Bbbk \leftrightarrows \mathsf{Vect}_\Bbbk^{basis} \rightleftarrows \mathsf{Vect}_\Bbbk^{fd}$$

defines an equivalence of categories. For instance, the morphisms in the category $\mathsf{Vect}_\Bbbk^{basis}$ are defined so that $U\colon \mathsf{Vect}_\Bbbk^{basis} \to \mathsf{Vect}_\Bbbk^{fd}$ is fully faithful.   □

A category is **connected** if any pair of objects can be connected by a finite zig-zag of morphisms.

PROPOSITION 1.5.12. *Any connected groupoid is equivalent, as a category, to the automorphism group of any of its objects.*

PROOF. Choose any object g of a connected groupoid G and let $G = \mathsf{G}(g, g)$ denote its automorphism group. The inclusion $BG \hookrightarrow \mathsf{G}$ mapping the unique object of BG to $g \in \mathsf{G}$ is full and faithful, by definition, and essentially surjective, since G was assumed to be connected. Apply Theorem 1.5.9.   □

As a special case, we obtain the following result:

COROLLARY 1.5.13. *In a path-connected space X, any choice of basepoint $x \in X$ yields an isomorphic fundamental group $\pi_1(X, x)$.*

PROOF. Recall from Example 1.1.12(ii) that any space X has a fundamental groupoid $\Pi_1(X)$ whose objects are points in X and whose morphisms are endpoint-preserving homotopy classes of paths in X. Picking any point x, the group of automorphisms of the object $x \in \Pi_1(X)$ is exactly the fundamental group $\pi_1(X, x)$. Proposition 1.5.12 implies that any pair of automorphism groups are equivalent, as categories, to the fundamental groupoid

$$\pi_1(X, x) \overset{\sim}{\longrightarrow} \Pi_1(X) \overset{\sim}{\longleftarrow} \pi_1(X, x')$$

and thus to each other. An equivalence between 1-object categories is an isomorphism. Exercise 1.3.i reveals that an isomorphism of groups, regarded as 1-object categories, is exactly an isomorphism of groups in the usual sense (a bijective homomorphism). Thus, all of the fundamental groups defined by choosing a basepoint in a path-connected space are isomorphic.   □

REMARK 1.5.14. Frequently, one functor of an equivalence of categories can be defined canonically, while the inverse equivalence requires the axiom of choice. In the case of the equivalence between the fundamental group and fundamental groupoid of a path-connected space, a more precise statement is that one comparison equivalence is natural while the other, making use of the axiom of choice, is not. Write Top_*^{pc} for the category of path-connected based topological spaces. We regard the fundamental group π_1 and fundamental groupoid Π_1 as a parallel pair of functors:

$$\pi_1\colon \mathsf{Top}_*^{pc} \overset{\pi_1}{\longrightarrow} \mathsf{Group} \hookrightarrow \mathsf{Cat} \qquad \text{and} \qquad \Pi_1\colon \mathsf{Top}_*^{pc} \overset{U}{\longrightarrow} \mathsf{Top} \overset{\Pi_1}{\longrightarrow} \mathsf{Groupoid} \hookrightarrow \mathsf{Cat}.$$

The inclusion of the fundamental group into the fundamental groupoid defines a natural transformation $\pi_1 \Rightarrow \Pi_1$ such that each component $\pi_1(X, x) \to \Pi_1(X)$, itself a functor, is furthermore an equivalence of categories. The definition of the inverse equivalence $\Pi_1(X) \to \pi_1(X, x)$ requires the choice, for each point $p \in X$, of a path-connecting p to the basepoint x. These (path homotopy classes of) chosen paths need not be preserved by maps in $\mathsf{Top}_*^{\mathrm{pc}}$. Thus, the inverse equivalences $\Pi_1(X) \to \pi_1(X, x)$ do not assemble into a natural transformation.

The group of automorphisms of any object in a connected groupoid, considered in Proposition 1.5.12, is one example of a *skeleton* of a category.

DEFINITION 1.5.15. A category C is **skeletal** if it contains just one object in each isomorphism class. The **skeleton** skC of a category C is the unique (up to isomorphism) skeletal category that is equivalent to C.

REMARK 1.5.16. The category skC may be constructed by choosing one object in each isomorphism class in C and defining skC to be the full subcategory on this collection of objects. By Theorem 1.5.9, the inclusion skC $\hookrightarrow$ C defines an equivalence of categories. This construction, however, fails to define a functor sk(−): CAT → CAT because there is no reason that a functor $F\colon$ C → D would necessarily restrict to define a functor between the chosen skeletal subcategories. It is possible to choose a functor sk$F\colon$ skC → skD whose inclusion into D is naturally isomorphic to the restriction of F to skC, but these choices will not be strictly functorial.[39]

Note than an equivalence between skeletal categories is necessarily an isomorphism of categories. Thus, two categories are equivalent if and only if their skeletons are isomorphic. For this reason, we feel free to speak of *the* skeleton of a category, even though its construction is not canonical.

EXAMPLE 1.5.17.

(i) The skeleton of a connected groupoid is the group of automorphisms of any of its objects (see Proposition 1.5.12).
(ii) The skeleton of the category defined by a preorder, as described in Example 1.1.4(iii), is a poset.
(iii) The skeleton of the category $\mathsf{Vect}_{\Bbbk}^{\mathrm{fd}}$ is the category $\mathsf{Mat}_{\Bbbk}$.
(iv) The skeleton of the category $\mathsf{Fin}_{\mathrm{iso}}$ is the category whose objects are positive integers and with $\mathrm{Hom}(n, n) = \Sigma_n$, the group of permutations of n elements. The hom-sets between distinct natural numbers are all empty.

EXAMPLE 1.5.18 (a categorification of the orbit-stabilizer theorem). Let $X\colon \mathsf{B}G \to \mathsf{Set}$ be a left G-set. Its **translation groupoid** $\mathsf{T}_G X$ has elements of X as objects. A morphism $g\colon x \to y$ is an element $g \in G$ so that $g \cdot x = y$. The objects in the skeleton $\mathsf{sk}\mathsf{T}_G X$ are the connected components in the translation groupoid. These are precisely the **orbits** of the group action, which partition X in precisely this manner.

Consider $x \in X$ as a representative of its orbit O_x. Because the translation groupoid is equivalent to its skeleton, we must have

$$\mathrm{Hom}_{\mathsf{sk}\mathsf{T}_G X}(O_x, O_x) \cong \mathrm{Hom}_{\mathsf{T}_G X}(x, x) =: G_x,$$

the set of automorphisms of x. This group consists of precisely those $g \in G$ so that $g \cdot x = x$. In other words, the group $\mathrm{Hom}_{\mathsf{T}_G X}(x, x)$ is the **stabilizer** G_x of x with respect to the G-action. Note that this argument implies that any pair of elements in the same orbit must have isomorphic stabilizers. As is always the case for a skeletal groupoid, there are no morphisms between distinct objects. In summary, the skeleton of the translation groupoid, as a category, is the disjoint union of the stabilizer groups, indexed by the orbits of the action of G on X.

---

[39]There is, however, a *pseudofunctor* sk(−): CAT → CAT, but a precise definition of this concept is beyond the scope of this text.

The set of morphisms in the translation groupoid with domain x is isomorphic to G. This set may be expressed as a disjoint union of hom-sets $\mathrm{Hom}_{T_G X}(x, y)$, where y ranges over the orbit O_x. Each of these hom-sets is isomorphic to $\mathrm{Hom}_{T_G X}(x, x) = G_x$. In particular, $|G| = |O_x| \cdot |G_x|$, proving the **orbit-stabilizer theorem**.

A guiding principle in category theory is that categorically-defined concepts should be equivalence invariant. Some category theorists go so far as to call a definition "evil" if it is not invariant under equivalence of categories. The only evil definitions that have been introduced thus far are smallness and discreteness. A category is **essentially small** if it is equivalent to a small category or, equivalently, if its skeleton is a small category. A category is **essentially discrete** if it is equivalent to a discrete category.

The following constructions and definitions are equivalence invariant:

- If a category is locally small, any category equivalent to it is again locally small.
- If a category is a groupoid, any category equivalent to it is again a groupoid.
- If $\mathsf{C} \simeq \mathsf{D}$, then $\mathsf{C}^{\mathrm{op}} \simeq \mathsf{D}^{\mathrm{op}}$.
- The product of a pair of categories is equivalent to the product of any pair of equivalent categories.
- An arrow in C is an isomorphism if and only if its image under an equivalence $\mathsf{C} \xrightarrow{\simeq} \mathsf{D}$ is an isomorphism.

The last of these properties can be generalized. By Theorem 1.5.9, a full and faithful functor $F: \mathsf{C} \to \mathsf{D}$ defines an equivalence onto its **essential image**, the full subcategory of objects isomorphic to Fc for some $c \in \mathsf{C}$. Fully faithful functors have a useful property stated as Exercise 1.5.iv: if F is full and faithful and Fc and Fc' are isomorphic in D, then c and c' are isomorphic in C. We will introduce what are easily the most important fully faithful functors in category theory in Chapter 2: the covariant and contravariant Yoneda embeddings.

Exercises.

EXERCISE 1.5.i. Prove Lemma 1.5.1.

EXERCISE 1.5.ii. Segal defined a category Γ in [Seg74] as follows:

Γ is the category whose objects are all finite sets, and whose morphisms from S to T are the maps $\theta: S \to P(T)$ such that $\theta(\alpha)$ and $\theta(\beta)$ are disjoint when $\alpha \neq \beta$. The composite of $\theta: S \to P(T)$ and $\phi: T \to P(U)$ is $\psi: S \to P(U)$, where $\psi(\alpha) = \bigcup_{\beta \in \theta(\alpha)} \phi(\beta)$.

Prove that Γ is equivalent to the opposite of the category Fin_* of finite pointed sets. In particular, the functors introduced in Example 1.3.2(xi) define presheaves on Γ.

EXERCISE 1.5.iii. Prove Lemma 1.5.10.

EXERCISE 1.5.iv. Show that a full and faithful functor $F: \mathsf{C} \to \mathsf{D}$ both **reflects** and **creates isomorphisms**. That is, show:

(i) If f is a morphism in C so that Ff is an isomorphism in D, then f is an isomorphism.
(ii) If x and y are objects in C so that Fx and Fy are isomorphic in D, then x and y are isomorphic in C.

By Lemma 1.3.8, the converses of these statements hold for any functor.

EXERCISE 1.5.v. Find an example to show that a faithful functor need not reflect isomorphisms.

EXERCISE 1.5.vi.

(i) Prove that the composite of a pair of full, faithful, or essentially surjective functors again has the same properties.

(ii) Prove that if C ≃ D and D ≃ E, then C ≃ E. Conclude that equivalence of categories is an equivalence relation.[40]

EXERCISE 1.5.vii. Let G be a connected groupoid and let G be the group of automorphisms at any of its objects. The inclusion BG ↪ G defines an equivalence of categories. Construct an inverse equivalence G → BG.

EXERCISE 1.5.viii. Klein's Erlangen program studies groupoids of geometric spaces of various kinds. Prove that the groupoid Affine of affine planes is equivalent to the groupoid Proj^{l} of projective planes with a distinguished line, called the "line at infinity." The morphisms in each groupoid are bijections on both points and lines (preserving the distinguished line in the case of projective planes) that preserve and reflect the incidence relation. The functor Proj^{l} → Affine removes the line at infinity and the points it contains. Explicitly describe an inverse equivalence.

EXERCISE 1.5.ix. Show that any category that is equivalent to a locally small category is locally small.

EXERCISE 1.5.x. Characterize the categories that are equivalent to discrete categories. A category that is connected and essentially discrete is called **chaotic**.

EXERCISE 1.5.xi. Consider the functors Ab → Group (inclusion), Ring → Ab (forgetting the multiplication), $(-)^{\times}$: Ring → Group (taking the group of units), Ring → Rng (dropping the multiplicative unit), Field → Ring (inclusion), and Mod$_R$ → Ab (forgetful). Determine which functors are full, which are faithful, and which are essentially surjective. Do any define an equivalence of categories? (Warning: A few of these questions conceal research-level problems, but they can be fun to think about even if full solutions are hard to come by.)

1.6. The art of the diagram chase

> The diagrams incorporate a large amount of information. Their use provides extensive savings in space and in mental effort. In the case of many theorems, the setting up of the correct diagram is the major part of the proof. We therefore urge that the reader stop at the end of each theorem and attempt to construct for himself the relevant diagram before examining the one which is given in the text. Once this is done, the subsequent demonstration can be followed more readily; in fact, the reader can usually supply it himself.
>
> Samuel Eilenberg and Norman Steenrod,
> *Foundations of Algebraic Topology* [ES52]

The proof of Theorem 1.5.9 used a technique called "diagram chasing," also called *abstract nonsense*, that we now consider more formally. A **diagram** is typically presented informally as a directed graph of morphisms in a category. In this informal presentation, the diagram **commutes** if any two paths of composable arrows in the directed graph with

---

[40]A second, more direct proof of this result appears as Exercise 1.7.vi.

common source and target have the same composite. For example, a commutative triangle

(1.6.1)
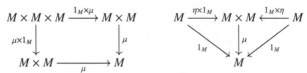

asserts that the hypotenuse h equals the composite gf of the two legs. Commutative diagrams in a category can be used to define more complicated mathematical objects. For example:

DEFINITION 1.6.2. A **monoid** is an object $M \in \mathsf{Set}$ together with a pair of morphisms $\mu\colon M \times M \to M$ and $\eta\colon 1 \to M$ so that the following diagrams commute:

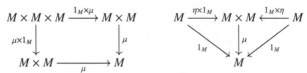

The morphism $\mu\colon M \times M \to M$ defines a binary "multiplication" operation on M. The morphism $\eta\colon 1 \to M$, whose domain is a singleton set, identifies an element $\eta \in M$. The three axioms demand that multiplication is associative and that multiplication on the left or right by the element η acts as the identity. The advantage of the commutative diagrams approach to this definition is that it readily generalizes to other categories. For example:

DEFINITION 1.6.3. A **topological monoid** is an object $M \in \mathsf{Top}$ together with morphisms $\mu\colon M \times M \to M$ and $\eta\colon 1 \to M$ so that the following diagrams commute:

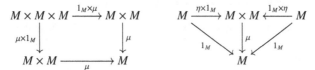

A **unital ring**[41] is an object $R \in \mathsf{Ab}$ together with morphisms $\mu\colon R \otimes_{\mathbb{Z}} R \to R$ and $\eta\colon \mathbb{Z} \to R$ so that the following diagrams commute:

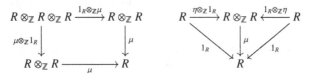

A $\Bbbk$-**algebra** is an object $R \in \mathsf{Vect}_{\Bbbk}$ together with morphisms $\mu\colon R \otimes_{\Bbbk} R \to R$ and $\eta\colon \Bbbk \to R$ so that the following diagrams commute:

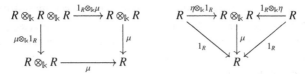

There are evident formal similarities in each of these four definitions; they are all special cases of a general notion of a monoid in a *monoidal category*, as defined in §E.2.

---

[41]A not-necessarily unital ring may be defined by ignoring the morphism η and the pair of commutative triangles.

The morphisms $\eta\colon 1 \to M$ in the case of topological monoids, $\eta\colon \mathbb{Z} \to R$ in the case of unital rings, and $\eta\colon \Bbbk \to R$ in the case of $\Bbbk$-algebras do no more and no less than specify an element of M or R to serve as the multiplicative unit. We will introduce language to describe the role played in each case by the topological space 1, the abelian group $\mathbb{Z}$, and the vector space $\Bbbk$ in Example 2.1.5(iii).

In the case of a topological monoid, the condition that $\mu\colon M \times M \to M$ is a morphism in Top demands that the multiplication function is continuous; for instance, the circle $S^1 \subset \mathbb{C}$ defines a topological monoid with addition of angles. For unital rings, the morphism $\mu\colon R \otimes_\mathbb{Z} R \to R$ represents a bilinear homomorphism of abelian groups from $R \times R$ to R; in particular, multiplication distributes over addition in R. The role of the tensor product in the definition of a $\Bbbk$-algebra is similar.

Let us now give precise meaning to the term "diagram."

DEFINITION 1.6.4. A **diagram** in a category C is a functor $F\colon \mathsf{J} \to \mathsf{C}$ whose domain, the **indexing category**, is a small category.

A diagram is typically depicted by drawing the objects and morphisms in its image, with the domain category left implicit, particularly in the case where the indexing category J is a preorder, so that any two paths of composable arrows have a common composite. Nonetheless, the indexing category J plays an important role. Functoriality requires that any composition relations that hold in J must hold in the image of the diagram, which is what it means to say that the directed graph defined by the image of the diagram in C is **commutative**. An immediate consequence of the form of our definition of a commutative diagram is the following result.

LEMMA 1.6.5. *Functors preserve commutative diagrams.*

PROOF. A diagram in C is given by a functor $F\colon \mathsf{J} \to \mathsf{C}$, whose domain is a small category. Given any functor $G\colon \mathsf{C} \to \mathsf{D}$, the composite $GF\colon \mathsf{J} \to \mathsf{D}$ defines the image of the diagram in D.   □

A few examples will help to illustrate the connection between commutative diagrams, as formalized in Definition 1.6.4, and their informal directed graph presentations.

EXAMPLE 1.6.6. Consider 2×2, the category with four objects and the displayed non-identity morphisms

In 2×2, the diagonal morphism is the composite of both the top and right morphisms and the left and bottom morphisms; in particular, these composites are equal. A diagram indexed by 2×2, typically drawn without the diagonal composite, is a **commutative square**.

REMARK 1.6.7. In practice, one thinks of the indexing category as a directed graph, defining the **shape** of the diagram, together with specified commutativity relations. For example, to define a functor with domain 2×2, it suffices to specify the images of the four objects together with four morphisms

$$
\begin{array}{ccc}
a & \xrightarrow{\ f\ } & b \\
{\scriptstyle g}\big\downarrow & & \big\downarrow{\scriptstyle h} \\
c & \xrightarrow[\ k\ ]{} & d
\end{array}
$$

subject to the relation that $hf = kg$. When indexing categories are represented in this way, the commutativity relations become an essential part of the data. They distinguish between the category 2×2 that indexes a commutative square and the category

that indexes a not-necessarily commutative square; here the two diagonals represent distinct composites of the two paths along the edges of the square.

EXAMPLE 1.6.8. The category 4 has four objects and the six displayed non-identity morphisms

This category is a preorder; each of the four displayed triangular "faces" commutes. These commutativity relations in 4 imply that a diagram of shape 4 in a category C is given by a sequence of three composable morphisms together with their composites

Associativity of composition implies that the bottom and back faces commute in C. Example 4.1.13 introduces terminology to explain why a diagram of shape 4 is determined by a sequence of three composable morphisms: the category 4 is *free on the directed graph* $\bullet \to \bullet \to \bullet \to \bullet$.

EXAMPLE 1.6.9. Consider 2×3, the category with six objects and the displayed non-identity morphisms

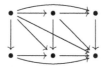

The long diagonal asserts that the outer triples of composable morphisms have a common composite, i.e., that the outer rectangle commutes. The short diagonals assert, respectively, that the left-hand and right-hand squares commute. The inner parallelogram also commutes. A diagram indexed by 2×3, typically drawn without any of the diagonals, is a **commutative rectangle**.

Two commutative squares define a commutative rectangle: a collection of morphisms with the indicated sources and targets

(1.6.10)

$$
\begin{array}{ccc}
a & \xrightarrow{\ f\ } & b & \xrightarrow{\ j\ } & c \\
\downarrow{\scriptstyle g} & & \downarrow{\scriptstyle h} & & \downarrow{\scriptstyle \ell} \\
a' & \xrightarrow[\ k\ ]{} & b' & \xrightarrow[\ m\ ]{} & c'
\end{array}
$$

define a 2×3-shaped diagram provided that $hf = kg$ and $\ell j = mh$. This is a special case of the following more general result, which describes the induced relations in the "algebra of composition" encoded by the arrows in a category.

LEMMA 1.6.11. *Suppose $f_1, \ldots, f_n$ is a composable sequence—a "path"—of morphisms in a category. If the composite $f_k f_{k-1} \cdots f_{i+1} f_i$ equals $g_m \cdots g_1$, for another composable sequence of morphisms $g_1, \ldots, g_m$, then $f_n \cdots f_1 = f_n \cdots f_{k+1} g_m \cdots g_1 f_{i-1} \cdots f_1$.*

PROOF. Composition is well-defined: if the composites $g_m \cdots g_1$ and $f_k f_{k-1} \cdots f_{i+1} f_i$ define the same arrow, then the results of pre- or post-composing with other sequences of arrows must also be the same.   □

This very simple result underlies most proofs by "diagram chasing." When a diagram is depicted by a simple (meaning there is at most one edge between any two vertices) acyclic directed graph, the most common convention is to include commutativity relations that assert that any two paths in the diagram with a common source and target commute, i.e., that the directed graph represents a poset category. For example, the category $2 \times 2 \times 2$ indexes the **commutative cube**, which is typically depicted as follows:

In such cases, Lemma 1.6.11 and transitivity of equality implies that commutativity of the entire diagram may be checked by establishing commutativity of each minimal subdiagram in the directed graph. Here, a minimal subdiagram corresponds to a composition relation $h_n \cdots h_1 = k_m \cdots k_1$ that cannot be factored into a relation between shorter paths of composable morphisms. The graph corresponding to a minimal relation is a "directed polygon"

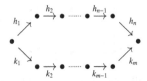

with a commutative triangle, as in (1.6.1), being the simplest case. This sort of argument is called "equational reasoning" in [**Sim11a**, 2.1], which provides an excellent short introduction to diagram chasing.

The following results have simple proofs by diagram chasing.

LEMMA 1.6.12. *If the triangle displayed on the left commutes*

and if f is an isomorphism, then the triangle displayed on the right also commutes. Dually, for any triple of morphisms with domains and codomains as displayed with k an isomorphism

the left-hand triangle commutes if and only if the right-hand triangle commutes.

PROOF. Pre-compose the composition relation $h = gf$ with f^{-1} to yield $hf^{-1} = g$. □

LEMMA 1.6.13. *For any commutative square $\beta\alpha = \delta\gamma$ in which each of the morphisms is an isomorphism, then the inverses define a commutative square $\alpha^{-1}\beta^{-1} = \gamma^{-1}\delta^{-1}$.*

PROOF. Apply Lemma 1.6.12 four times.   □

In certain special cases, commutativity of diagrams can be automatic. For instance, any parallel sequences of composable morphisms in a preorder must have a common composite precisely because any hom-set in a preorder has at most one element. Parallel sequences of composable morphisms also necessarily have a common composite when the domain or codomain objects have a certain special property:

DEFINITION 1.6.14. An object $i \in \mathsf{C}$ is **initial** if for every $c \in \mathsf{C}$ there is a unique morphism $i \to c$. Dually, an object $t \in \mathsf{C}$ is **terminal** if for every $c \in \mathsf{C}$ there is a unique morphism $c \to t$.

EXAMPLE 1.6.15. Many of the categories of our acquaintance have initial and terminal objects.

   (i) The empty set is an initial object in Set and any singleton set is terminal.
   (ii) In Top, the empty and singleton spaces are, respectively, initial and terminal.
   (iii) In Set$_*$, any singleton set is both initial and terminal.
   (iv) In Mod$_R$, the zero module is both initial and terminal. Similarly, the trivial group is both initial and terminal in Group.
   (v) The zero ring is the terminal ring. To identify an initial object, we must clarify what sort of rings and ring homomorphisms are being considered. Recall that Ring denotes the category of unital rings and ring homomorphisms that preserve the multiplicative identity. This is a non-full subcategory of the larger category Rng of rings that do not necessarily have a multiplicative identity and homomorphisms that need not preserve one if it happens to exist. The integers define an initial object in Ring but not in Rng, in which the zero ring is initial.
   (vi) The category Field of fields has neither initial nor terminal objects. Indeed, there are no homomorphisms between fields of different characteristic.
   (vii) The empty category defines an initial object in Cat, and the category $\mathbb{1}$ is terminal.

(viii) An initial object in a preorder is a global minimal element, and a terminal object is a global maximal element. These initial and terminal objects may or may not exist in each particular instance.

LEMMA 1.6.16. *Let $f_1, \ldots, f_n$ and $g_1, \ldots, g_m$ be composable sequences of morphisms so that the domain of f_1 equals the domain of g_1 and the codomain of f_n equals the codomain of g_m. If this common codomain is a terminal object, or if this common domain is an initial object, then $f_n \cdots f_1 = g_m \cdots g_1$.*

PROOF. The two dual statements are immediate consequences of the uniqueness part of Definition 1.6.14.   □

In certain cases, one can prove that a diagram commutes by appealing to "elements" of the objects. For instance, this is possible in any concrete category.

DEFINITION 1.6.17. A **concrete category** is a category C equipped with a faithful functor $U\colon C \to$ Set.

The functor U typically carries an object of C to its "underlying set." The faithfulness condition asserts that any parallel pair of morphisms $f, g\colon c \rightrightarrows c'$ that induce the same function $Uf = Ug$ between the underlying sets must be equal in C. The idea is that the question of whether a map between the underlying sets of objects in a concrete category is a map in the category is a condition (e.g., continuity). By contrast, the functor $U\colon C \to$ Set is not faithful if the maps in C have extra structure that is not visible at the level of the underlying sets (e.g., homotopy classes of maps [**Fre04**]).

EXAMPLE 1.6.18. Each category listed in Example 1.1.3 is concrete, although care must be taken in the cases of Graph and Ch_R, whose objects are "multi-sorted" sets. The most obvious forgetful functors, which send a graph to its set of vertices or its set of edges, are not faithful. However, the functor $V \sqcup E\colon$ Graph $\to$ Set considered in Example 1.3.2(ii) that sends a graph to the union of its set of vertices and edges is faithful.

Because faithful functors reflect identifications between parallel morphisms:

LEMMA 1.6.19. *If $U\colon C \to D$ is faithful, then any diagram in C whose image commutes in D also commutes in C.*

PROOF. Here a "diagram" in C is a directed graph of morphisms together with a collection of desired composition relations between composable paths of morphisms. If $f_1, \ldots, f_n$ and $g_1, \ldots, g_m$ are parallel sequences of composable morphisms in C so that

$$Uf_n \cdots Uf_1 = Ug_m \cdots Ug_1$$

in D, then by faithfulness (and functoriality) of U, $f_n \cdots f_1 = g_m \cdots g_1$ in C.   □

In particular, to prove that a diagram in a concrete category commutes, it suffices to prove commutativity of the induced diagram of underlying sets. This amounts to showing that certain composite functions between underlying sets are the same, and this can be checked by considering the actions of these functions on the elements of their domain.

We close with a word of warning.

REMARK 1.6.20 (on cancellability). We have seen that commutativity of a pair of adjacent squares as in (1.6.10) implies commutativity of the exterior rectangle, but the converse need not hold, as illustrated by the following diagram in Ab, in which the outer rectangle commutes but neither square does:

$$\begin{array}{ccccc}
\mathbb{Z} & \xrightarrow{1_\mathbb{Z}} & \mathbb{Z} & \longrightarrow & 0 \\
\downarrow & & \downarrow{\scriptstyle 1_\mathbb{Z}} & & \downarrow \\
0 & \longrightarrow & \mathbb{Z} & \xrightarrow{1_\mathbb{Z}} & \mathbb{Z}
\end{array}$$

Neither does commutativity of one of the two squares, plus the outer rectangle, imply commutativity of the other in general. The issue is that a composition relation of the form

$$gh_n \cdots h_1 f = gk_m \cdots k_1 f$$

need not imply that $h_n \cdots h_1 = k_m \cdots k_1$ unless f is "right cancellable" and g is "left cancellable," i.e., unless f is an epimorphism and g is a monomorphism; see Definition 1.2.7.

LEMMA 1.6.21. *Consider morphisms with the indicated sources and targets*

$$\begin{array}{ccccc}
a & \xrightarrow{f} & b & \xrightarrow{j} & c \\
{\scriptstyle g}\downarrow & & \downarrow{\scriptstyle h} & & \downarrow{\scriptstyle \ell} \\
a' & \xrightarrow{k} & b' & \xrightarrow{m} & c'
\end{array}$$

and suppose that the outer rectangle commutes. This data defines a commutative rectangle if either:

(i) *the left-hand square commutes and m is a monomorphism; or*
(ii) *the right-hand square commutes and f is an epimorphism.*

PROOF. The statements are dual. Assuming (i), $mkg = \ell jf = mhf$ by commutativity of the outer rectangle and right-hand square. Since m is a monomorphism, it follows that $kg = hf$ and thus that the diagram commutes.   □

Exercises.

EXERCISE 1.6.i. Show that any map from a terminal object in a category to an initial one is an isomorphism. An object that is both initial and terminal is called a **zero object**.

EXERCISE 1.6.ii. Show that any two terminal objects in a category are connected by a unique isomorphism.

EXERCISE 1.6.iii. Show that any faithful functor reflects monomorphisms. That is, if $F\colon \mathsf{C} \to \mathsf{D}$ is faithful, prove that if Ff is a monomorphism in D, then f is a monomorphism in C. Argue by duality that faithful functors also reflect epimorphisms. Conclude that in any concrete category, any morphism that defines an injection of underlying sets is a monomorphism and any morphism that defines a surjection of underlying sets is an epimorphism.

EXERCISE 1.6.iv. Find an example to show that a faithful functor need not preserve epimorphisms. Argue by duality, or by another counterexample, that a faithful functor need not preserve monomorphisms.

EXERCISE 1.6.v. More specifically, find a concrete category that contains a monomorphism whose underlying function is not injective. Find a concrete category that contains an epimorphism whose underlying function is not surjective. Exercise 4.5.v explains why the latter examples may seem less familiar than the former.

EXERCISE 1.6.vi. A **coalgebra** for an endofunctor $T: \mathsf{C} \to \mathsf{C}$ is an object $C \in \mathsf{C}$ equipped with a map $\gamma: C \to TC$. A morphism $f: (C, \gamma) \to (C', \gamma')$ of coalgebras is a map $f: C \to C'$ so that the square

$$
\begin{array}{ccc}
C & \xrightarrow{\ f\ } & C' \\
\gamma \downarrow & & \downarrow \gamma' \\
TC & \xrightarrow[Tf]{} & TC'
\end{array}
$$

commutes. Prove that if (C, γ) is a **terminal coalgebra**, that is a terminal object in the category of coalgebras, then the map $\gamma: C \to TC$ is an isomorphism.

1.7. The 2-category of categories

A number of important facts about natural transformations are proven by diagram chasing. In this section, we define "vertical" and "horizontal" composition operations for natural transformations. The upshot is that categories, functors, and natural transformations assemble into a 2-dimensional categorical structure called a *2-category*, a definition that is stated at the conclusion.

In French, a natural transformation is called a *morphisme de foncteurs*. Indeed, for any fixed pair of categories C and D, there is a **functor category** D^{C} whose objects are functors $\mathsf{C} \to \mathsf{D}$ and whose morphisms are natural transformations. Given a functor $F: \mathsf{C} \to \mathsf{D}$, its identity natural transformation $1_F: F \Rightarrow F$ is the natural transformation whose components $(1_F)_c := 1_{Fc}$ are identities. The following lemma defines the composition of morphisms in D^{C}.

LEMMA 1.7.1 (vertical composition). *Suppose $\alpha: F \Rightarrow G$ and $\beta: G \Rightarrow H$ are natural transformations between parallel functors $F, G, H: \mathsf{C} \to \mathsf{D}$. Then there is a natural transformation $\beta \cdot \alpha: F \Rightarrow H$ whose components*

$$(\beta \cdot \alpha)_c := \beta_c \cdot \alpha_c$$

are defined to be the composites of the components of α and β.

PROOF. Naturality of α and β implies that for any $f: c \to c'$ in the domain category, each square, and thus also the composite rectangle, commutes:

$$
\begin{array}{ccc}
Fc & \xrightarrow{\alpha_c} Gc \xrightarrow{\beta_c} & Hc \\
Ff \downarrow & \quad Gf \downarrow \qquad & \downarrow Hf \\
Fc' & \xrightarrow[\alpha_{c'}]{} Gc' \xrightarrow[\beta_{c'}]{} & Hc'
\end{array}
$$
□

COROLLARY 1.7.2. *For any pair of categories C and D, the functors from C to D and natural transformations between them define a category D^{C}.*

PROOF. It remains only to verify that the composition operation defined by Lemma 1.7.1 is associative and unital. It suffices to verify these properties componentwise, and they follow immediately from the associativity and unitality of composition in D. □

REMARK 1.7.3 (sizes of functor categories). Care should be taken with size when discussing functor categories. If C and D are small, then D^{C} is again a small category, but if C and D are locally small, then D^{C} need not be. This is only guaranteed if D is locally small and C is small; see Exercise 1.7.i. In summary, the formation of functor categories defines a bifunctor $\mathsf{Cat}^{op} \times \mathsf{Cat} \to \mathsf{Cat}$ or $\mathsf{Cat}^{op} \times \mathsf{CAT} \to \mathsf{CAT}$, but the category of functors between two non-locally small categories may be even larger than these categories are.

The composition operation defined in Lemma 1.7.1 is called **vertical composition**. Drawing the parallel functors horizontally, a composable pair of natural transformations in the category D^{C} fits into a *pasting diagram*

$$
\begin{array}{c}
F \\
C \overset{\Downarrow \alpha}{\underset{\Downarrow \beta}{\xrightarrow{\;\;G\;\;}}} D \\
H
\end{array}
\quad = \quad
\begin{array}{c}
F \\
C \;\; \Downarrow \beta \cdot \alpha \;\; D \\
H
\end{array}
$$

As the terminology suggests, there is also a **horizontal composition** operation

$$
\begin{array}{c}
F \qquad H \\
C \;\; \Downarrow \alpha \;\; D \;\; \Downarrow \beta \;\; E \\
G \qquad K
\end{array}
\quad = \quad
\begin{array}{c}
HF \\
C \;\; \Downarrow \beta * \alpha \;\; E \\
KG
\end{array}
$$

defined by the following lemma.

LEMMA 1.7.4 (horizontal composition). *Given a pair of natural transformations*

$$
\begin{array}{c}
F \qquad H \\
C \;\; \Downarrow \alpha \;\; D \;\; \Downarrow \beta \;\; E \\
G \qquad K
\end{array}
$$

there is a natural transformation $\beta * \alpha \colon HF \Rightarrow KG$ whose component at $c \in \mathsf{C}$ is defined as the composite of the following commutative square

(1.7.5)
$$
\begin{array}{ccc}
HFc & \xrightarrow{\;\beta_{Fc}\;} & KFc \\
{\scriptstyle H\alpha_c}\Big\downarrow & {\scriptstyle (\beta * \alpha)_c} & \Big\downarrow{\scriptstyle K\alpha_c} \\
HGc & \xrightarrow[\;\beta_{Gc}\;]{} & KGc
\end{array}
$$

PROOF. The square (1.7.5) commutes by naturality of $\beta \colon H \Rightarrow K$ applied to the morphism $\alpha_c \colon Fc \to Gc$ in D. To prove that the components $(\beta * \alpha)_c \colon HFc \to KGc$ so-defined are natural, we must show that $KGf \cdot (\beta * \alpha)_c = (\beta * \alpha)_{c'} \cdot HFf$ for any $f \colon c \to c'$ in C. This relation holds on account of the commutative rectangle

$$
\begin{array}{ccccc}
HFc & \xrightarrow{\;H\alpha_c\;} & HGc & \xrightarrow{\;\beta_{Gc}\;} & KGc \\
{\scriptstyle HFf}\Big\downarrow & & {\scriptstyle HGf}\Big\downarrow & & \Big\downarrow{\scriptstyle KGf} \\
HFc' & \xrightarrow[\;H\alpha_{c'}\;]{} & HGc' & \xrightarrow[\;\beta_{Gc'}\;]{} & KGc'
\end{array}
$$

The right-hand square commutes by naturality of β. The left-hand square commutes by naturality of α and Lemma 1.6.5, which states that functors, in this case the functor H, preserve commutative diagrams.[42]   □

REMARK 1.7.6. The natural transformations

$$H\alpha\colon HF \Rightarrow HG, \quad K\alpha\colon KF \Rightarrow KG, \quad \beta F\colon HF \Rightarrow KF, \quad \text{and} \quad \beta G\colon HG \Rightarrow KG$$

appearing in Lemma 1.7.4 are defined by **whiskering** the natural transformations α and β with the functors H and K or F and G, respectively. A precise definition of this construction is given in Exercise 1.7.ii. The terminology is on account of the following graphical depiction of the whiskered composite

$$L\beta F\colon LHF \Rightarrow LKF \qquad C \xrightarrow{F} D \Downarrow\beta E \xrightarrow{L} F$$

of the natural transformation $\beta\colon H \Rightarrow K$ with the functors F and L. Exercise 1.7.iii explains the particular interest in the case where either L or F is an identity.

Importantly, vertical and horizontal composition can be performed in either order, satisfying the rule of **middle four interchange**:

LEMMA 1.7.7 (middle four interchange). *Given functors and natural transformations*

$$C \xrightarrow{G} D \xrightarrow{K} E$$

the natural transformation $JF \Rightarrow LH$ defined by first composing vertically and then composing horizontally equals the natural transformation defined by first composing horizontally and then composing vertically:

$$C \Downarrow\beta\cdot\alpha\ D \Downarrow\delta\cdot\gamma\ E \quad = \quad C \xrightarrow{KG} E$$

PROOF. Exercise 1.7.iv.   □

Lemmas 1.7.1, 1.7.4, and 1.7.7 prove that categories, functors, and natural transformations assemble into a *2-category*. Aside from this example, we will not meet any other 2-categories in this text. Nonetheless, the following definition is useful as an axiomatization of the composition operations for natural transformations that are available. A succinct introduction to 2-categories and *pasting diagrams*, which are used to display composite natural transformations, can be found in [**KS74**].

DEFINITION 1.7.8. A **2-category** is comprised of:

---

[42]Naturality of $\beta * \alpha$ could also be deduced from a second commutative rectangle that defines the component $(\beta * \alpha)_c$ as the top-right composite of (1.7.5). The point is that the squares (1.7.5) and the morphisms obtained by applying the four functors HF, HG, KF, and KG to a morphism in C define a commutative cube.

- objects, for example the categories C,
- 1-morphisms between pairs of objects, for example, the functors $C \xrightarrow{F} D$, and
- 2-morphisms between parallel pairs of 1-morphisms, for example, the natural transformations $C \overset{F}{\underset{G}{\Downarrow \alpha}} D$

so that:

- The objects and 1-morphisms form a category, with identities $1_C \colon C \to C$.
- For each fixed pair of objects C and D, the 1-morphisms $F \colon C \to D$ and 2-morphisms between such form a category under an operation called vertical composition, as described in Lemma 1.7.1, with identities $C \overset{F}{\underset{F}{\Downarrow 1_F}} D$.
- There is also a category whose objects are the objects in which a morphism from C to D is a 2-cell $C \overset{F}{\underset{G}{\Downarrow \alpha}} D$ under an operation called horizontal composition, with identities $C \overset{1_C}{\underset{1_C}{\Downarrow 1_{1_C}}} C$. The source and target 1-morphisms of a horizontal composition must have the form described in Lemma 1.7.4.
- The law of middle four interchange described in Lemma 1.7.7 holds.

The reader who has taken the categorical philosophy to heart might ask: What is a morphism between 2-categories? *2-functors* will make an appearance in §4.4.

Exercises.

EXERCISE 1.7.i. Prove that if C is small and D is locally small, then D^C is locally small by defining a monomorphism from the collection of natural transformations between a fixed pair of functors $F, G \colon C \rightrightarrows D$ into a set. (Hint: Think about the function that sends a natural transformation to its collection of components.)

EXERCISE 1.7.ii. Given a natural transformation $\beta \colon H \Rightarrow K$ and functors F and L as displayed in

$$C \xrightarrow{F} D \overset{H}{\underset{K}{\Downarrow \beta}} E \xrightarrow{L} F$$

define a natural transformation $L\beta F \colon LHF \Rightarrow LKF$ by $(L\beta F)_c = L\beta_{Fc}$. This is the **whiskered composite** of β with L and F. Prove that $L\beta F$ is natural.

EXERCISE 1.7.iii. Redefine the horizontal composition of natural transformations introduced in Lemma 1.7.4 using vertical composition and whiskering.

EXERCISE 1.7.iv. Prove Lemma 1.7.7.

EXERCISE 1.7.v. Show that for any category C, the collection of natural endomorphisms of the identity functor 1_C defines a commutative monoid, called the **center of the category**.

The proof of Proposition 1.4.4 demonstrates that the center of Ab_{fg} is the multiplicative monoid $(\mathbb{Z}, \times, 1)$.

EXERCISE 1.7.vi. Suppose the functors and natural isomorphisms

$$\mathsf{C} \underset{G}{\overset{F}{\rightleftarrows}} \mathsf{D} \qquad \eta: 1_\mathsf{C} \cong GF \qquad \epsilon: FG \cong 1_\mathsf{D}$$

$$\mathsf{D} \underset{G'}{\overset{F'}{\rightleftarrows}} \mathsf{E} \qquad \eta': 1_\mathsf{D} \cong G'F' \qquad \epsilon': F'G' \cong 1_\mathsf{E}$$

define equivalences of categories $\mathsf{C} \simeq \mathsf{D}$ and $\mathsf{D} \simeq \mathsf{E}$. Prove (again) that there is a composite equivalence of categories $\mathsf{C} \simeq \mathsf{E}$ by defining composite natural isomorphisms $1_\mathsf{C} \cong GG'F'F$ and $F'FGG' \cong 1_\mathsf{E}$.

EXERCISE 1.7.vii. Prove that a bifunctor $F: \mathsf{C} \times \mathsf{D} \to \mathsf{E}$ determines and is uniquely determined by:

(i) A functor $F(c, -): \mathsf{D} \to \mathsf{E}$ for each $c \in \mathsf{C}$.
(ii) A natural transformation $F(f, -): F(c, -) \Rightarrow F(c', -)$ for each $f: c \to c'$ in C, defined functorially in C.

In other words, prove that there is a bijection between functors $\mathsf{C} \times \mathsf{D} \to \mathsf{E}$ and functors $\mathsf{C} \to \mathsf{E}^\mathsf{D}$. By symmetry of the product of categories, these classes of functors are also in bijection with functors $\mathsf{D} \to \mathsf{E}^\mathsf{C}$.

CHAPTER 2

Universal Properties, Representability, and the Yoneda Lemma

> ... a mathematical object X is best thought of in
> the context of a category surrounding it, and is
> determined by the network of relations it enjoys
> with *all* the objects of that category. Moreover, to
> understand X it might be more germane to deal
> directly with the functor representing it.
>
> Barry Mazur, "Thinking about Grothendieck"
> [Maz16]

The aim in this chapter is to explain what it means to say that the natural numbers is the universal discrete dynamical system, that the Sierpinski space is the universal space with an open subset, or that the complete graph on n-vertices is the universal n-colored graph. Universal properties expressed in this plain language manner might appear somewhat ad hoc, but this is a false impression. By the chapter's end, we will see several equivalent ways in which the notion of universal property can be made precise. The key input to the theory developed here is something that is a priori non-obvious: a functor valued in the category of sets. The category of sets plays a special role because traditional approaches to the foundations of mathematics are based on set theory, with a wide variety of mathematical objects defined to be sets with additional structure.

To illustrate, consider the set of vertex colorings of a graph subject to the requirement that adjacent vertices are assigned distinct colors. There is a contravariant functor from the category of graphs to the category of sets that takes a graph to the set of n-colorings of its vertices. To explain the contravariance, note that an n-coloring of a graph G and a graph homomorphism $G' \to G$ induce an n-coloring of G' that colors each vertex of G' to match the color of its image: as graph homomorphisms preserve the incidence relation on vertices, any two adjacent vertices of G' are assigned distinct colors. This defines the action of the functor $n\text{-Color} \colon \mathsf{Graph}^{\mathrm{op}} \to \mathsf{Set}$ on morphisms.[1]

The graph with the fewest vertices and fewest edges that can be colored with no less than n colors is the complete graph on n vertices K_n. Indeed, the functor $n\text{-Color}$ encodes a *universal property* of the graph K_n in the sense that the graph K_n *represents* the functor $n\text{-Color}$. Specifically, this means that there is a natural bijection between the set $n\text{-Color}(G)$ of n-colorings of a graph G and the set of graph homomorphisms $G \to K_n$. In §2.1, we introduce the notion of a *representable functor* along with a plethora of examples, intended to convey the ubiquity of this abstract notion.

---

[1] The **four color theorem** states that if G is a simple planar graph, then the set $4\text{-Color}(G)$ is non-empty. Functoriality implies that any graph admitting a graph homomorphism to a planar graph also admits a 4-coloring. For instance, complete bipartite graphs, which are typically non-planar, will admit homomorphisms of this type.

By the Yoneda lemma, which we prove in §2.2, this universal property uniquely characterizes the graph K_n. The Yoneda lemma also establishes a correspondence between natural endomorphisms of the functor n-Color—"color permutations"—and symmetries of the graph K_n and proves that there is no uniform way to convert an m-coloring into an n-coloring if $m > n$, this result deduced from the fact that there are no graph homomorphisms $K_m \to K_n$.

The Yoneda lemma is arguably the most important result in category theory, although it takes some time to explore the depths of the consequences of this simple statement. In §2.3, we define the notion of *universal element* that witnesses a universal property of some object in a locally small category. The universal element witnessing the universal property of the complete graph is an n-coloring of K_n, an element of the set n-Color(K_n). In §2.4, we use the Yoneda lemma to show that the pair comprised of an object characterized by a universal property and its universal element defines either an initial or a terminal object in the *category of elements* of the functor that it represents. This gives precise meaning to the term **universal**: it is a synonym for either "initial" or "terminal," with context disambiguating between the two cases. For instance, K_n is the **terminal n-colored graph**: the terminal object in the category of n-colored graphs and graph homomorphisms that preserve the coloring of vertices.

2.1. Representable functors

> The further you go in mathematics, especially
> pure mathematics, the more universal properties
> you will meet.
>
> ———————————————————————
> Tom Leinster, *Basic Category Theory* [**Lei14**]

The most basic formulation of a universal property is to say that a particular object defines an initial or terminal object in its ambient category. The problem with this paradigm is that the most familiar categories—for instance, of sets, spaces, groups, modules, and so on—tend to have uninteresting initial and terminal objects. To express the universal properties of more complicated objects, one has to cook up a less familiar category. For example:

EXAMPLE 2.1.1. A set X with an endomorphism $f \colon X \to X$ and a distinguished element x_0 is called a **discrete dynamical system**. This data allows one to consider the discrete-time evolution of the initial element x_0, a sequence defined by $x_{n+1} := f(x_n)$. The principle of mathematical recursion asserts that the natural numbers $\mathbb{N}$, the successor function $s \colon \mathbb{N} \to \mathbb{N}$, and the element $0 \in \mathbb{N}$ define the **universal discrete dynamical system**: which is to say, there is a unique function $r \colon \mathbb{N} \to X$ so that $r(n) = x_n$ for each n, i.e., so that $r(0) = x_0$ and so that the diagram

$$
\begin{array}{ccc}
\mathbb{N} & \xrightarrow{\ s\ } & \mathbb{N} \\
{\scriptstyle r}\big\downarrow & & \big\downarrow{\scriptstyle r} \\
X & \xrightarrow[\ f\ ]{} & X
\end{array}
$$

(2.1.2)

commutes.

In this section, we introduce a vehicle for expressing universal properties that is less-contrived than spontaneously generating an appropriate category in which the universal object is initial or terminal. The notion that we introduce can be understood as a generalization of the universal properties describing initial or terminal objects. In §2.4, we will show that it is not a true generalization.

To say that an object $c \in C$ is initial is to say that for all other objects x, the set $C(c, x)$ is a singleton. Dually, $c \in C$ is terminal, if and only if for all $x \in C$ the set $C(x, c)$ is a singleton. These properties can be expressed more categorically as characterizations of the co- and contravariant functors $C(c, -)$ and $C(-, c)$ represented by c that were introduced in Definition 1.3.11:

$$
\begin{array}{ccc}
C \xrightarrow{\;C(c,-)\;} Set & \qquad & C^{op} \xrightarrow{\;C(-,c)\;} Set \\[2mm]
x \;\longmapsto\; C(c, x) & & x \;\longmapsto\; C(x, c) \\
f\downarrow \quad\longmapsto\quad \downarrow f_* & & f\downarrow \quad\longmapsto\quad \uparrow f^* \\
y \;\longmapsto\; C(c, y) & & y \;\longmapsto\; C(y, c)
\end{array}
$$

DEFINITION 2.1.3 (a representable characterization of initial or terminal objects).

(i) An object c in a category C is **initial** if and only if the functor $C(c, -): C \to Set$ is naturally isomorphic to the constant functor $*: C \to Set$ that sends every object to the singleton set.

(ii) An object $c \in C$ is **terminal** if and only if the functor $C(-, c): C^{op} \to Set$ is naturally isomorphic to the constant functor $*: C^{op} \to Set$ that sends every object to the singleton set.

In terminology we now introduce, Definition 2.1.3 asserts that C has an initial object if and only if the constant functor $*: C \to Set$ is representable, and dually, C has a terminal object if and only if $*: C^{op} \to Set$ is representable.

DEFINITION 2.1.4.

(i) A covariant or contravariant functor F from a locally small category C to Set is **representable** if there is an object $c \in C$ and a natural isomorphism between F and the functor of appropriate variance[2] represented by c, in which case one says that the functor F is **represented by** the object c.

(ii) A **representation** for a covariant functor F is a choice of object $c \in C$ together with a specified natural isomorphism $C(c, -) \cong F$, if F is covariant, or $C(-, c) \cong F$, if F is contravariant.

The domain C of a representable functor is required to be locally small so that the hom-functors $C(c, -)$ and $C(-, c)$ are valued in the category of sets.

As in the special case of Definition 2.1.3, a representable functor $F: C \to Set$ or $F: C^{op} \to Set$ encodes a **universal property** of its representing object. Put colloquially, a universal property of an object X is a description of the covariant functor $Hom(X, -)$ or of the contravariant functor $Hom(-, X)$ associated to that object. Many examples occur "in nature."[3]

EXAMPLE 2.1.5. The following covariant functors are representable.

(i) The identity functor $1_{Set}: Set \to Set$ is represented by the singleton set 1. That is, for any set X, there is a natural isomorphism $Set(1, X) \cong X$ that defines a bijection

---

[2]Some use the term "corepresentable" for covariant representable functors, reserving "representable" for the contravariant case. We argue that this distinction is unnecessary since the variance of the functor F ought to be evident from its definition.

[3]If a friend or oracle is nearby, we suggest first encountering the examples listed in 2.1.5 and 2.1.6 as exercises. Have the friend list the functors that are mentioned and challenge yourself to find the representing objects.

between elements $x \in X$ and functions $x: 1 \to X$ carrying the singleton element to x. Naturality says that for any $f: X \to Y$, the diagram

$$
\begin{array}{ccc}
\mathsf{Set}(1,X) & \xrightarrow{\;\cong\;} & X \\
{\scriptstyle f_*}\big\downarrow & & \big\downarrow{\scriptstyle f} \\
\mathsf{Set}(1,Y) & \xrightarrow[\;\cong\;]{} & Y
\end{array}
$$

commutes, i.e., that the composite function $1 \xrightarrow{x} X \xrightarrow{f} Y$ corresponds to the element $f(x) \in Y$, as is evidently the case.

(ii) The forgetful functor $U: \mathsf{Group} \to \mathsf{Set}$ is represented by the group $\mathbb{Z}$. That is, for any group G, there is a natural isomorphism $\mathsf{Group}(\mathbb{Z}, G) \cong UG$ that associates, to every element $g \in UG$, the unique homomorphism $\mathbb{Z} \to G$ that maps the integer 1 to g. This defines a bijection because every homomorphism $\mathbb{Z} \to G$ is determined by the image of the generator 1; that is to say, $\mathbb{Z}$ is the **free group on a single generator**.

This bijection is natural because the composite group homomorphism $\mathbb{Z} \xrightarrow{g} G \xrightarrow{\phi} H$ carries the integer 1 to $\phi(g) \in H$.

(iii) For any unital ring R, the forgetful functor $U: \mathsf{Mod}_R \to \mathsf{Set}$ is represented by the R-module R. That is, there is a natural bijection between R-module homomorphisms $R \to M$ and elements of the underlying set of M, in which $m \in UM$ is associated to the unique R-module homomorphism that carries the multiplicative identity of R to m; this is to say, R is the **free R-module on a single generator**. This explains the appearance of the abelian group $\mathbb{Z}$ and the vector space $\Bbbk$ in Definition 1.6.3, where maps with these domains were used to specify elements in the codomains.

(iv) The functor $U: \mathsf{Ring} \to \mathsf{Set}$ is represented by the unital ring $\mathbb{Z}[x]$, the polynomial ring in one variable with integer coefficients. A unital ring homomorphism $\mathbb{Z}[x] \to R$ is uniquely determined by the image of x; put another way, $\mathbb{Z}[x]$ is the **free unital ring on a single generator**.

(v) The functor $U(-)^n: \mathsf{Group} \to \mathsf{Set}$ that sends a group G to the set of n-tuples of elements of G is represented by the **free group F_n on n generators**. Similarly, the functor $U(-)^n: \mathsf{Ab} \to \mathsf{Set}$ is represented by the **free abelian group $\bigoplus_n \mathbb{Z}$ on n generators**.

(vi) More generally, any group presentation, such as

$$
S_3 := \big\langle s, t \,\big|\, s^2 = t^2 = 1,\, sts = tst \big\rangle,
$$

defines a functor $\mathsf{Group} \to \mathsf{Set}$ that carries a group G to the set

$$
\big\{(g_1, g_2) \in G^2 \,\big|\, g_1^2 = g_2^2 = e,\, g_1 g_2 g_1 = g_2 g_1 g_2 \big\}.
$$

The functor is represented by the group admitting the given presentation, in this case by the symmetric group S_3 on three elements: the presentation tells us that homomorphisms $S_3 \to G$ are classified by pairs of elements $g_1, g_2 \in G$ satisfying the listed relations.

(vii) The functor $(-)^\times: \mathsf{Ring} \to \mathsf{Set}$ that sends a unital ring to its set of units is represented by the ring $\mathbb{Z}[x, x^{-1}]$ of Laurent polynomials in one variable. That is to say, a ring homomorphism $\mathbb{Z}[x, x^{-1}] \to R$ may be defined by sending x to any unit of R and is completely determined by this assignment, and moreover there are no ring homomorphisms that carry x to a non-unit.

(viii) The forgetful functor U: Top → Set is represented by the singleton space: there is a natural bijection between elements of a topological space and continuous functions from the one-point space.

(ix) The functor ob: Cat → Set that takes a small category to its set of objects is represented by the terminal category $\mathbb{1}$: a functor $\mathbb{1}$ → C is no more and no less than a choice of object in C.

(x) The functor mor: Cat → Set that takes a small category to its set of morphisms is represented by the category 2: a functor 2 → C is no more and no less than a choice of morphism in C. In this sense, the category 2 is the **free** or **walking arrow**.

(xi) The functor iso: Cat → Set that takes a small category to its set of isomorphisms (pointing in a specified direction) is represented by the category $\mathbb{I}$, with two objects and exactly one morphism in each hom-set. In this sense, the category $\mathbb{I}$ is the **free** or **walking isomorphism**.

(xii) The functor comp: Cat → Set that takes a small category to the set of composable pairs of morphisms in it is represented by the category 3. Generalizing, the ordinal $\mathbb{n} + \mathbb{1} = 0 → 1 → \cdots → n$ represents the functor that takes a small category to the set of paths of n composable morphisms in it.

(xiii) The forgetful functor U: Set$_*$ → Set is represented by the two-element based set: based functions out of this set correspond naturally and bijectively to elements of the target based set, the element in question being the image of the non-basepoint element.

(xiv) The functor Path: Top → Set that carries a topological space to its set of paths and the functor Loop: Top$_*$ → Set that carries a based space to its set of based loops are each representable by definition, by the unit interval I and the based circle S^1, respectively. A **path** in X is a continuous function $I → X$ while a (based) **loop** in X is a based continuous function $S^1 → X$.

The adjective "free" is reserved for universal properties expressed by covariant represented functors. It could be applied to any of the objects listed in Example 2.1.5: 2 is the free category with an arrow, S^1 is the free space containing a loop. The dual term "cofree," for universal properties expressed by contravariant represented functors, is less commonly used.

EXAMPLE 2.1.6. The following contravariant functors are representable.

(i) The contravariant power set functor P: Set^{op} → Set is represented by the set $\Omega = \{\top, \bot\}$ with two elements. The natural isomorphism Set$(A, \Omega) \cong PA$ is defined by the bijection that associates a function $A → \Omega$ with the subset that is the preimage of $\top$; reversing perspectives, a subset $A' \subset A$ is identified with its **classifying function** $\chi_{A'} : A → \Omega$, which sends exactly the elements of A' to the element $\top$. The naturality condition stipulates that for any function $f: A → B$, the diagram

$$
\begin{array}{ccc}
\mathrm{Set}(B, \Omega) & \xrightarrow{\cong} & PB \\
\scriptstyle{f^*} \downarrow & & \downarrow \scriptstyle{f^{-1}} \\
\mathrm{Set}(A, \Omega) & \xrightarrow[\cong]{} & PA
\end{array}
$$

commutes. That is, naturality asserts that given a function $\chi_{B'} : B → \Omega$ classifying the subset $B' \subset B$, the composite function $A \xrightarrow{f} B \xrightarrow{\chi_{B'}} \Omega$ classifies the subset $f^{-1}(B') \subset A$.

(ii) The functor $O\colon \mathsf{Top}^{\mathrm{op}} \to \mathsf{Set}$ that sends a space to its set of open subsets is represented by the **Sierpinski space** S, the topological space with two points, one closed and one open. The natural bijection $\mathsf{Top}(X, S) \cong O(X)$ associates a continuous function $X \to S$ to the preimage of the open point. This bijection is natural because a composite function $Y \to X \to S$ classifies the preimage of the open subset of X under the function $Y \to X$.

(iii) The Sierpinski space also represents the functor $C\colon \mathsf{Top}^{\mathrm{op}} \to \mathsf{Set}$ that sends a space to its set of closed subsets. Composing the natural isomorphisms $O \cong \mathsf{Top}(-, S) \cong C$ we see that the closed set and open set functors are naturally isomorphic. The composite natural isomorphism carries an open subset to its complement, which is closed. This recovers the natural isomorphism described in Example 1.4.3(v).

(iv) The functor $\mathrm{Hom}(- \times A, B)\colon \mathsf{Set}^{\mathrm{op}} \to \mathsf{Set}$ that sends a set X to the set of functions $X \times A \to B$ is represented by the set B^A of functions from A to B. That is, there is a natural bijection between functions $X \times A \to B$ and functions $X \to B^A$. This natural isomorphism is referred to as **currying** in computer science; by fixing a variable in a two-variable function, one obtains a family of functions in a single variable.

(v) The functor $U(-)^*\colon \mathsf{Vect}_{\Bbbk}^{\mathrm{op}} \to \mathsf{Set}$ that sends a vector space to the set of vectors in its dual space is represented by the vector space $\Bbbk$, i.e., linear maps $V \to \Bbbk$ are, by definition, precisely the vectors in the dual space V^*.

(vi) For any fixed abelian group A and any $n \geq 0$, *singular cohomology with coefficients in A* defines a functor $H^n(-; A)\colon \mathsf{Top}^{\mathrm{op}} \to \mathsf{Ab}$. As in Example 1.3.2(iii), this functor is a homotopy invariant, factoring through the quotient $\mathsf{Top} \to \mathsf{Htpy}$ to define a functor $H^n(-; A)\colon \mathsf{Htpy}^{\mathrm{op}} \to \mathsf{Ab}$. Passing to underlying sets and restricting to a subcategory of "nice" spaces, such as the *CW complexes*, the resulting functor $H^n(-; A)\colon \mathsf{Htpy}_{\mathrm{CW}}^{\mathrm{op}} \to \mathsf{Set}$ is represented by the **Eilenberg–MacLane space** $K(A, n)$. That is, for any CW complex X, homotopy classes of maps $X \to K(A, n)$ stand in bijection with elements of the nth singular cohomology group $H^n(X; A)$ of X with coefficients in A.

(vii) A **classifying space** for a topological group G is a CW complex BG that represents the functor $\mathsf{Htpy}_{\mathrm{CW}}^{\mathrm{op}} \to \mathsf{Set}$ that takes a CW complex to the set of isomorphism classes of *principle G-bundles* over it.[4]

Our language asserts that a representation encodes some sort of universal property of its representing object, but many questions remain:

- How unique are these universal properties? If two objects represent the same functor, are they isomorphic?
- What data is involved in the construction of a natural isomorphism between a representable functor F and the functor represented by an object c?
- How do the universal properties expressed by representable functors relate to initial and terminal objects—our first paradigm for "universality"?

Answers to all of these questions will make use of the Yoneda lemma, to which we now turn.

Exercises.

---

[4]If G is a discrete group, the CW complex BG is built from a collection of combinatorial data that defines the *nerve* of the category BG (see Exercise 6.5.iv), hence our use of this notation for 1-object groupoids.

EXERCISE 2.1.i. For each of the three functors

$$1 \underset{1}{\overset{0}{\rightleftarrows}} 2$$

between the categories 1 and 2, describe the corresponding natural transformations between the covariant functors $\mathsf{Cat} \Rrightarrow \mathsf{Set}$ represented by the categories 1 and 2.

EXERCISE 2.1.ii. Prove that if $F: \mathsf{C} \to \mathsf{Set}$ is representable, then F preserves monomorphisms, i.e., sends every monomorphism in C to an injective function. Use the contrapositive to find a covariant set-valued functor defined on your favorite concrete category that is not representable.

EXERCISE 2.1.iii. Suppose $F: \mathsf{C} \to \mathsf{Set}$ is equivalent to $G: \mathsf{D} \to \mathsf{Set}$ in the sense that there is an equivalence of categories $H: \mathsf{C} \to \mathsf{D}$ so that GH and F are naturally isomorphic.

(i) If G is representable, then is F representable?
(ii) If F is representable, then is G representable?

EXERCISE 2.1.iv. A functor F defines a **subfunctor** of G if there is a natural transformation $\alpha: F \Rightarrow G$ whose components are monomorphisms. In the case of $G: \mathsf{C}^{\mathrm{op}} \to \mathsf{Set}$, a subfunctor is given by a collection of subsets $Fc \subset Gc$ so that each $Gf: Gc \to Gc'$ restricts to a function $Ff: Fc \to Fc'$. Characterize those subsets that assemble into a subfunctor of the representable functor $\mathsf{C}(-, c)$.

EXERCISE 2.1.v. The functor of Example 2.1.5(xi) that sends a category to its collection of isomorphisms is a subfunctor of the functor of Example 2.1.5(x) that sends a category to its collection of morphisms. Define a functor between the representing categories 1 and 2 that induces the corresponding monic natural transformation between these representable functors.

2.2. The Yoneda lemma

> Yoneda enjoyed relating the story of the origins of this lemma, as follows ... Yoneda spent a year in France (apparently in 1954 and 1955). There he met Saunders Mac Lane. Mac Lane, then visiting Paris, was anxious to learn from Yoneda, and commenced an interview with Yoneda in a café at Gare du Nord. The interview was continued on Yoneda's train until its departure. In its course, Mac Lane learned about the lemma and subsequently baptized it.
>
> Saunders Mac Lane, "The Yoneda lemma"
> **[ML98b]**

To gain a better understanding of the representations considered in Examples 2.1.5 and 2.1.6 above, it would help to answer the following question: Given a functor $F: \mathsf{C} \to \mathsf{Set}$, what data is needed to define a natural isomorphism $\mathsf{C}(c, -) \cong F$? More generally, what data is needed to define a natural transformation $\mathsf{C}(c, -) \Rightarrow F$?

Before addressing the general question, let us cut our teeth on a few warm-up examples.

EXAMPLE 2.2.1. Consider diagrams indexed by the ordinal category ω. A functor $F: \omega \to \mathsf{Set}$ is given by a family of sets $(F_n)_{n \in \omega}$ together with functions $f_{n,n+1}: F_n \to F_{n+1}$. The functor $\omega(k, -): \omega \to \mathsf{Set}$ represented by the object $k \in \omega$ corresponds to a family whose first k sets, indexed by objects $n < k$, are empty and whose remaining sets, indexed by

$n \geq k$ are singletons. The data of a natural transformation $\alpha\colon \omega(k, -) \Rightarrow F$ is given by its components $(\alpha_n\colon \omega(k, n) \to F_n)_{n\in\omega}$, which must satisfy a naturality condition. Using the fact that the sets $\omega(k, n)$ are either empty or singletons, this data is displayed by the diagram:

$$
\begin{array}{ccccccccccccc}
\emptyset & \longrightarrow & \emptyset & \longrightarrow & \cdots & \longrightarrow & \emptyset & \longrightarrow & * & \longrightarrow & * & \longrightarrow & \cdots & \longrightarrow & * & \longrightarrow & \cdots \\
\downarrow{\scriptstyle\alpha_0} & & \downarrow{\scriptstyle\alpha_1} & & & & \downarrow{\scriptstyle\alpha_{k-1}} & & \downarrow{\scriptstyle\alpha_k} & & \downarrow{\scriptstyle\alpha_{k+1}} & & & & \downarrow{\scriptstyle\alpha_n} & & \\
F_0 & \underset{f_{0,1}}{\longrightarrow} & F_1 & \longrightarrow & \cdots & \longrightarrow & F_{k-1} & \underset{f_{k-1,k}}{\longrightarrow} & F_k & \underset{f_{k,k+1}}{\longrightarrow} & F_{k+1} & \longrightarrow & \cdots & \longrightarrow & F_n & \longrightarrow & \cdots
\end{array}
$$

Evidently, the components α_n contain no information for $n < k$. For $n \geq k$, the components determine elements $\alpha_n \in F_n$ (see Example 2.1.5(i)). The naturality condition demands that $\alpha_{n+1} = f_{n,n+1}(\alpha_n)$. By an inductive argument, the natural transformation $\alpha\colon \omega(k, -) \Rightarrow F$ is completely and uniquely determined by the choice of the first element $\alpha_k \in F_k$, which is the image of the identity morphism $1_k \in \omega(k, k)$ under the kth component map $\alpha_k\colon \omega(k, k) \to F_k$. Moreover, any element of F_k may be chosen as this image; naturality places no further restrictions.

As a second example, we consider diagrams whose indexing categories are groups.

EXAMPLE 2.2.2. There is a single object in the category BG whose endomorphisms define a group G. Thus, there is a unique covariant represented functor $BG \to$ Set and a unique contravariant represented functor $BG^{\mathrm{op}} \to$ Set. Example 1.3.9 characterized the functors from BG to Set; in the covariant case, they correspond to left G-sets, and in the contravariant case, they correspond to right G-sets. Recalling that the elements of G define the set of automorphisms of the unique object in the category BG, we see that the covariant represented functor is the G-set G, with its action by left multiplication, while the contravariant represented functor is again G, but with its right action by right multiplication.

Now consider a G-set $X\colon BG \to$ Set. Example 1.4.3(iv) observed that a natural transformation $\phi\colon G \Rightarrow X$ is exactly a G-equivariant map $\phi\colon G \to X$. Here $\phi\colon G \to X$ is the unique component of the natural transformation, and equivariance of this map expresses the naturality condition. To define ϕ we must specify elements $\phi(g) \in X$ for each $g \in G$. Equivariance demands that $\phi(g \cdot h) = g \cdot \phi(h)$. Taking h to be the identity element, we see that $\phi(g) = g \cdot \phi(e)$. In other words, the choice of $\phi(e) \in X$ forces us to define $\phi(g)$ to be $g \cdot \phi(e)$. Moreover, any choice of $\phi(e) \in X$ is permitted, because the left action of G on G is free.[5]

This argument proves:

PROPOSITION 2.2.3. *G-equivariant maps $G \to X$ correspond bijectively to elements of X, identified as the image of the identity element $e \in G$.*

For diagrams indexed by the poset ω and by a group BG, Examples 2.2.1 and 2.2.2 reveal that natural transformations whose domain is a represented functor are determined by the choice of a single element, which lives in the set defined by evaluating the codomain functor at the representing object, and moreover any choice is permitted. In each case, this element is the image of the identity morphism at the representing object.[6] This is no

---

[5]The action of a group G on a set X is **free** if every stabilizer group is trivial. In the context of the left action of G on itself, if there were distinct elements k, h so that $g = k \cdot e = h \cdot e$, we might be forced to make contradictory definitions of $\phi(g)$.

[6]Recall, the identity element in a group corresponds to the identity morphism in the one-object category.

coincidence. Indeed, the same is true for diagrams of any shape by what is arguably the most important result in category theory: the Yoneda lemma.[7]

THEOREM 2.2.4 (Yoneda lemma). *For any functor $F\colon \mathsf{C} \to \mathsf{Set}$, whose domain C is locally small and any object $c \in \mathsf{C}$, there is a bijection*

$$\mathrm{Hom}(\mathsf{C}(c, -), F) \cong Fc$$

that associates a natural transformation $\alpha\colon \mathsf{C}(c, -) \Rightarrow F$ to the element $\alpha_c(1_c) \in Fc$. Moreover, this correspondence is natural in both c and F.

As C is locally small but not necessarily small, a priori the collection of natural transformations $\mathrm{Hom}(\mathsf{C}(c, -), F)$ might be large. However, the bijection in the Yoneda lemma proves that this particular collection of natural transformations indeed forms a set.

The statement of the dual form of Theorem 2.2.4, for contravariant functors, is left to Exercise 2.2.i. We divide the proof into two parts, first demonstrating the bijection and then turning our attention to the naturality statement.

PROOF OF THE BIJECTION. There is clearly a function $\Phi\colon \mathrm{Hom}(\mathsf{C}(c, -), F) \to Fc$ that maps a natural transformation $\alpha\colon \mathsf{C}(c, -) \Rightarrow F$ to the image of 1_c under the component function $\alpha_c\colon \mathsf{C}(c, c) \to Fc$, i.e.,

$$\Phi\colon \mathrm{Hom}(\mathsf{C}(c, -), F) \to Fc \qquad \Phi(\alpha) := \alpha_c(1_c).$$

Our first aim is to define an inverse function $\Psi\colon Fc \to \mathrm{Hom}(\mathsf{C}(c, -), F)$ that constructs a natural transformation $\Psi(x)\colon \mathsf{C}(c, -) \Rightarrow F$ from any $x \in Fc$. To this end, we must define components $\Psi(x)_d\colon \mathsf{C}(c, d) \to Fd$ so that naturality squares, such as displayed for $f\colon c \to d$ in C, commute:

$$
\begin{array}{ccc}
\mathsf{C}(c, c) & \xrightarrow{\;\Psi(x)_c\;} & Fc \\
{\scriptstyle f_*}\downarrow & & \downarrow{\scriptstyle Ff} \\
\mathsf{C}(c, d) & \xrightarrow[\;\Psi(x)_d\;]{} & Fd
\end{array}
$$

The image of the identity element $1_c \in \mathsf{C}(c, c)$ under the left-bottom composite is $\Psi(x)_d(f) \in Fd$, the value of the component $\Psi(x)_d$ at the element $f \in \mathsf{C}(c, d)$. The image under the top-right composite is $Ff(\Psi(x)_c(1_c))$. For Ψ to define an inverse for Φ, we must define $\Psi(x)_c(1_c) = x$. Now, naturality forces us to define

$$(2.2.5) \qquad \Psi\colon Fc \to \mathrm{Hom}(\mathsf{C}(c, -), F) \qquad \Psi(x)_d(f) := Ff(x).$$

This condition completely determines the components $\Psi(x)_d$ of $\Psi(x)$.

It remains to verify that $\Psi(x)$ is natural. To that end, we must show for a generic morphism $g\colon d \to e$ in C (one whose domain is not necessarily the distinguished object c), that the square

$$
\begin{array}{ccc}
\mathsf{C}(c, d) & \xrightarrow{\;\Psi(x)_d\;} & Fd \\
{\scriptstyle g_*}\downarrow & & \downarrow{\scriptstyle Fg} \\
\mathsf{C}(c, e) & \xrightarrow[\;\Psi(x)_e\;]{} & Fe
\end{array}
$$

commutes. The image of $f \in \mathsf{C}(c, d)$ along the left-bottom composite is $\Psi(x)_e(gf) := F(gf)(x)$. The image along the top-right composite is $Fg(\Psi(x)_d(f)) := Fg(Ff(x))$. By

---

[7]For the oral history of the Yoneda lemma, which first appeared in print in a paper of Grothendieck [**Gro60**], see [**ML98b**], quoted at the beginning of this section.

functoriality of F, $F(gf) = Fg \cdot Ff$, so these elements agree. Thus, the formula (2.2.5) defines the function $\Psi\colon Fc \to \mathrm{Hom}(\mathrm{C}(c, -), F)$.

By construction, $\Phi\Psi(x) = \Psi(x)_c(1_c) = x$, so Ψ is a right inverse to Φ. We wish to show that $\Psi\Phi(\alpha) = \alpha$, i.e., that the natural transformation $\Psi(\alpha_c(1_c))$ is α. It suffices to show that these natural transformations have the same components. By definition,

$$\Psi(\alpha_c(1_c))_d(f) = Ff(\alpha_c(1_c)).$$

By naturality of α, the square

(2.2.6)
$$\begin{array}{ccc} \mathrm{C}(c,c) & \xrightarrow{\alpha_c} & Fc \\ {\scriptstyle f_*}\downarrow & & \downarrow{\scriptstyle Ff} \\ \mathrm{C}(c,d) & \xrightarrow[\alpha_d]{} & Fd \end{array}$$

commutes, from which we see that $Ff(\alpha_c(1_c)) = \alpha_d(f)$. Thus, $\Psi(\alpha_c(1_c))_d = \alpha_d$, proving that Ψ is also a left inverse to Φ. The explicit bijections Φ and Ψ prove that evaluation of a natural transformation at the identity of the representing object defines an isomorphism

$$\mathrm{Hom}(\mathrm{C}(c, -), F) \xrightarrow{\cong} Fc,$$

as claimed.   □

PROOF OF NATURALITY. The naturality in the statement of the Yoneda lemma amounts to the following pair of assertions. Naturality in the functor asserts that, given a natural transformation $\beta\colon F \Rightarrow G$, the element of Gc representing the composite natural transformation $\beta\alpha\colon \mathrm{C}(c, -) \Rightarrow F \Rightarrow G$ is the image under $\beta_c\colon Fc \to Gc$ of the element of Fc representing $\alpha\colon \mathrm{C}(c, -) \Rightarrow F$, i.e., the diagram

$$\begin{array}{ccc} \mathrm{Hom}(\mathrm{C}(c,-), F) & \xrightarrow{\ \Phi_F\ } & Fc \\ {\scriptstyle \beta_*}\downarrow\ {\scriptstyle\cong} & & {\scriptstyle\cong}\ \downarrow{\scriptstyle \beta_c} \\ \mathrm{Hom}(\mathrm{C}(c,-), G) & \xrightarrow[\ \Phi_G\ ]{\cong} & Gc \end{array}$$

commutes in Set. By definition, $\Phi_G(\beta \cdot \alpha) = (\beta \cdot \alpha)_c(1_c)$, which is $\beta_c(\alpha_c(1_c))$ by the definition of vertical composition of natural transformations given in Lemma 1.7.1, and this is $\beta_c(\Phi_F(\alpha))$.

Naturality in the object asserts that, given a morphism $f\colon c \to d$ in C, the element of Fd representing the composite natural transformation $\alpha f^*\colon \mathrm{C}(d, -) \Rightarrow \mathrm{C}(c, -) \Rightarrow F$ is the image under $Ff\colon Fc \to Fd$ of the element of Fc representing α, i.e., the diagram

$$\begin{array}{ccc} \mathrm{Hom}(\mathrm{C}(c,-), F) & \xrightarrow{\ \Phi_c\ } & Fc \\ {\scriptstyle (f^*)^*}\downarrow & & \downarrow{\scriptstyle Ff} \\ \mathrm{Hom}(\mathrm{C}(d,-), F) & \xrightarrow[\ \Phi_d\ ]{\cong} & Fd \end{array}$$

commutes. Here, the image of $\alpha \in \mathrm{Hom}(\mathrm{C}(c, -), F)$ along the top-right composite is $Ff(\alpha_c(1_c))$, and the image along the left-bottom composite is $(\alpha \cdot f^*)_d(1_d)$. By the definition of vertical composition, the dth component of the composite natural transformation $\alpha \cdot f^*$

is the function

$$C(d, d) \xrightarrow{\;f^*\;} C(c, d) \xrightarrow{\;\alpha_d\;} Fd$$

$$1_d \quad \mapsto \quad f \quad \mapsto \quad \alpha_d(f)$$

As demonstrated by the commutative square (2.2.6), $\alpha_d(f) = Ff(\alpha_c(1_c))$, which proves that this second naturality square commutes. □

REMARK 2.2.7. Were it not for size issues, we could express Theorem 2.2.4 more concisely as saying that the maps Φ define the components of a natural isomorphism between two functors that we now introduce. The pair c and F in the statement of the Yoneda lemma define an object in the product category $C \times \mathsf{Set}^C$; recall Set^C is the category of functors $C \to \mathsf{Set}$ and natural transformations between them defined in Corollary 1.7.2. There is a bifunctor $\mathrm{ev} \colon C \times \mathsf{Set}^C \to \mathsf{Set}$ that maps (c, F) to the set Fc, i.e., to the set obtained when evaluating the functor F at the object c. This functor defines the codomain of the natural isomorphism Φ.

The definition of the domain of Φ makes use of a functor

$$C^{\mathrm{op}} \xrightarrow{\;y\;} \mathsf{Set}^C$$

$$
\begin{array}{ccc}
c & \mapsto & C(c, -) \\
f \downarrow & \mapsto & \uparrow f^* \\
d & \mapsto & C(d, -)
\end{array}
$$

This is the functor obtained by applying Exercise 1.7.vii to the bifunctor introduced in Definition 1.3.13. Using Exercise 1.3.v to regard y as a functor $C \to (\mathsf{Set}^C)^{\mathrm{op}}$, the domain of Φ is defined to be the composite

$$\mathrm{Hom}(y(-), -) := C \times \mathsf{Set}^C \xrightarrow{\;y \times 1_{\mathsf{Set}^C}\;} (\mathsf{Set}^C)^{\mathrm{op}} \times \mathsf{Set}^C \xrightarrow{\;\mathrm{Hom}\;} \mathsf{SET}$$

$$(c, F) \quad \mapsto \quad (C(c, -), F) \quad \mapsto \quad \mathrm{Hom}(C(c, -), F)$$

Here is where the size issues arise. If C is small, then Set^C is locally small, and the hom functor is valued in the category of sets, as in Definition 1.3.13. However, if C is only locally small, Set^C need not be locally small; see Remark 1.7.3. We write SET to indicate that the collection of natural transformations between a pair of functors $F, G \colon C \rightrightarrows \mathsf{Set}$ might not be a set.

The composite functor $\mathrm{Hom}(y(-), -) \colon C \times \mathsf{Set}^C \to \mathsf{Set}$ is valued in the usual category of sets by the proof of the bijection in Theorem 2.2.4 just given. So the Yoneda lemma asserts that evaluating at the representing object's identity morphism defines a natural isomorphism

$$
C \times \mathsf{Set}^C \underset{\mathrm{ev}}{\overset{\mathrm{Hom}(y(-),-)}{\cong \Downarrow \; \Phi}} \mathsf{Set}
$$

An easy application of the Yoneda lemma completely characterizes natural transformations between representable functors: any locally small category C is isomorphic to the full subcategory of $\mathsf{Set}^{C^{\mathrm{op}}}$ spanned by the contravariant represented functors, and C^{op} is isomorphic to the full subcategory of Set^C spanned by the covariant represented functors, via the canonically-defined **covariant** and **contravariant Yoneda embeddings**.

COROLLARY 2.2.8 (Yoneda embedding). *The functors*

$$C \overset{y}{\longleftrightarrow} \mathsf{Set}^{C^{op}} \qquad C^{op} \overset{y}{\longleftrightarrow} \mathsf{Set}^{C}$$

$$
\begin{array}{ccc}
c & \mapsto & C(-,c) \\
f\downarrow & \mapsto & \downarrow f_* \\
d & \mapsto & C(-,d)
\end{array}
\qquad
\begin{array}{ccc}
c & \mapsto & C(c,-) \\
f\downarrow & \mapsto & \uparrow f^* \\
d & \mapsto & C(d,-)
\end{array}
$$

define full and faithful embeddings.

Via Exercise 1.7.vii, the co- and contravariant Yoneda embeddings are really just two different incarnations of a common bifunctor

$$C(-,-) \colon C^{op} \times C \to \mathsf{Set}$$

defined for any locally small category C, namely the two-sided represented functor introduced in Definition 1.3.13. The natural transformations $f_* \colon C(-,c) \Rightarrow C(-,d)$ and $f^* \colon C(d,-) \Rightarrow C(c,-)$ are defined in Example 1.4.7, by post- and pre-composition with the morphism f, respectively.

PROOF. By Definition 1.5.7, the functors $y \colon C \hookrightarrow \mathsf{Set}^{C^{op}}$ and $y \colon C^{op} \hookrightarrow \mathsf{Set}^{C}$ are fully faithful if they define local bijections between hom-sets

$$C(c,d) \overset{\cong}{\to} \mathrm{Hom}(C(-,c), C(-,d)) \qquad C(c,d) \overset{\cong}{\to} \mathrm{Hom}(C(d,-), C(c,-)).$$

From the construction of the functors y, it is easy to see that these maps are injections: distinct morphisms $c \rightrightarrows d$ induce distinct natural transformations (see Exercise 1.4.iv). The Yoneda lemma implies that every natural transformation between represented functors arises in this way. By Theorem 2.2.4, natural transformations

$$\alpha \colon C(d,-) \Rightarrow C(c,-)$$

correspond to elements of $C(c,d)$, i.e., to morphisms $f \colon c \to d$ in C, where the element f is $\alpha_d(1_d)$. The natural transformation $f^* \colon C(d,-) \Rightarrow C(c,-)$ defined by pre-composition by f sends 1_d to f. Thus, the bijection implies that $\alpha = f^*$. □

Corollary 2.2.8, that natural transformations between represented functors correspond to morphisms between the representing objects, is an extraordinarily powerful result, as we shall gradually discover. Common convention refers to both Theorem 2.2.4 and Corollary 2.2.8 as "the Yoneda lemma."

We close this section with two non-categorical applications of the fullness of the Yoneda embedding. A third even more paradigmatic application appears as Proposition 2.3.9.

COROLLARY 2.2.9. *Every row operation on matrices with n rows is defined by left multiplication by some n × n matrix, namely the matrix obtained by performing the row operation on the identity matrix.*

PROOF. Recall Mat_R is the category whose objects are positive integers and in which a morphism $m \to n$ is an $n \times m$ matrix, with n rows and m columns, whose entries are elements of the unital ring R. The elements in the image of the represented functor $\mathrm{Hom}(-,n)$ are matrices with n rows. The row operations of elementary linear algebra—for instance, replacing the ith row with the sum of the ith and jth row—define natural endomorphisms of $\mathrm{Hom}(-,n)$; naturality here follows from linearity of (right) matrix multiplication. Thus, by Corollary 2.2.8, every row operation must be definable by left multiplication by a suitable

$n \times n$ matrix. Moreover, Theorem 2.2.4 allows us to identify this matrix: it is obtained by applying the row operation in question to the $n \times n$ identity matrix. □

Another corollary of the Yoneda lemma is a classical result known as **Cayley's theorem**, stating that any abstract group may be realized as a subgroup of a permutation group.

COROLLARY 2.2.10. *Any group is isomorphic to a subgroup of a permutation group.*

PROOF. Regarding a group G as a one-object category BG, Example 2.2.2 identifies the image of the covariant Yoneda embedding $BG \hookrightarrow \mathsf{Set}^{BG^{\mathrm{op}}}$ as the right G-set G, with G acting by right multiplication. Corollary 2.2.8 tells us that the only G-equivariant endomorphisms of the right G-set G are those maps defined by left multiplication with a fixed element of G. In particular, any G-equivariant endomorphism of G must be an automorphism, a fact that is not otherwise obvious.

In this way, the Yoneda embedding defines an isomorphism between G and the automorphism group of the right G-set G, an object in $\mathsf{Set}^{BG^{\mathrm{op}}}$. Composing with the faithful forgetful functor $\mathsf{Set}^{BG^{\mathrm{op}}} \to \mathsf{Set}$, we obtain an isomorphism between G and a subgroup of the automorphism group $\mathrm{Sym}(G)$ of the set G. □

Exercises.

EXERCISE 2.2.i. State and prove the dual to Theorem 2.2.4, characterizing natural transformations $\mathsf{C}(-, c) \Rightarrow F$ for a contravariant functor $F \colon \mathsf{C}^{\mathrm{op}} \to \mathsf{Set}$.

EXERCISE 2.2.ii. Explain why the Yoneda lemma does not dualize to classify natural transformations from an arbitrary set-valued functor to a represented functor.

EXERCISE 2.2.iii. As discussed in Section 2.2, diagrams of shape ω are determined by a countably infinite family of objects and a countable infinite sequence of morphisms. Describe the Yoneda embedding $y \colon \omega \hookrightarrow \mathsf{Set}^{\omega^{\mathrm{op}}}$ in this manner (as a family of ω^{op}-indexed functors and natural transformations). Prove directly, without appealing to the Yoneda lemma, that y is full and faithful.

EXERCISE 2.2.iv. Prove the following strengthening of Lemma 1.2.3, demonstrating the equivalence between an isomorphism in a category and a **representable isomorphism** between the corresponding co- or contravariant represented functors: the following are equivalent:

(i) $f \colon x \to y$ is an isomorphism in C.
(ii) $f_* \colon \mathsf{C}(-, x) \Rightarrow \mathsf{C}(-, y)$ is a natural isomorphism.
(iii) $f^* \colon \mathsf{C}(y, -) \Rightarrow \mathsf{C}(x, -)$ is a natural isomorphism.

EXERCISE 2.2.v. By the Yoneda lemma, natural endomorphisms of the contravariant power set functor $P \colon \mathsf{Set}^{\mathrm{op}} \to \mathsf{Set}$ correspond bijectively to endomorphisms of its representing object $\Omega = \{\bot, \top\}$. Describe the natural endomorphisms of P that correspond to each of the four elements of $\mathrm{Hom}(\Omega, \Omega)$. Do these functions induce natural endomorphisms of the covariant power set functor?

EXERCISE 2.2.vi. Do there exist any non-identity natural endomorphisms of the category of spaces? That is, does there exist any family of continuous maps $X \to X$, defined for all spaces X and not all of which are identities, that are natural in all maps in the category Top?

EXERCISE 2.2.vii. Use the Yoneda lemma to explain the connection between homeomorphisms of the standard unit interval $I = [0, 1] \subset \mathbb{R}$ and natural automorphisms of the path functor $\mathrm{Path} \colon \mathsf{Top} \to \mathsf{Set}$ of Example 2.1.5(xiv), which we call **re-parameterizations**.

2.3. Universal properties and universal elements

The Yoneda lemma enables us to answer the questions posed at the end of §2.1. In this section, we study the data associated to a representation and prove that a representable functor *defines* its representing object.

In a locally small category C, any pair of isomorphic objects $x \cong y$ are **representably isomorphic**, meaning that $C(-, x) \cong C(-, y)$ and $C(x, -) \cong C(y, -)$. This is an immediate consequence of the functoriality of the Yoneda embeddings $C \hookrightarrow \mathsf{Set}^{C^{op}}$ and $C^{op} \hookrightarrow \mathsf{Set}^{C}$; see Exercise 2.2.iv. The Yoneda lemma supplies the converse:

PROPOSITION 2.3.1. *Consider a pair of objects x and y in a locally small category C.*

(i) *If either the co- or contravariant functors represented by x and y are naturally isomorphic, then x and y are isomorphic.*

(ii) *In particular, if x and y represent the same functor, then x and y are isomorphic.*

PROOF. The full and faithful Yoneda embeddings $C \hookrightarrow \mathsf{Set}^{C^{op}}$ and $C^{op} \hookrightarrow \mathsf{Set}^{C}$ create isomorphisms (see Exercise 1.5.iv). So an isomorphism between represented functors is induced by a unique isomorphism between their representing objects, which in particular must be isomorphic, proving (i). Now (ii) is an immediate consequence: given a functor represented by both x and y, the representing natural isomorphisms compose to demonstrate that x and y are representably isomorphic. □

There may be many isomorphisms between the objects x and y appearing in the proof of Proposition 2.3.1, but there is a unique natural isomorphism commuting with the chosen representations. On account of this, one typically refers to *the* representing object of a representable functor. Category theorists often use the definite article "the" in contexts where the object in question is well-defined up to canonical isomorphism. For instance:

COROLLARY 2.3.2. *The full subcategory of C spanned by its terminal objects is either empty or is a contractible groupoid. In particular, any two terminal objects in C are uniquely isomorphic.* [8]

A **contractible groupoid** is a category that is equivalent to the terminal category $\mathbb{1}$. Explicitly, a contractible groupoid is a category with a unique morphism in each hom-set.

PROOF. By Theorem 2.2.4, natural transformations $C(-, t) \Rightarrow C(-, t')$ between the functors represented by a pair of terminal objects $t, t' \in C$ stand in bijection with elements of the singleton set $C(t, t')$. So there can be at most one morphism between any pair of terminal objects. Proposition 2.3.1 implies that this morphism is an isomorphism: the objects t and t' are terminal if and only if they represent the functor $*: C^{op} \to \mathsf{Set}$ that is constant at the singleton set. □

The Yoneda lemma also provides a more explicit characterization of the data encoded by a representation.

DEFINITION 2.3.3. A **universal property** of an object $c \in C$ is expressed by a representable functor F together with a **universal element** $x \in Fc$ that defines a natural isomorphism $C(c, -) \cong F$ or $C(-, c) \cong F$, as appropriate, via the Yoneda lemma.

---

[8]Corollary 2.3.2 proves that the so-called "universal cover" of a topological space X is not an initial object in the category of connected covering spaces over X, unless X is simply connected. If it were, the Deck transformation group, which is isomorphic to the fundamental group $\pi_1(X)$, would necessarily be trivial.

EXAMPLE 2.3.4. Recall from Example 2.1.5(iv) that the forgetful functor U: Ring $\to$ Set is represented by the ring $\mathbb{Z}[x]$. The universal element, which defines the natural isomorphism

(2.3.5) $\text{Ring}(\mathbb{Z}[x], R) \cong UR,$

is the element $x \in \mathbb{Z}[x]$. As in the proof of the Yoneda lemma, the bijection (2.3.5) is implemented by evaluating a ring homomorphism $\phi: \mathbb{Z}[x] \to R$ at the element $x \in \mathbb{Z}[x]$ to obtain an element $\phi(x) \in R$.

EXAMPLE 2.3.6. A functor $E: BG \to$ Set, i.e., a left G-set E, is representable if and only if there is an isomorphism $G \cong E$ of left G-sets. This implies that the action of G on E is **free** (every stabilizer group is trivial) and **transitive** (the orbit of any point is the entire set), and that E is non-empty. Conversely, any non-empty free and transitive left G-set is representable; a proof is given in Example 2.4.10. By the Yoneda lemma, the data of a representation for E, i.e., a specific isomorphism $G \cong E$, is determined by the choice of an element $e \in E$, the universal element, serving as the image of the identity $e \in G$. In other words, a representable left G-set E is a group that has forgotten its identity element. A representable G-set is called a G-**torsor**.

For example, a torsor for the group $\mathbb{R}^n$ under addition is n-dimensional **affine space**, commonly denoted by $\mathbb{A}^n$ (or $\mathbb{A}^n_{\mathbb{R}}$ to emphasize the ground field). We think of $\mathbb{A}^n$ as being the collection of points in n-dimensional real space, but without a chosen origin. Thinking of points in $\mathbb{R}^n$ as vectors, the free and transitive action of $\mathbb{R}^n$ on $\mathbb{A}^n$ is easy to describe: a point in $\mathbb{A}^n$ is transported along the vector in question. This action is free, because every non-zero vector acts on every point non-trivially, and transitive, because any two points in affine space can be "subtracted" to define a unique vector in $\mathbb{R}^n$ that carries the one to the other. A choice of an origin point o in $\mathbb{A}^n$ determines coordinates for all of the other points, hence an isomorphism $\mathbb{R}^n \cong \mathbb{A}^n$. The point $o \in \mathbb{A}^n$ is the universal element: the coordinates for another point $p \in \mathbb{A}^n$ are defined to be the unique vector $\vec{x} \in \mathbb{R}^n$ so that $\vec{x} \cdot o = p$.

EXAMPLE 2.3.7. Fix $\Bbbk$-vector spaces V and W and consider the functor

$$\text{Bilin}(V, W; -): \text{Vect}_{\Bbbk} \to \text{Set}$$

that sends a vector space U to the set of $\Bbbk$-bilinear maps $V \times W \to U$. A **bilinear map** $f: V \times W \to U$ is a function of two variables so that for all $v \in V$, $f(v, -): W \to U$ is a linear map and for all $w \in W$, $f(-, w): V \to U$ is a linear map. Equivalently, by "currying," a bilinear map may be defined to be a linear map $V \to \text{Hom}(W, U)$ or $W \to \text{Hom}(V, U)$, where the codomains are vector spaces of linear maps.

A representation for the functor $\text{Bilin}(V, W; -)$ defines a vector space $V \otimes_{\Bbbk} W$, the **tensor product** of V and W. That is, the tensor product is defined by an isomorphism

(2.3.8) $\text{Vect}_{\Bbbk}(V \otimes_{\Bbbk} W, U) \cong \text{Bilin}(V, W; U),$

between the set of linear maps $V \otimes_{\Bbbk} W \to U$ and the set of bilinear maps $V \times W \to U$ that is natural in U.[9]

Theorem 2.2.4 tells us that the natural isomorphism (2.3.8) is determined by a universal element of $\text{Bilin}(V, W; V \otimes_{\Bbbk} W)$, i.e., by a bilinear map $\otimes: V \times W \to V \otimes_{\Bbbk} W$. The tensor product $V \otimes_{\Bbbk} W$ is the **universal vector space equipped with a bilinear map** from $V \times W$. The Yoneda lemma can be used to unpack what this means. The natural isomorphism (2.3.8) identifies any bilinear map $f: V \times W \to U$ with a linear map $\bar{f}: V \otimes_{\Bbbk} W \to U$. To

---

[9]This isomorphism is also natural in V and W; see Proposition 4.3.6.

understand this identification, consider the naturality square induced by $\bar{f}$.

$$
\begin{array}{ccc}
\mathsf{Vect}_{\Bbbk}(V \otimes_{\Bbbk} W, V \otimes_{\Bbbk} W) & \xrightarrow{\;\cong\;} & \mathsf{Bilin}(V, W; V \otimes_{\Bbbk} W) \\
\bar{f}_* \downarrow & & \downarrow \bar{f}_* \\
\mathsf{Vect}_{\Bbbk}(V \otimes_{\Bbbk} W, U) & \xrightarrow[\;\cong\;]{} & \mathsf{Bilin}(V, W; U)
\end{array}
$$

Tracing $1_{V \otimes_{\Bbbk} W}$ around this commutative square reveals that the bilinear map f factors uniquely through the bilinear map $\otimes$ along the linear map $\bar{f}$.

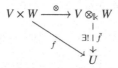

In other words, the bijection (2.3.8) is implemented by composing a linear map $V \otimes_{\Bbbk} W \to U$ with the universal bilinear map $\otimes$. This universal property tells us that the bilinear map $\otimes \colon V \times W \to V \otimes_{\Bbbk} W$ is initial in a category that will be described in Example 2.4.12(iii).

Moreover, the defining universal property of the tensor product gives a recipe for its construction as a vector space. Supposing the vector space $V \otimes_{\Bbbk} W$ exists, consider its quotient by the vector space spanned by[10] the image of the bilinear map $- \otimes -$. By definition, the quotient map $V \otimes_{\Bbbk} W \to V \otimes_{\Bbbk} W / \langle v \otimes w \rangle$ restricts along $- \otimes -$ to yield the zero bilinear map. But the zero map $V \otimes_{\Bbbk} W \to V \otimes_{\Bbbk} W / \langle v \otimes w \rangle$ also has this property, so by the universal property of $V \otimes W$, these linear maps must agree. Because the quotient map is surjective, this implies that $V \otimes_{\Bbbk} W$ is isomorphic to the span of the vectors $v \otimes w$ for all $v \in V$ and $w \in W$ modulo the bilinearity relations satisfied by $- \otimes -$, recovering the usual constructive definition.

The universal property of the tensor product of a pair of $\Bbbk$-vector spaces, more than simply characterizing the vector space up to isomorphism, allows one to prove useful properties about it without ever appealing to a specific basis.

PROPOSITION 2.3.9. *For any $\Bbbk$-vector spaces V and W, $V \otimes_{\Bbbk} W \cong W \otimes_{\Bbbk} V$.*

PROOF. By Proposition 2.3.1, it suffices to show that the functors $\mathsf{Bilin}(V, W; -)$ and $\mathsf{Bilin}(W, V; -)$ are naturally isomorphic. The natural isomorphism $\mathsf{Bilin}(V, W; U) \cong \mathsf{Bilin}(W, V; U)$ sends a bilinear map $f \colon V \times W \to U$ to the bilinear map $f^\sharp \colon W \times V \to U$ defined by $f^\sharp(w, v) := f(v, w)$. Composing natural isomorphisms, the represented functors

(2.3.10) $\mathsf{Vect}_{\Bbbk}(V \otimes_{\Bbbk} W, -) \cong \mathsf{Bilin}(V, W; -) \cong \mathsf{Bilin}(W, V; -) \cong \mathsf{Vect}_{\Bbbk}(W \otimes_{\Bbbk} V, -)$

are naturally isomorphic. By Proposition 2.3.1, a corollary of the Yoneda lemma, this natural isomorphism must arise from an isomorphism $V \otimes_{\Bbbk} W \cong W \otimes_{\Bbbk} V$ in $\mathsf{Vect}_{\Bbbk}$ between the representing objects.   □

REMARK 2.3.11. Moreover, the Yoneda lemma provides an explicit isomorphism $V \otimes_{\Bbbk} W \cong W \otimes_{\Bbbk} V$. The image of the identity linear transformation under the composite isomorphism

$\mathsf{Vect}_{\Bbbk}(V \otimes_{\Bbbk} W, V \otimes_{\Bbbk} W) \cong \mathsf{Bilin}(V, W; V \otimes_{\Bbbk} W)$

$\cong \mathsf{Bilin}(W, V; V \otimes_{\Bbbk} W) \cong \mathsf{Vect}_{\Bbbk}(W \otimes_{\Bbbk} V, V \otimes_{\Bbbk} W)$

defines an isomorphism $\phi \colon W \otimes_{\Bbbk} V \xrightarrow{\;\cong\;} V \otimes_{\Bbbk} W$ so that the natural isomorphism (2.3.10) is defined by precomposing with ϕ. In Example 2.3.7, we saw that the identity linear map

---

[10]The image of a bilinear map is not itself a sub-vector space, so closing under span is necessary.

is sent under the bijection $\mathsf{Vect}_\Bbbk(V \otimes_\Bbbk W, V \otimes_\Bbbk W) \cong \mathrm{Bilin}(V, W; V \otimes_\Bbbk W)$ to the universal bilinear map $\otimes \colon V \times W \to V \otimes_\Bbbk W$. This in turn is sent to the bilinear map

$$W \times V \xrightarrow{(w,v) \mapsto (v,w)} V \times W \xrightarrow{\otimes} V \otimes_\Bbbk W.$$

Appealing to the universal property of $\otimes \colon W \times V \to W \otimes_\Bbbk V$, which defines the natural isomorphism $\mathrm{Bilin}(W, V; V \otimes_\Bbbk W) \cong \mathsf{Vect}_\Bbbk(W \otimes_\Bbbk V, V \otimes_\Bbbk W)$, this composite bilinear map is sent to the unique linear map ϕ that makes the diagram

$$
\begin{array}{ccc}
W \times V & \xrightarrow{\;\otimes\;} & W \otimes_\Bbbk V \\
{\scriptstyle (w,v) \mapsto (v,w)} \downarrow & & \Big\downarrow {\scriptstyle \exists! \mid \phi} \\
V \times W & \xrightarrow{\;\otimes\;} & V \otimes_\Bbbk W
\end{array}
$$

commute. This ϕ is the desired linear isomorphism.

EXAMPLE 2.3.12. Fixing a group G, there is a functor

$$\mathsf{Ring} \xrightarrow{\mathrm{Group}(G,(-)^\times)} \mathsf{Set}$$

that sends a unital ring R to the set of group homomorphisms from G to the group of units $R^\times$. This functor expresses the universal property of the **group ring**[11] $\mathbb{Z}[G]$. By the Yoneda lemma, the natural isomorphism

$$\mathsf{Ring}(\mathbb{Z}[G], -) \cong \mathsf{Group}(G, (-)^\times),$$

is encoded by a group homomorphism $G \to \mathbb{Z}[G]^\times$. As in the proof of the Yoneda lemma, there is natural bijection between ring homomorphisms $\mathbb{Z}[G] \to R$ and group homomorphisms $G \to R^\times$ constructed as follows: a ring homomorphism $\mathbb{Z}[G] \to R$ induces a group homomorphism $\mathbb{Z}[G]^\times \to R^\times$ and thus a group homomorphism $G \to R^\times$ by precomposing with $G \to \mathbb{Z}[G]^\times$.

Exercises.

EXERCISE 2.3.i. What are the universal elements for the representations:

(i) defining the category 2 in Example 2.1.5(x)?
(ii) defining the Sierpinski space in Example 2.1.6(ii)?
(iii) defining the Sierpinski space in Example 2.1.6(iii)?

EXERCISE 2.3.ii. Use the defining universal property of the tensor product to prove that

(i) $\Bbbk \otimes_\Bbbk V \cong V$ for any $\Bbbk$-vector space V; and
(ii) $U \otimes_\Bbbk (V \otimes_\Bbbk W) \cong (U \otimes_\Bbbk V) \otimes_\Bbbk W$ for any $\Bbbk$-vector spaces U, V, W.

EXERCISE 2.3.iii. The set B^A of functions from a set A to a set B represents the contravariant functor $\mathsf{Set}(- \times A, B) \colon \mathsf{Set}^{\mathrm{op}} \to \mathsf{Set}$. The universal element for this representation is a function

$$\mathrm{ev} \colon B^A \times A \to B$$

called the **evaluation map**. Define the evaluation map and describe its universal property, in analogy with the universal bilinear map $\otimes$ of Example 2.3.7.

---

[11]There is a more general notion of group ring in which $\mathbb{Z}$ is replaced by an arbitrary commutative unital ring K. The universal property of $K[G]$ can be described analogously by replacing the category Ring with the category of K-algebras.

2.4. The category of elements

A **universal property** for an object $c \in C$ is expressed either by a contravariant functor F together with a representation $C(-, c) \cong F$ or by a covariant functor F together with a representation $C(c, -) \cong F$. The representations define a natural characterization of the maps into (in the contravariant case) or out of (in the covariant case) the object c. Proposition 2.3.1 implies that a universal property characterizes the object $c \in C$ up to isomorphism. More precisely, there is a unique isomorphism between c and any other object representing F that commutes with the chosen representations.

In such contexts, the phrase "c is the universal object in C with an x" assets that $x \in Fc$ is a **universal element** in the sense of Definition 2.3.3, i.e., x is the element of Fc that classifies the natural isomorphism that defines the representation by the Yoneda lemma. In this section, we prove that the term "universal" is being used in the precise sense alluded to at the beginning of this chapter: the universal element is either initial or terminal in an appropriate category. The category in question, called the *category of elements*, can be constructed in a canonical way from the data of the representable functor F. The main result of this section, Proposition 2.4.8, proves that any universal property can be understood as defining an initial or terminal object, as variance dictates, completing the promise made in §2.1.

The category of elements is also used to answer another question inspired by consideration of representable functors: namely, how to determine whether a given set-valued functor F is representable. And if this is the case, is there a way to identify when an element $x \in Fc$ determines a natural isomorphism $C(-, c) \cong F$, rather than simply a natural transformation $C(-, c) \Rightarrow F$?

DEFINITION 2.4.1. The **category of elements** $\int F$ of a covariant functor $F: C \to \text{Set}$ has

- as objects, pairs (c, x) where $c \in C$ and $x \in Fc$, and
- a morphism $(c, x) \to (c', x')$ is a morphism $f: c \to c'$ in C so that $Ff(x) = x'$.

The category of elements has an evident forgetful functor $\Pi: \int F \to C$.

If $Ff(x) = x'$, then

$$(c, x) \xrightarrow{f} (c', x') \qquad \in \qquad \int F$$

$$\Big\downarrow{\Pi} \qquad\qquad\qquad\qquad\qquad \Big\downarrow{\Pi}$$

$$c \xrightarrow[f]{} c' \qquad\qquad \in \qquad C$$

For a contravariant functor $F: C^{\text{op}} \to \text{Set}$, the category of elements $\int F$ is defined to be the opposite of the category of elements of F, regarded as covariant functor of C^{op}. For both co- and contravariant functors on C, our convention is that the category of elements should have a canonical forgetful functor $\int F \to C$. This dictates the direction for the morphisms in $\int F$ in both Definitions 2.4.1 and 2.4.2.

Explicitly:

DEFINITION 2.4.2. The **category of elements** $\int F$ of a contravariant functor $F: C^{\text{op}} \to \text{Set}$ has

- as objects, pairs (c, x) where $c \in C$ and $x \in Fc$, and
- a morphism $(c, x) \to (c', x')$ is a morphism $f: c \to c'$ in C so that $Ff(x') = x$.

The category of elements has an evident forgetful functor $\Pi\colon \int F \to \mathsf{C}$.

If $Ff(x') = x$, then
$$(c, x) \xrightarrow{\ f\ } (c', x') \qquad \in \qquad \int F$$
$$\downarrow \Pi \qquad\qquad\qquad\qquad\qquad \downarrow \Pi$$
$$c \xrightarrow{\ f\ } c' \qquad\qquad \in \qquad \mathsf{C}$$

EXAMPLE 2.4.3. Recall the functor n-Color: $\mathsf{Graph}^{\mathrm{op}} \to \mathsf{Set}$ that carries a graph to the set of n-colorings of its vertices with the property that no adjacent vertices are assigned the same color. An object in the category of elements of n-Color is a graph together with a chosen n-coloring; that is, objects are n-colored graphs. A morphism $\phi\colon G \to G'$ between a pair of n-colored graphs is a graph homomorphism $\phi\colon G \to G'$ so that the induced function n-Color$(\phi)\colon n$-Color$(G') \to n$-Color(G) carries the chosen coloring of G' to the chosen coloring of G. In other words, the graph homomorphism ϕ preserves the chosen colorings, in the sense that each purple vertex of G is carried to a purple vertex of G'. In summary, $\int n$-Color is the category of n-colored graphs and color-preserving graph homomorphisms.

EXAMPLE 2.4.4. For a concrete category C, the objects in the category of elements for the forgetful functor $U\colon \mathsf{C} \to \mathsf{Set}$ are elements $x \in Uc$ in the underlying set of a specified object $c \in \mathsf{C}$. Morphisms are maps in C whose underlying functions preserve the chosen elements. Extending previous notation introduced in certain special cases, it would be reasonable to write C_* for $\int U$ and refer to the category of elements as the category of **based objects** in C.

EXAMPLE 2.4.5. When C is a discrete category, a functor $F\colon \mathsf{C} \to \mathsf{Set}$ encodes an indexed family of sets $(F_c)_{c \in \mathsf{C}}$. The category of elements of F is again a discrete category whose set of objects is $\coprod_{c \in \mathsf{C}} F_c$. The canonical projection $\Pi\colon \coprod_{c \in \mathsf{C}} F_c \to \mathsf{C}$ is defined so that the fiber over $c \in \mathsf{C}$ is the set F_c. The set $\coprod_{c \in \mathsf{C}} F_c$ is called the **dependent sum** of the indexed family of sets $(F_c)_{c \in \mathsf{C}}$. The related **dependent product** $\prod_{c \in \mathsf{C}} F_c$ is the set of sections of the functor $\Pi\colon \coprod_{c \in \mathsf{C}} F_c \to \mathsf{C}$.

EXAMPLE 2.4.6. Objects in the category of elements of $\mathsf{C}(c, -)$ are morphisms $f\colon c \to x$ in C, i.e., the objects of $\int \mathsf{C}(c, -)$ are morphisms in C with domain c. A morphism from $f\colon c \to x$ to $g\colon c \to y$ is a morphism $h\colon x \to y$ so that $g = hf$; we say that h is a morphism under c:

$$\begin{array}{ccc} & c & \\ {}^{f}\swarrow & & \searrow^{g} \\ x & \xrightarrow[h]{} & y \end{array}$$

This category has another name: it is the **slice category** c/C **under** the object $c \in \mathsf{C}$ introduced in Exercise 1.1.iii. The forgetful functor $c/\mathsf{C} \to \mathsf{C}$ sends a morphism $f\colon c \to x$ to its codomain and a commutative triangle to the leg opposite the object c.

Dually, $\int \mathsf{C}(-, c)$ is the **slice category** C/c **over** the object $c \in \mathsf{C}$. Objects are morphisms $f\colon x \to c$ with codomain c, and a morphism from f to $g\colon y \to c$ is a morphism $h\colon x \to y$ so that $gh = f$; h is a morphism over c:

$$\begin{array}{ccc} x & \xrightarrow{\ h\ } & y \\ {}_{f}\searrow & & \swarrow_{g} \\ & c & \end{array}$$

The forgetful functor $\mathsf{C}/c \to \mathsf{C}$ projects onto the domain.

Note that $C(-, c) \cong C^{op}(c, -) \colon C^{op} \to \mathsf{Set}$. And indeed $c/(C^{op})$, the category of elements of $C(-, c)$ regarded as a covariant functor of C^{op}, is isomorphic to $(C/c)^{op}$, the opposite of the category of elements of the contravariant functor $C(-, c)$; see Exercise 1.2.i.

The Yoneda lemma supplies an alternative definition of the category of elements of F. In the contravariant case, objects $(c, x) \in \int F$ are in bijection with natural transformations $\Psi(x) \colon C(-, c) \Rightarrow F$ whose domain is a represented functor and whose codomain is F. A morphism from $\Psi(x)$ to a second element $\Psi(x') \colon C(-, c') \Rightarrow F$ is a natural transformation $C(-, c) \Rightarrow C(-, c')$—i.e., by Corollary 2.2.8, is a morphism $f \colon c \to c'$—so that the triangle of natural transformations

$$
\begin{array}{ccc}
C(-, c) \overset{\Psi(x)}{\Longrightarrow} F & & x \in Fc \\
f_* \Big\Downarrow \quad \nearrow_{\Psi(x')} & \rightsquigarrow & \uparrow \;\; \Big\uparrow Ff \\
C(-, c') & & x' \in Fc'
\end{array}
$$

commutes. This proves that the category of elements can be reconstructed as a comma category, as defined in Exercise 1.3.vi.

LEMMA 2.4.7. *For $F \colon C^{op} \to \mathsf{Set}$, the category of elements is isomorphic to the comma category*

$$\textstyle\int F \cong y \downarrow F$$

defined relative to the Yoneda embedding $y \colon C \to \mathsf{Set}^{C^{op}}$ and the object $F \colon \mathbb{1} \to \mathsf{Set}^{C^{op}}$.

Similarly, in the covariant case, the Yoneda lemma defines a bijection between objects $(c, x) \in \int F$ and natural transformations $\Psi(x) \colon C(c, -) \Rightarrow F$, and a morphism from $\Psi(x)$ to $\Psi(x') \colon C(c', -) \Rightarrow F$ is a natural transformation $C(c', -) \Rightarrow C(c, -)$, i.e., a morphism $f \colon c \to c'$, so that

$$
\begin{array}{ccc}
C(c, -) \overset{\Psi(x)}{\Longrightarrow} F & & x \in Fc \\
f^* \Big\Uparrow \quad \nearrow_{\Psi(x')} & \rightsquigarrow & \uparrow \;\; \Big\downarrow Ff \\
C(c', -) & & x' \in Fc'
\end{array}
$$

commutes.

Importantly, the category of elements can be used to show that all universal properties can be encoded by saying that certain data, the universal element of Definition 2.3.3, defines an initial or terminal object in an appropriate category, namely the category of elements of the representable functor.

PROPOSITION 2.4.8 (universal elements are universal elements). *A covariant set-valued functor is representable if and only if its category of elements has an initial object. Dually, a contravariant set-valued functor is representable if and only if its category of elements has a terminal object.*

It is easy to see that this condition is necessary. A natural isomorphism $C(c, -) \cong F$ induces an isomorphism of categories $\int F \cong \int C(c, -) \cong c/C$ (see Exercise 2.4.vii), and the latter has an initial object: the identity $1_c \in c/C$. Thus, if F is representable, then $\int F$ has an initial object. The surprise is that the presence of an initial object in $\int F$ is sufficient to establish representability of the functor F. The key to the proof, unsurprisingly, is the Yoneda lemma in the form of Theorem 2.2.4.[12]

---

[12] In fact, now that the statement and proof strategy are known, we invite the reader to first try to come up with the argument on their own.

PROOF. Consider a functor $F: \mathsf{C} \to \mathsf{Set}$ and suppose $(c, x) \in \int F$ is initial. We will show that the natural transformation $\Psi(x): \mathsf{C}(c, -) \Rightarrow F$ defined by the Yoneda lemma is a natural isomorphism. For any $y \in Fd$, initiality of $x \in Fc$ says there exists a unique morphism $(c, x) \to (d, y)$, i.e., a unique morphism $f: c \to d$ in C so that $Ff(x) = y$. As y runs through the elements of the set Fd, this says exactly that the component $\Psi(x)_d: \mathsf{C}(c, d) \to Fd$ is an isomorphism: in the proof of the Yoneda lemma, $\Psi(x)_d(f)$ was defined to be $Ff(x)$. Existence of the morphism $(c, x) \to (d, y)$ in $\int F$ asserts that the function $\Psi(x)_d$ is surjective while uniqueness asserts that it is injective.

Reversing this argument, a natural isomorphism $\alpha: \mathsf{C}(c, -) \cong F$ defines an object $\alpha_c(1_c) \in Fc$ in the category of elements, which we prove is initial. The bijection $\alpha_d: \mathsf{C}(c, d) \xrightarrow{\cong} Fd$ says that for each object $y \in Fd$, there is a unique morphism $f: c \to d$ so that $Ff(\alpha_c(1_c)) = y$. Thus, for each $(d, y) \in \int F$, there is a unique morphism $f: (c, \alpha_c(1_c)) \to (d, y)$ in $\int F$, which says exactly that $(c, \alpha_c(1_c))$ defines an initial object in the category of elements. □

Recall that a representation for a functor $F: \mathsf{C} \to \mathsf{Set}$ consists of an object $c \in \mathsf{C}$ together with a natural isomorphism $\mathsf{C}(c, -) \cong F$. Equivalently by Proposition 2.4.8, a representation for F is an initial object in $\int F$. Representations are not strictly unique: if c' is any object isomorphic to c, then a representation $\mathsf{C}(c, -) \cong F$ induces a representation $\mathsf{C}(c', -) \cong \mathsf{C}(c, -) \cong F$ defined by composing with the chosen isomorphism. However, representations are unique in an appropriate category-theoretic sense.

PROPOSITION 2.4.9. *For any functor $F: \mathsf{C} \to \mathsf{Set}$, the full subcategory of $\int F$ spanned by its representations is either empty or a contractible groupoid.*

PROOF. If F is not representable, then it has no representations and this subcategory is empty. Otherwise, Proposition 2.4.8 implies that any representation defines an initial object in $\int F$. By Corollary 2.3.2, the subcategory spanned by the initial objects is either empty or is a contractible groupoid: there exists a unique (iso)morphism between any two initial objects. □

EXAMPLE 2.4.10. The category of elements of a G-set $X: BG \to \mathsf{Set}$ is the translation groupoid $T_G X$ introduced in Example 1.5.18: its objects are elements $x \in X$ and its morphisms $g: x \to y$ are elements $g \in G$ so that $g \cdot x = y$. By Example 2.3.6, if X is representable, then any element of X may be chosen as a universal element. By Proposition 2.4.9, it follows that X is representable if and only if $\int X \cong T_G X$ is a contractible groupoid.

Assuming X is non-empty, this groupoid is contractible if and only if for each pair of elements $x, y \in X$, there is a unique $g \in G$ so that $g \cdot x = y$. Existence tells us that the action of G on X is transitive. Uniqueness, in the case $x = y$, tells us that it is free. This proves the claim made in Example 2.3.6 that any free and transitive left G-set is representable.

Trivially, any category E arises as a category of elements, namely for the constant functor $*: \mathsf{E} \to \mathsf{Set}$ valued at the singleton set, a representation for which defines an initial object in E. More interesting is when E arises as the category of elements for a functor indexed by a different category C, related by a forgetful functor $\mathsf{E} \to \mathsf{C}$.[13] For instance:

EXAMPLE 2.4.11. The category of discrete dynamical systems, introduced in Example 2.1.1, is the category of elements for the functor $U: \mathsf{End} \to \mathsf{Set}$ whose domain is the category of sets equipped with an endomorphism and whose maps are functions so that the diagram

---

[13] An elementary characterization of the functors $\mathsf{E} \to \mathsf{C}$ arising as categories of elements for some $\mathsf{C} \to \mathsf{Set}$ is suggested by Exercise 2.4.viii.

analogous to (2.1.2) commutes. This functor is represented by the object $(\mathbb{N}, s\colon \mathbb{N} \to \mathbb{N})$ and the representation is given by the universal element $0 \in \mathbb{N} = U(\mathbb{N}, s)$. Proposition 2.4.8 then implies the universal property observed in Example 2.1.1: that $(\mathbb{N}, s\colon \mathbb{N} \to \mathbb{N}, 0 \in \mathbb{N})$ is the universal discrete dynamical system.

EXAMPLE 2.4.12.

(i) The objects in the category of elements of the contravariant power set functor are pairs $A' \subset A$, a set with a chosen subset. A morphism $f\colon (A' \subset A) \to (B' \subset B)$ is a function $f\colon A \to B$ so that $f^{-1}(B') = A'$. The terminal object, corresponding to the representation described in Example 2.1.6(i), is the set $\Omega := \{\top, \bot\}$ of two elements with the distinguished singleton subset $\{\top\} \subset \Omega$. This is the universal set equipped with a subset.[14] For any $A' \subset A$, there is a unique function $h\colon A \to \Omega$ so that $h^{-1}(\top) = A'$. Note that even if we begin with a skeletal category of sets, $\{\top\} \subset \Omega$ is not the unique terminal object in the category of elements. It is isomorphic to $\{\bot\} \subset \Omega$.

(ii) An object in the category of elements of the forgetful functor $U\colon \mathsf{Vect}_{\Bbbk} \to \mathsf{Set}$ is a vector v in some $\Bbbk$-vector space V. A morphism $(V, v) \to (W, w)$ is a linear map $T\colon V \to W$ that carries v to w. The vector space $\Bbbk$, which represents U, together with its universal element $1 \in \Bbbk$, defines an initial object in the category of elements but it is not the only one. Any non-zero scalar $c \in \Bbbk$ defines a linear isomorphism $c \cdot -\colon \Bbbk \to \Bbbk$. Thus, the pairs $(\Bbbk, c)$ are also initial in the category of elements. More generally, any 1-dimensional vector space V and non-zero vector v define an initial object. Fixing such a pair (V, v), linear maps $V \to W$ are in bijection with vectors $w \in W$, taken to be the image of v.

(iii) Objects in the category of elements of $\mathrm{Bilin}(V, W; -)$ are bilinear maps $f\colon V \times W \to U$, with target some $\Bbbk$-vector space U. Morphisms are linear maps $T\colon U \to U'$ so that the diagram of functions

commutes, i.e., so that the bilinear map f' is the composite of the bilinear map f and the linear map T. The universal property of the universal bilinear map $\otimes\colon V \times W \to V \otimes_{\Bbbk} W$ described in Example 2.3.7 says exactly that $\otimes$ is initial in this category.

(iv) An object in the category of elements of the functor $U(-)^n\colon \mathsf{Group} \to \mathsf{Set}$ consists of a group G together with an n-tuple of elements $g_1, \ldots, g_n \in G$. A morphism $(g_1, \ldots, g_n \in G) \to (h_1, \ldots, h_n \in H)$ is a homomorphism $\phi\colon G \to H$ so that $\phi(g_i) = h_i$ for all i. The universal group with n elements, denoted by F_n, is the group with specified elements $x_1, \ldots, x_n \in F_n$ with the universal property that for any $g_1, \ldots, g_n \in G$ there is a unique group homomorphism $\phi\colon F_n \to G$ so that $\phi(x_i) = g_i$. The free group on n generators $x_1, \ldots, x_n$ has this property. Every element in F_n is a product of the x_i and their inverses, so a choice of n elements $\phi(x_i) \in G$ determines the entire map $\phi\colon F_n \to G$. Moreover, there are no relations between the x_i in F_n so any choices are permitted.

---

[14]In the *topos* of sets, $\{\top\} \subset \Omega$ is called the **subobject classifier**; see §E.4.

(v) An object in the category of elements of the functor $U(-)^*\colon \mathsf{Vect}_{\Bbbk}^{\mathrm{op}} \to \mathsf{Set}$ is a vector space with a dual vector. That is, an object is simply a linear map $f\colon V \to \Bbbk$ with codomain $\Bbbk$. A morphism from $f\colon V \to \Bbbk$ to $g\colon W \to \Bbbk$ is a linear map $T\colon V \to W$ so that $f = gT$. Immediately from this definition it is clear that the category of elements of $U(-)^*$ is the slice category $\mathsf{Vect}_{\Bbbk}/\Bbbk$. The identity on $\Bbbk$ is terminal, i.e., $1\colon \Bbbk \to \Bbbk$ is the universal dual vector.

(vi) An object in the category of elements of $U\colon \mathsf{Ring} \to \mathsf{Set}$ is a unital ring R with an element $r \in R$. Maps are ring homomorphisms preserving the chosen elements. The initial object is $x \in \mathbb{Z}[x]$. In the category of elements, the object $x \in \mathbb{Z}[x]$ has no non-identity endomorphisms: initial objects never do. But the ring $\mathbb{Z}[x]$ has many endomorphisms in the category of rings: the universal property $\mathsf{Ring}(\mathbb{Z}[x], -) \cong U$ tells us that maps $\mathbb{Z}[x] \to \mathbb{Z}[x]$ are classified by polynomials with integer coefficients, i.e., by elements $p(x) \in \mathbb{Z}[x]$. By Corollary 2.2.8, all natural endomorphisms of $U\colon \mathsf{Ring} \to \mathsf{Set}$ must have components $R \to R$ defined in this manner: i.e., the component of any natural endomorphism of rings is a map $p(x)\colon R \to R$ defined by $r \mapsto p(r)$, where p is an integer polynomial.

Definitions via universal properties emerge as an important theme in the coming chapters. To whet the reader's appetite, let us pose the following puzzle. Fixing two objects A, B in a locally small category C, define a functor

$$\mathsf{C}(A, -) \times \mathsf{C}(B, -)\colon \mathsf{C} \to \mathsf{Set}$$

that carries an object X to the set $\mathsf{C}(A, X) \times \mathsf{C}(B, X)$ whose elements are pairs of maps $a\colon A \to X$ and $b\colon B \to X$ in C. What would it mean for this functor to be representable?

Exercises.

EXERCISE 2.4.i. Given $F\colon \mathsf{C} \to \mathsf{Set}$, show that $\int F$ is isomorphic to the comma category $* \downarrow F$ of the singleton set $*\colon \mathbb{1} \to \mathsf{Set}$ over the functor $F\colon \mathsf{C} \to \mathsf{Set}$.

EXERCISE 2.4.ii. Characterize the terminal objects of C/c.

EXERCISE 2.4.iii. Use the principle of duality to convert the proof that a covariant functor is representable if and only if its category of elements has an initial object into a proof that a contravariant functor is representable if and only if its category of elements has a terminal object.

EXERCISE 2.4.iv. Explain the sense in which the Sierpinski space is the universal topological space with an open subset.

EXERCISE 2.4.v. Define a contravariant functor $F\colon \mathsf{Set}^{\mathrm{op}} \to \mathsf{Set}$ that carries a set to the set of preorders on it. What is its category of elements? Is F representable?

EXERCISE 2.4.vi. For a locally small category C, regard the two-sided represented functor $\mathrm{Hom}(-, -)\colon \mathsf{C}^{\mathrm{op}} \times \mathsf{C} \to \mathsf{Set}$ as a covariant functor of its domain $\mathsf{C}^{\mathrm{op}} \times \mathsf{C}$. The category of elements of Hom is called the **twisted arrow category**. Justify this name by describing its objects and morphisms.

EXERCISE 2.4.vii. Prove that the construction of the category of elements defines the action on objects of a functor

$$\int(-)\colon \mathsf{Set}^{\mathsf{C}} \to \mathsf{CAT}/\mathsf{C}.$$

Conclude that if $F, G\colon \mathsf{C} \to \mathsf{Set}$ are naturally isomorphic, then $\int F \cong \int G$ over C.

EXERCISE 2.4.viii. Prove that for any $F\colon \mathsf{C} \to \mathsf{Set}$, the canonical forgetful functor $\Pi\colon \int F \to \mathsf{C}$ has the following property: for any morphism $f\colon c \to d$ in the "base category" C and any object (c, x) in the fiber over c, there is a unique lift of the morphism f to a morphism in $\int F$ with domain (c, x) that projects along Π to f. A functor with this property is called a **discrete left fibration**.

EXERCISE 2.4.ix. Formulate the dual definition of a **discrete right fibration** satisfied by the canonical functor $\Pi \colon \int F \to \mathsf{C}$ associated to a contravariant functor $F \colon \mathsf{C}^{\mathrm{op}} \to \mathsf{Set}$.

EXERCISE 2.4.x. Answer the question posed at the end of this section: fixing two objects A, B in a locally small category C, we define a functor

$$\mathsf{C}(A, -) \times \mathsf{C}(B, -) \colon \mathsf{C} \to \mathsf{Set}$$

that carries an object X to the set $\mathsf{C}(A, X) \times \mathsf{C}(B, X)$ whose elements are pairs of maps $a \colon A \to X$ and $b \colon B \to X$ in C. What would it mean for this functor to be representable?

CHAPTER 3

Limits and Colimits

> Many properties of mathematical constructions may be represented by universal properties of diagrams.
>
> Saunders Mac Lane, *Categories for the Working Mathematician* [**ML98a**]

From a very simple topological space, the real line $\mathbb{R}$ with its usual metric, one can build a wide variety of new topological spaces. Taking *products* of $\mathbb{R}$ with itself, one defines the Euclidean spaces $\mathbb{R}^n$ in finite or infinite dimensions. The space $\mathbb{R}^n$ has interesting *subspaces*, defined to be the solutions to certain continuous polynomial functions $\mathbb{R}^n \to \mathbb{R}$, including the $(n-1)$-sphere S^{n-1}, which bounds the closed n-disk D^n. From the sphere S^n, one can define real projective space $\mathbb{R}P^n$ as a *quotient* by identifying each pair of antipodal points. From spheres and disks one can build tori, the Möbius band, the Klein bottle, and indeed any *cell complex* through a sequence of *gluing* constructions, in which disks are attached to an existing space along their boundary spheres. In each case, the newly constructed object is a particular set equipped with a specific topology. In this chapter, we will explain how all of these topologies can be defined in a uniform way, via a universal property that characterizes the newly constructed space either as a *limit* or a *colimit* of a particular diagram in the category of topological spaces.

Limits and colimits can be defined in any category. Special cases include constructions of the infimum and supremum, free products, cartesian products, direct sums, kernels, cokernels, fiber products, amalgamated free products, inverse limits of sequences, unions, and even the category of elements, among many others. This chapter begins in §3.1 by introducing the abstract notions of limits and colimits and the special terminology used for particular diagram shapes. We then turn our focus to "computational" results, which describe how more complicated limits and colimits can be built out of simpler ones. To ground intuition for limits and colimits, these abstract constructions are first introduced in §3.2 in the special case of limits of diagrams valued in the category of sets. Indeed, as a consequence of the Yoneda lemma, the set-theoretical constructions of limits suffice to prove general formulae for limits and colimits in any category. To understand this, we consider a variety of possible interactions between functors and limits in §3.3, presenting functors that *preserve*, *reflect*, or *create* limits and colimits. Such functors can be used to recognize cases in which limits or colimits in one category can be constructed from limits or colimits in another. This vocabulary is used in §3.4 to describe the representable nature of limits and colimits and extend the constructions of general limits from simple limits from Set to any category.

In §3.5, we apply the computational theorems from §3.4 to prove that the categories of sets, spaces, and categories are each **complete** and **cocomplete**, meaning that any diagram indexed by a small category has both a limit and a colimit. Most of the "algebraic" categories introduced in Example 1.1.3 are also complete and cocomplete—the category of fields being a notable exception—but we defer the proof of these facts to §5.6 where they can be deduced simultaneously as special cases of a general theorem. In §3.6, we introduce limit and colimit functors defined on categories of diagrams and use the functoriality of the limit and the colimit to prove the result that motivated Eilenberg and Mac Lane to introduce the concept of naturality: that naturally isomorphic diagrams have isomorphic limits.

In §3.7, we explain why few categories admit limits and colimits of diagrams of unbounded size. In §3.8, we study the interactions between limits and colimits. While limits commute with limits and colimits commute with colimits, in general, limits and colimits fail to commute with each other, with a few notable exceptions that depend both on the indexing shapes and the ambient category in which the diagrams are valued. To illustrate, we prove the most famous of these: that finite limits commute with *filtered* colimits in the category of sets.

3.1. Limits and colimits as universal cones

Recall that a **diagram** of **shape** J in a category C is a functor $F: J \to C$. Our aim in this section is to introduce the dual notions of *limit* and *colimit* of a diagram, which are defined, respectively, to be the universal *cones* over and under the diagram. The first task is to explain what is meant by a cone.

DEFINITION 3.1.1. For any object $c \in C$ and any category J, the **constant functor** $c: J \to C$ sends every object of J to c and every morphism in J to the identity morphism 1_c. The constant functors define an embedding $\Delta: C \to C^J$ that sends an object c to the constant functor at c and a morphism $f: c \to c'$ to the **constant natural transformation**, in which each component is defined to be the morphism f.

DEFINITION 3.1.2. A **cone over** a diagram $F: J \to C$ with **summit** or **apex** $c \in C$ is a natural transformation $\lambda: c \Rightarrow F$ whose domain is the constant functor at c. The components $(\lambda_j: c \to Fj)_{j \in J}$ of the natural transformation are called the **legs** of the cone. Explicitly:

- The data of a cone over $F: J \to C$ with summit c is a collection of morphisms $\lambda_j: c \to Fj$, indexed by the objects $j \in J$.
- A family of morphisms $(\lambda_j: c \to F_j)_{j \in J}$ defines a cone over F if and only if, for each morphism $f: j \to k$ in J, the following triangle commutes in C:

(3.1.3)
$$
\begin{array}{ccc}
 & c & \\
\lambda_j \swarrow & & \searrow \lambda_k \\
Fj & \xrightarrow[Ff]{} & Fk
\end{array}
$$

Dually, a **cone under** F with **nadir** c is a natural transformation $\lambda: F \Rightarrow c$, whose **legs** are the components $(\lambda_j: F_j \to c)_{j \in J}$. The naturality condition asserts that, for each morphism $f: j \to k$ of J, the triangle

$$
\begin{array}{ccc}
Fj & \xrightarrow{Ff} & Fk \\
\lambda_j \searrow & & \swarrow \lambda_k \\
 & c &
\end{array}
$$

commutes in C.

Cones under a diagram are also called **cocones**—a cone under $F: J \to C$ is precisely a cone over $F: J^{op} \to C^{op}$—but we find the terminology "under" and "over" to be more evocative. To illustrate, if F is a diagram indexed by the poset category $(\mathbb{Z}, \leq)$, then a cone over F with summit c is comprised of a family of morphisms $(\lambda_n: c \to Fn)_{n \in \mathbb{Z}}$ so that, for each $n \leq m$, the triangle defined by the morphisms λ_n, λ_m, and $Fn \to Fm$ commutes:

(3.1.4)

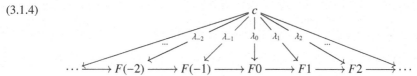

A cone under F with nadir c is comprised of a family of morphisms $(\lambda_n: Fn \to c)_{n \in \mathbb{Z}}$ so that, for each $n \leq m$, the triangle defined by the morphisms λ_n, λ_m, and $F_n \to F_m$ commutes:

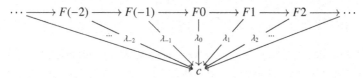

The limit of F is most simply described as the universal cone over the diagram F, while the colimit is the universal cone under F. As discussed in Chapter 2, there are two equivalent ways to understand this universal property. A limit may be defined to be a representation for a particular contravariant functor or as a terminal object in its category of elements; dually, a colimit is a representation for a particular covariant functor and, equivalently, is an initial object in its category of elements. As is always the case for representable definitions, limits and colimits can be contemplated for any diagram valued in any category, but in any particular case these universal objects may or may not exist.

DEFINITION 3.1.5 (limits and colimits I). For any diagram $F: \mathsf{J} \to \mathsf{C}$, there is a functor

$$\mathrm{Cone}(-, F): \mathsf{C}^{\mathrm{op}} \to \mathsf{Set}$$

that sends $c \in \mathsf{C}$ to the set[1] of cones over F with summit c; we leave a description of its action on morphisms as Exercise 3.1.i. A **limit** of F is a representation for $\mathrm{Cone}(-, F)$. By the Yoneda lemma, a limit consists of an object $\lim F \in \mathsf{C}$ together with a universal cone $\lambda: \lim F \Rightarrow F$, called the **limit cone**, defining the natural isomorphism

$$\mathsf{C}(-, \lim F) \cong \mathrm{Cone}(-, F).$$

Dually, there is a functor

$$\mathrm{Cone}(F, -): \mathsf{C} \to \mathsf{Set}$$

that sends c to the set of cones under F with nadir c. A **colimit** of F is a representation for $\mathrm{Cone}(F, -)$. By the Yoneda lemma, a colimit consists of an object $\mathrm{colim}\, F \in \mathsf{C}$ together with a universal cone $\lambda: F \Rightarrow \mathrm{colim}\, F$, called the **colimit cone**, defining the natural isomorphism

$$\mathsf{C}(\mathrm{colim}\, F, -) \cong \mathrm{Cone}(F, -).$$

Applying Proposition 2.4.8, limits and colimits may also be defined to be terminal and initial objects, respectively, in the appropriate categories of elements.

DEFINITION 3.1.6 (limits and colimits II). For any diagram $F: \mathsf{J} \to \mathsf{C}$, a **limit** is a terminal object in the category of cones over F, i.e., in the category $\int \mathrm{Cone}(-, F)$. An object in the category of cones over F is a cone over F, with any summit. In particular, the data

---

[1]To guarantee that there is a *set* of cones with fixed summit over the diagram F, it suffices to assume that J is small and C is locally small, so that C^{J} is locally small. Until §3.7, all of the limits and colimits under consideration are indexed by small categories.

of a terminal object in the category of cones consists of a limit object in C together with a specified **limit cone**. A morphism from a cone $\lambda: c \Rightarrow F$ to a cone $\mu: d \Rightarrow F$ is a morphism $f: c \to d$ in C so that for each $j \in J$, $\mu_j \cdot f = \lambda_j$. In other words, a morphism of cones is a map between the summits so that each leg of the domain cone factors through the corresponding leg of the codomain cone along this map.

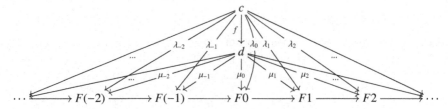

The forgetful functor $\int \mathrm{Cone}(-, F) \to \mathsf{C}$ takes a cone to its summit.

Dually, a **colimit** is an initial object in the category of cones under F, i.e., in the category $\int \mathrm{Cone}(F, -)$. An object in the category of cones under F is a cone under F, with any nadir. In particular, the data of an initial object is comprised of the colimit object in C together with a specified **colimit cone**. A morphism from a cone $\lambda: F \Rightarrow c$ to a cone $\mu: F \Rightarrow d$ is a morphism $f: c \to d$ so that for each $j \in J$, $\mu_j = f \cdot \lambda_j$. In other words, a morphism of cones is a map between the nadirs so that each leg of the codomain cone factors through the corresponding leg of the domain cone along this map. The forgetful functor $\int \mathrm{Cone}(F, -) \to \mathsf{C}$ takes a cone to its nadir.

Limits and colimits of a diagram F of shape J might be denoted by $\lim_J F$ or $\mathrm{colim}_J F$ for emphasis, particularly when multiple diagram shapes are involved. On other occasions, no special notation is used: to say that a cone is a "limit cone" is to assert that its summit is a limit, however it is denoted.

On account of the following proposition, it is common to use the definite article "the" to refer to a limit or a colimit of a fixed diagram because these objects are well-defined up to a unique isomorphism:

PROPOSITION 3.1.7 (essential uniqueness of limits and colimits). *Given any two limit cones $\lambda: \ell \Rightarrow F$ and $\lambda': \ell' \Rightarrow F$ over a common diagram F, there is a unique isomorphism $\ell \cong \ell'$ that commutes with the legs of the limit cones.*

Note that this result does not prohibit limit objects from having non-trivial automorphisms. If $\lambda: \ell \Rightarrow F$ is a limit cone and $\sigma: \ell \cong \ell$ is any automorphism of ℓ, then $\lambda\sigma: \ell \Rightarrow F$ is another limit cone. However, Proposition 3.1.7 implies that the only automorphism of ℓ that commutes with the specified limit cone λ is the identity.

PROOF. The limit cones $\lambda: \ell \Rightarrow F$ and $\lambda': \ell' \Rightarrow F$ each define terminal objects in the category of cones over F. By Corollary 2.3.2, the terminal objects in any category define a contractible groupoid. This means, in particular, that these objects are uniquely isomorphic in the category of cones, i.e., that there is a unique isomorphism between ℓ and ℓ' commuting with the cone legs.   □

REMARK 3.1.8 (limit and colimit diagrams). The data of a diagram, together with a limit cone over it, is called a **limit diagram** and the data of a diagram, together with a colimit cone under it, is called a **colimit diagram**. A J-indexed diagram and a cone over it—such as displayed in (3.1.4)—combine to define a diagram indexed by the category $J^\triangleleft$, which has J as a full subcategory together with a freely-adjoined initial object. Dually, a J-indexed diagram and a cone under it combine to define a diagram indexed by the category $J^\triangleright$, which is built from the category J together with a freely-adjoined terminal object. Exercise 3.5.iv provides a more explicit construction of this category.

There are special names for limit and colimit diagrams of certain shapes.

DEFINITION 3.1.9. A **product** is a limit of a diagram indexed by a discrete category, with only identity morphisms. A diagram in C indexed by a discrete category J is simply a collection of objects $F_j \in C$ indexed by $j \in J$. A cone over this diagram is a J-indexed family of morphisms $(\lambda_j \colon c \to F_j)_{j \in J}$, subject to no further constraints. The limit is typically denoted by $\prod_{j \in J} F_j$ and the legs of the limit cone are maps

$$\left(\pi_k \colon \prod_{j \in J} F_j \to F_k \right)_{k \in J}$$

called **(product) projections**. The universal property asserts that composition with the product projections defines a natural isomorphism

$$C(c, \textstyle\prod_{j \in J} F_j) \xrightarrow[\cong]{(\pi_k)_*} \prod_{k \in J} C(c, F_k) \cong \mathrm{Cone}(c, F).$$

EXAMPLE 3.1.10. The product of a pair of spaces X and Y is a space $X \times Y$ equipped with continuous projection functions

$$X \xleftarrow{\ \pi_X\ } X \times Y \xrightarrow{\ \pi_Y\ } Y$$

satisfying the following universal property: for any other space A with continuous maps $f \colon A \to X$ and $g \colon A \to Y$, there is a unique continuous function $h \colon A \to X \times Y$ so that the diagram

$$
\begin{array}{ccc}
 & A & \\
f \swarrow & \downarrow \exists! h & \searrow g \\
X \xleftarrow[\pi_X]{} & X \times Y & \xrightarrow[\pi_Y]{} Y
\end{array}
$$

commutes.

Taking A to be the point, which represents the underlying set functor $U \colon \mathsf{Top} \to \mathsf{Set}$, the bijection

$$\mathsf{Top}(*, X \times Y) \cong \mathsf{Top}(*, X) \times \mathsf{Top}(*, Y)$$

tells us that the points of $X \times Y$ correspond to points in the cartesian product of the underlying sets of X and Y.

It remains to define the topology on the set $X \times Y$. Taking A to be the set $X \times Y$ equipped with various topologies, this universal property forces $X \times Y$ to be defined to be the coarsest topology on the cartesian product of the underlying sets of X and Y so that the projection functions π_X and π_Y are continuous.[2]

Similarly, the product of an arbitrary (possibly infinite) family of spaces $(X_j)_{j \in J}$ is defined by assigning the product of the underlying sets the coarsest topology so that each of the projection functions $\pi_k \colon \prod_j X_j \to X_k$ is continuous. The more intuitive "box topology" on the set $\prod_j X_j$, whose basic open sets are products of basic opens in each X_j, coincides with the product topology for finite indexing sets, but fails to satisfy the universal property of Definition 3.1.9 when the indexing set J is infinite.

---

[2]**Coarsest** means "having the fewest open and closed sets"; the antonym is **finest**, meaning "having the most open and closed sets."

DEFINITION 3.1.11. A **terminal object** is often regarded as a trivial special case of a product, where the indexing category is empty. A cone over an empty diagram is just an object in the target category, its summit, together with no additional data. Any morphism between the summits defines a map of empty cones. Thus, the category of cones in this special case is just the ambient category itself and so the limit, a terminal object in this category, is just a terminal object in the sense of Definition 1.6.14.

EXAMPLE 3.1.12. The category $\mathbb{1}$ with one object and only its identity morphism is the **terminal category**, the terminal object in Cat or in CAT.

DEFINITION 3.1.13. An **equalizer** is a limit of a diagram indexed by the **parallel pair**, the category $\bullet \rightrightarrows \bullet$ with two objects and two parallel non-identity morphisms. A diagram of this shape is simply a parallel pair of morphisms $f, g: A \rightrightarrows B$ in the target category C. A cone over this diagram with summit C consists of a pair of morphisms $a: C \to A$ and $b: C \to B$ so that $fa = b$ and $ga = b$; these two equations correspond to the naturality conditions (3.1.3) imposed by each of the two non-identity morphisms in the indexing category. Together, they assert that $fa = ga$; the morphism b is necessarily equal to this common composite. Thus, a cone over a parallel pair $f, g: A \rightrightarrows B$ is represented by a single morphism $a: C \to A$ so that $fa = ga$.

The equalizer $h: E \to A$ is the universal arrow with this property. The limit diagram

$$E \xrightarrow{\ h\ } A \underset{g}{\overset{f}{\rightrightarrows}} B$$

is called an **equalizer diagram**. The universal property asserts that given any $a: C \to A$ so that $fa = ga$, there exists a unique factorization $k: C \to E$ of a through h.

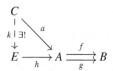

EXAMPLE 3.1.14. For instance, the equalizer of a group homomorphism $\phi: G \to H$ and the trivial homomorphism $e: G \to H$, sending every element of G to the identity in H, is the kernel of ϕ, and the leg of the limit cone is the inclusion $\ker \phi \hookrightarrow G$.

The equalizer of a pair of homomorphisms $\phi, \psi: G \rightrightarrows H$ is the subgroup of elements $g \in G$ so that $\phi(g) = \psi(g)$. If H is abelian, this is the kernel of the homomorphism $\phi - \psi: G \to H$.

The fact that the equalizer defines a subgroup of G is no coincidence: the map from an equalizer into the domain of the parallel pair that it equalizes is always a monomorphism; see Exercise 3.1.vi.

DEFINITION 3.1.15. A **pullback** is a limit of a diagram indexed by the poset category

$$\bullet \to \bullet \leftarrow \bullet$$

comprised of two non-identity morphisms with common codomain. Writing f and g for the morphisms defining the image of a diagram of this shape in a category C, a cone with summit D consists of a triple of morphisms, one for each object in the indexing category, so that both triangles in the diagram

(3.1.16)

$$
\begin{array}{ccc}
D & \xrightarrow{\ c\ } & C \\
{\scriptstyle b}\big\downarrow & \searrow{\scriptstyle a} & \big\downarrow{\scriptstyle g} \\
B & \xrightarrow[\ f\ ]{} & A
\end{array}
$$

commute; the two triangles represent the two naturality conditions (3.1.3) imposed by the non-identity morphisms in the indexing category. The leg a asserts that gc and fb have a common composite. Thus, the data of a cone over $B \xrightarrow{f} A \xleftarrow{g} C$ may be described more simply as a pair of morphisms $B \xleftarrow{b} D \xrightarrow{c} C$ defining a commutative square.

The pullback, the universal cone over f and g, is a commutative square $fh = gk$ with the following universal property: given any commutative square (3.1.16), there is a unique factorization of its legs through the summit of the pullback cone:

(3.1.17)

$$
\begin{array}{ccc}
D & & \\
& \searrow{\scriptstyle \exists!} \quad \searrow{\scriptstyle c} & \\
{\scriptstyle b}& P \xrightarrow{\ k\ } C & \\
& {\scriptstyle h}\big\downarrow \quad\ \ \big\downarrow{\scriptstyle g} & \\
& B \xrightarrow[\ f\ ]{} A &
\end{array}
$$

The symbol "$\lrcorner$" indicates that the square $gk = fh$ is a pullback, i.e., a limit diagram, and not simply a commutative square. The pullback P is also called the **fiber product** and is frequently denoted by $B \times_A C$. The precise relationship between pullbacks and products is explored in §3.2 and §3.4.

Particularly when the map $f\colon 1 \to A$ represents an "element" of the object A, such as when its domain is an object $1 \in C$ that represents an "underlying set" functor $C \to Set$, the pullback P

$$
\begin{array}{ccc}
P & \longrightarrow & C \\
\big\downarrow & & \big\downarrow{\scriptstyle g} \\
1 & \xrightarrow[\ f\ ]{} & A
\end{array}
$$

defines the **fiber** of the map $g\colon C \to A$ over f.

EXAMPLE 3.1.18. For instance, consider the continuous quotient map $\rho\colon \mathbb{R} \to S^1$ in the category of spaces that sends a real number t to the point $e^{2\pi i t}$ of S^1, thought of as the unit circle in the complex plane. The pullback of the diagram

$$
\begin{array}{ccc}
\mathbb{Z} & \dashrightarrow & \mathbb{R} \\
\big\downarrow & & \big\downarrow{\scriptstyle \rho} \\
* & \xrightarrow[\ 1\ ]{} & S^1
\end{array}
$$

is the **fiber** of ρ over the point $1 \in S^1$. This is the discrete subspace $\mathbb{Z} \subset \mathbb{R}$.

The fact that the fiber defines a subspace of the domain of the map being pulled back is no coincidence: the pullback of a monomorphism $B \rightarrowtail A$ along any map $C \to A$ defines a monomorphism $B \times_A C \rightarrowtail C$; see Exercise 3.1.vii.

EXAMPLE 3.1.19. For another example of this flavor, recall that maps $\mathbb{Z} \to \mathbb{Z}$ of abelian groups correspond to elements $n \in \mathbb{Z}$ because $\mathbb{Z}$ represents the forgetful functor $U: \mathsf{Ab} \to \mathsf{Set}$, i.e., $\mathrm{Hom}(\mathbb{Z}, A) \cong UA$ for any abelian group A. Elements of the pullback of

$$\begin{array}{c} \mathbb{Z} \\ \downarrow{\scriptstyle n} \\ \mathbb{Z} \xrightarrow{\ m\ } \mathbb{Z} \end{array}$$

are pairs of integers x and y so that $nx = my$. From this description, assuming m and n are not both zero, it follows that the pullback is isomorphic to the abelian group $\mathbb{Z}$ and the legs of the pullback cone

(3.1.20)

$$\begin{array}{ccc} \mathbb{Z} & \overset{b}{\dashrightarrow} & \mathbb{Z} \\ {\scriptstyle a}\downarrow & & \downarrow{\scriptstyle n} \\ \mathbb{Z} & \xrightarrow{\ m\ } & \mathbb{Z} \end{array}$$

are defined to be the unique[3] integers so that $ma = nb$ is the least common multiple of n and m.

DEFINITION 3.1.21. The limit of a diagram indexed by the category ω^{op} is called an **inverse limit** of a tower or a sequence of morphisms. On account of this example, the term "inverse limit" is sometimes used to refer to limits of any shape. A diagram indexed by ω^{op} consists of a sequence of objects and morphisms

$$\cdots \longrightarrow F_3 \longrightarrow F_2 \longrightarrow F_1 \longrightarrow F_0$$

together with composites and identities, which are not displayed. A cone over this diagram is an extension of this data to a diagram of shape $(\omega + 1)^{\mathrm{op}}$. Explicitly, a cone consists of a new object "all the way to the left" together with morphisms making every triangle commute:

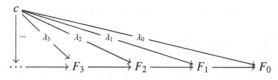

The inverse limit, frequently denoted by $\varprojlim F_n$, is the terminal cone. Similar remarks apply with any limit ordinal[4] α in place of ω.

EXAMPLE 3.1.22. The p-**adic integers** are defined to be the inverse limit of the following diagram of rings:

$$\mathbb{Z}_p = \varprojlim \mathbb{Z}/p^n \longrightarrow\!\!\!\!\rightarrow \cdots \longrightarrow\!\!\!\!\rightarrow \mathbb{Z}/p^4 \longrightarrow\!\!\!\!\rightarrow \mathbb{Z}/p^3 \longrightarrow\!\!\!\!\rightarrow \mathbb{Z}/p^2 \longrightarrow\!\!\!\!\rightarrow \mathbb{Z}/p$$

where the maps in the diagram are the canonical quotient maps.

DEFINITION 3.1.23. Definitions 3.1.9, 3.1.11, 3.1.13, 3.1.15, and 3.1.21 dualize to define:

---

[3]More accurately, the universal property described here only defines the pair of integers (a, b) up to a compatible choice of sign; see Exercise 3.1.x.

[4]Exercise 3.1.ix explains the limited interest of limits of diagrams indexed by opposites of successor ordinals.

- A **coproduct** $\coprod_{j \in J} A_j$ is the colimit of a diagram $(A_j)_{j \in J}$ indexed by a discrete category J. The legs of the colimit cone $\iota_{j'} : A_{j'} \to \coprod_{j \in J} A_j$ are referred to as **coproduct injections**, though in pathological cases these maps might not be monomorphisms (see Exercise 3.1.xi).
- An **initial object** is the colimit of the empty diagram.
- A **coequalizer** is a colimit of a diagram indexed by the parallel pair category $\bullet \rightrightarrows \bullet$. The coequalizer of a parallel pair of maps $f, g : A \rightrightarrows B$ is the universal map $h : B \to C$ with the property that $hf = hg$. The colimit cone

$$A \underset{g}{\overset{f}{\rightrightarrows}} B \overset{h}{\twoheadrightarrow} C$$

  is called a **coequalizer diagram**. Diagrams of this shape are also called **forks**.
- A **pushout** is a colimit of a diagram indexed by the poset category $\bullet \leftarrow \bullet \to \bullet$. Dualizing the convention introduced in Definition 3.1.15, the symbol "$\ulcorner$" indicates that a commutative square

$$
\begin{array}{ccc}
A & \overset{f}{\longrightarrow} & B \\
{\scriptstyle g}\downarrow & & \downarrow{\scriptstyle k} \\
C & \underset{h}{\longrightarrow} & P
\end{array}
$$

  is a pushout, i.e., is a colimit diagram. The pushout is the universal commutative square under the maps f and g.
- The colimit of a diagram indexed by the ordinal ω is called a **sequential colimit** or **direct limit**. The colimit of a diagram

$$F_0 \longrightarrow F_1 \longrightarrow F_2 \longrightarrow F_3 \longrightarrow \cdots$$

  frequently denoted by $\varinjlim F_n$, defines a diagram of shape $\omega + 1$. The term "direct limit" is sometimes also used to refer to colimits of any shape.

Each of the colimits introduced in Definition 3.1.23 is indexed by the *opposite* of the category used to index the corresponding limit notion. Exercise 3.1.ix explains why the category ω and the category $\bullet \leftarrow \bullet \to \bullet$ are used to index colimit diagrams, but not limit diagrams.

EXAMPLE 3.1.24. The figure eight or the wedge of two circles is a space $S^1 \vee S^1$ that can be defined to be the pushout of the diagram $S^1 \leftarrow * \to S^1$ of unbased spaces or as the coproduct of S^1 with itself in the category of based spaces. A further pushout diagram in Top

$$
\begin{array}{ccc}
S^1 & \overset{aba^{-1}b^{-1}}{\longrightarrow} & S^1 \vee S^1 \\
{\scriptstyle i}\downarrow & & \downarrow \\
D^2 & - - - - \to & T
\end{array}
$$

defines the torus $T \cong S^1 \times S^1$. Here i is the inclusion of the circle as the boundary of the disk and the map $aba^{-1}b^{-1}$ is the loop in $S^1 \vee S^1$ that wraps once around one circle, then once around the other, then again around the first but in the reversed orientation, and then again around the second but in the reversed orientation.

EXAMPLE 3.1.25. The coequalizer of a group homomorphism $\phi\colon G \to H$ paired with the trivial homomorphism $e\colon G \to H$ defines the **cokernel**, the quotient of H by the normal subgroup generated by the image of ϕ.

EXAMPLE 3.1.26. The colimit of a sequence of sets and inclusions

$$X_0 \hookrightarrow X_1 \hookrightarrow X_2 \hookrightarrow \cdots$$

can be understood to be their union $\bigcup_{n\geq 0} X_n$. For instance, a *CW complex* is a colimit of its *n-skeleta*, the subspaces built by attaching disks of dimension at most n.

The forms of the universal properties that characterize limits and colimits are quite useful when one is asked to define a map into a limit object or out of a colimit object. Moreover, the uniqueness statements in these universal properties can be used to prove the commutativity of a diagram whose initial vertex is a colimit object or whose terminal vertex is a limit object. We illustrate this in the simplest case, for products and coproducts, but these remarks apply more generally.

REMARK 3.1.27 (the universal properties of the product and the coproduct). In any category C with coproducts, maps $f\colon \coprod_{i\in I} A_i \to X$ correspond naturally and bijectively to I-indexed families of maps $(f_i\colon A_i \to X)_{i\in I}$. The i'-indexed component $f_{i'}$ of the map f is defined by restricting along the inclusion $\iota_{i'}\colon A_{i'} \to \coprod_{i\in I} A_i$ that defines the i'-indexed leg of the colimit cone.

$$\mathsf{C}(\textstyle\coprod_{i\in I} A_i, X) \xrightarrow{\;\cong\;} \prod_{i\in I}\mathsf{C}(A_i, X)$$

$$\textstyle\coprod_{i\in I} A_i \xrightarrow{f} X \quad\leadsto\quad (A_i \xrightarrow{f_i} X)_{i\in I}$$

Dually, in any category with products, maps $g\colon X \to \prod_{j\in J} B_j$ correspond naturally and bijectively to J-indexed families of maps $(g_j\colon X \to B_j)_{j\in J}$.

$$\mathsf{C}(X, \textstyle\prod_{j\in J} B_j) \xrightarrow{\;\cong\;} \prod_{j\in J}\mathsf{C}(X, B_j)$$

$$X \xrightarrow{g} \textstyle\prod_{j\in J} B_j \quad\leadsto\quad (X \xrightarrow{g_j} B_j)_{j\in J}$$

The j'-indexed component $g_{j'}$ of the map g is defined by composing with the projection $\pi_{j'}\colon \prod_{j\in J} B_j \to B_{j'}$ that defines the j'-indexed leg of the limit cone.

Combining these universal properties, to define a map from a coproduct to a product,

$$\coprod_{i\in I} A_i \xrightarrow{\;f\;} \prod_{j\in J} B_j$$
$$\iota_{i'}\uparrow \qquad\qquad \downarrow\pi_{j'}$$
$$A_{i'} \dashrightarrow[f_{(i',j')}] B_{j'}$$

it suffices, for each ordered pair $(i', j') \in I \times J$, to define a map $f_{(i',j')}\colon A_{i'} \to B_{j'}$. That is, a single map from a coproduct to a product is determined by a matrix of component maps indexed by the product of the indexing sets.[5]

In certain categories, this matrix representation can be pushed further. The categories Ab, Mod$_R$, or Ch$_R$ have a notion of a zero homomorphism between any pair of objects.

---

[5]In the future, we adopt the same notation for the bounded variable that indexes a product or coproduct and the variable that indexes a generic component so that we can avoid cluttering the text with primes: e.g., $f\colon \coprod_{i\in I} A_i \to \prod_{j\in J} B_j$ is given by component maps $f_{(i,j)} = \pi_j \cdot f \cdot \iota_i\colon A_i \to B_j$.

Given any finite collection of objects A_i, it makes sense then to consider the map

$$\coprod_{i \in I} A_i \xrightarrow{\begin{pmatrix} 1 & 0 & \cdots & 0 \\ 0 & \ddots & \ddots & \vdots \\ \vdots & \ddots & \ddots & 0 \\ 0 & \cdots & 0 & 1 \end{pmatrix}} \prod_{i \in I} A_i$$

encoded by the "identity matrix," and in each of these categories, this map turns out to be an isomorphism.[6] In such contexts, the isomorphic finite coproducts and products $\coprod_{i \in I} A_i \cong \prod_{i \in I} A_i$ are written as $\bigoplus_{i \in I} A_i$ and referred to as **direct sums**.

Combining the universal properties of the product and the coproduct, all maps between finite direct sums can be represented as matrices of component maps, and, as one might hope, the composite matrix is defined by "matrix multiplication":

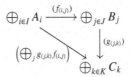

The examples mentioned above serve to illustrate the pervasiveness of limits and colimits among interesting mathematical objects, but how does one prove that a particular diagram has a limit or colimit and construct one if it exists? There are two general strategies. The first is to guess and check. A conjectured limit or colimit cone, such as displayed in (3.1.20), can often be shown to have the appropriate universal property via an elementary argument.[7]

A second strategy is to construct a limit of a more complicated diagram out of limits of simpler diagrams, whose constructions are easier to understand. In the next section, we begin to explore this viewpoint in the category of sets.

Exercises.

EXERCISE 3.1.i. For a fixed diagram $F \in C^J$, describe the actions of the cone functors $\text{Cone}(-, F) \colon C^{\text{op}} \to \text{Set}$ and $\text{Cone}(F, -) \colon C \to \text{Set}$ on morphisms in C.

EXERCISE 3.1.ii. For a fixed diagram $F \in C^J$, show that the cone functor $\text{Cone}(-, F)$ is naturally isomorphic to $\text{Hom}(\Delta(-), F)$, the restriction of the hom functor for the category C^J along the constant functor embedding defined in 3.1.1.

EXERCISE 3.1.iii. Prove that the category of cones over $F \in C^J$ is isomorphic to the comma category $\Delta \downarrow F$ formed from the constant functor $\Delta \colon C \to C^J$ and the functor $F \colon \mathbb{1} \to C^J$. Argue by duality that the category of cones under F is the comma category $F \downarrow \Delta$.

EXERCISE 3.1.iv. Give a second proof of Proposition 3.1.7 by using the universal properties of each of a pair of limit cones $\lambda \colon \ell \Rightarrow F$ and $\lambda' \colon \ell' \Rightarrow F$ to directly construct the unique isomorphism $\ell \cong \ell'$ between their summits.

---

[6]The "identity matrix" defines an isomorphism between the finite coproduct and the finite product in any *abelian category*; see Definition E.5.1.

[7]In the author's experience, most proofs that a desired universal property is fulfilled in a particular category proceed in this manner, by guess and check. The first step is to intuit that a particular construction might define the sought-for object; the next is to verify that the universal property holds.

EXERCISE 3.1.v. Consider a diagram $F: \mathsf{J} \to \mathsf{P}$ valued in a poset $(\mathsf{P}, \leq)$. Use order-theoretic language to characterize the limit and the colimit.

EXERCISE 3.1.vi. Prove that if

$$E \xrightarrow{\ h\ } A \underset{g}{\overset{f}{\rightrightarrows}} B$$

is an equalizer diagram, then h is a monomorphism.

EXERCISE 3.1.vii. Prove that if

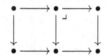

is a pullback square and f is a monomorphism, then k is a monomorphism.

EXERCISE 3.1.viii. Consider a commutative rectangle

$$\begin{array}{ccccc} \bullet & \longrightarrow & \bullet & \longrightarrow & \bullet \\ \downarrow & & \downarrow & \lrcorner & \downarrow \\ \bullet & \longrightarrow & \bullet & \longrightarrow & \bullet \end{array}$$

whose right-hand square is a pullback. Show that the left-hand square is a pullback if and only if the composite rectangle is a pullback.

EXERCISE 3.1.ix. Show that if J has an initial object, then the limit of any functor indexed by J is the value of that functor at an initial object. Apply the dual of this result to describe the colimit of a diagram indexed by a successor ordinal.

EXERCISE 3.1.x. If (a, b) are positive integers satisfying the universal property of (3.1.20), show that the pair $(-a, -b)$ also satisfies the same universal property. Explain why this observation does not imply that the pullback is ill-defined.

EXERCISE 3.1.xi. Use the universal property of the coproduct to prove that in any category with at least one morphism between each ordered pair of objects, the legs of a coproduct cone are split monomorphisms.

EXERCISE 3.1.xii. Suppose $E: \mathsf{I} \xrightarrow{\simeq} \mathsf{J}$ defines an equivalence between small categories and consider a diagram $F: \mathsf{J} \to \mathsf{C}$. Show that the category of J-shaped cones over F is equivalent to the category of I-shaped cones over FE, and use this equivalence to describe the relationship between limits of F and limits of FE.

EXERCISE 3.1.xiii. What is the coproduct in the category of commutative rings?

3.2. Limits in the category of sets

While a limit can be defined for any diagram of any shape valued in any category, the ability to state the definition is no guarantee that an object with that universal property exists. In this section, we restrict consideration to diagrams valued in the category of sets and give a constructive proof that a limit exists for any set-valued diagram that satisfies an important size constraint. Special cases of this general construction give computations of products, equalizers, pullbacks, and inverse limits of sequences of sets, the formulae for which provide intuition for general results that facilitate the computation of limits and colimits in more exotic categories.

DEFINITION 3.2.1. A diagram is **small** if its indexing category is a small category. A category C is **complete** if it admits limits of all small diagrams valued in C and is **cocomplete** if it admits colimits of all small diagrams valued in C.

A complete category is a category that "has all small limits." To prove that Set is complete, consider a small diagram $F: J \to$ Set. A limit is a representation

$$\mathrm{Set}(X, \lim F) \cong \mathrm{Cone}(X, F)$$

of the functor that sends a set X to the set of cones over F with summit X. Specializing to a fixed singleton set 1, which represents the identity functor Set $\to$ Set,

$$(3.2.2) \qquad \lim F \cong \mathrm{Set}(1, \lim F) \cong \mathrm{Cone}(1, F).$$

The idea is to take (3.2.2) as a *definition* of the limit of the diagram F:

DEFINITION 3.2.3. For any small diagram $F: J \to$ Set, define

$$\lim F := \mathrm{Cone}(1, F)$$

to be the set of cones over F with summit 1. Define the legs of the limit cone over F to be the functions $\lambda_j: \lim F \to Fj$, indexed by the objects $j \in J$, that are defined by

$$(3.2.4)$$
$$\lim F \xrightarrow{\quad \lambda_j \quad} Fj$$
$$\mu: 1 \Rightarrow F \quad \mapsto \quad \mu_j: 1 \to Fj$$

That is, λ_j is the function that carries a cone μ with summit 1 to its jth leg, regarded as an element of the set Fj.

REMARK 3.2.5. The size hypothesis in Definition 3.2.3 is essential. If J is small, then Set^J is locally small and, in particular, $\mathrm{Cone}(1, F)$, the collection of natural transformations from the constant diagram at 1 to F, is a set.

To prove that Set has all small limits, it remains to show that the data of Definition 3.2.3 has the desired universal property. The proof makes repeated use of the fact that 1 represents the identity functor on Set, allowing us to conflate elements $x \in X$ and morphisms $x: 1 \to X$.

THEOREM 3.2.6. *The category* Set *is complete, admitting all small limits.*

PROOF. Our first task is to show that the functions (3.2.4) define a cone over F, demonstrating that for each morphism $f: j \to k$ in J, the diagram

$$\begin{array}{ccc}
& \lim F & \\
{\scriptstyle \lambda_j}\swarrow & & \searrow{\scriptstyle \lambda_k} \\
Fj & \xrightarrow[Ff]{} & Fk
\end{array}$$

commutes. For any element $\mu: 1 \Rightarrow F$ of the set $\lim F$,

$$Ff(\lambda_j(\mu)) = Ff(\mu_j) = \mu_k = \lambda_k(\mu),$$

where the middle equality holds because $\mu: 1 \Rightarrow F$ defines a cone. This proves that the functions $(\lambda_j: \lim F \to Fj)_{j \in J}$ define a cone over F.

To prove that $\lambda: \lim F \Rightarrow F$ is the universal cone, consider a cone $\zeta: X \Rightarrow F$ with generic summit. We must show that ζ factors uniquely through $\lambda: \lim F \Rightarrow F$ along a function $r: X \to \lim F$. For each element $x \in X$, thought of as a function $x: 1 \to X$, there is a cone $\zeta x: 1 \Rightarrow F$ defined by restricting the cone ζ along the map x. Define

$r(x) \in \lim F = \mathrm{Cone}(1, F)$ to be this cone ζx. By the definition of the legs of the limit cone λ,

$$\lambda_j(r(x)) = \lambda_j(\zeta x) = (\zeta x)_j = \zeta_j x,$$

so the cone ζ indeed factors along r through the cone λ. This calculation also reveals that our definition $r(x) = \zeta x$ is necessary, proving uniqueness of the factorization and thereby verifying that $\lim F$ indeed defines a limit of F. □

We now apply Definition 3.2.3 to compute the limits of various diagrams of sets.

EXAMPLE 3.2.7. By Theorem 3.2.6, the product of sets A_j indexed by elements of some set $j \in J$ is the set of cones over this collection of sets with summit 1. A cone over a J-indexed discrete diagram of sets with summit 1 is just a J-tuple of elements in the sets. Thus, the categorical product in Set coincides with the usual cartesian product

$$\prod_{j \in J} A_j := \{(a_j \in A_j)_{j \in J}\}.$$

EXAMPLE 3.2.8. The terminal object in Set is the set of cones over the empty diagram with summit 1. There is a unique such cone, proving that the terminal object is again a singleton set.

EXAMPLE 3.2.9. Given a parallel pair of functions $f, g \colon X \rightrightarrows Y$, their equalizer is the set of maps $1 \xrightarrow{x} X$ so that $fx = gx$. In more familiar notation, the equalizer is the set E defined by

$$E := \{x \in X \mid f(x) = g(x)\}.$$

EXAMPLE 3.2.10. Elements of the limit of a diagram $F \colon \omega^{\mathrm{op}} \to$ Set are cones

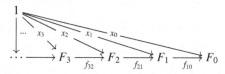

The data of such a cone is given by a tuple of elements $(x_n \in F_n)_{n \in \omega}$ making each triangle commute. Thus, we see that

$$\lim F := \left\{ (x_n)_n \in \prod_n F_n \;\middle|\; f_{n,n-1}(x_n) = x_{n-1} \right\}.$$

EXAMPLE 3.2.11. Elements of the pullback of a pair of functions $B \xrightarrow{f} A \xleftarrow{g} C$ are cones over this diagram with summit 1, i.e., are commutative squares

$$
\begin{array}{ccc}
1 & \xrightarrow{\ c\ } & C \\
{\scriptstyle b}\downarrow & & \downarrow{\scriptstyle g} \\
B & \xrightarrow[\ f\ ]{} & A
\end{array}
$$

The data of a cone of this form consists of a pair of elements $b \in B$ and $c \in C$ so that $f(b) = g(c)$. Thus, the pullback is defined to be the set

$$B \times_A C := \{(b, c) \in B \times C \mid f(b) = g(c)\}.$$

EXAMPLE 3.2.12. The limit of a left G-set $X\colon \mathsf{B}G \to \mathsf{Set}$ is the set of cones with summit 1. A map $x\colon 1 \to X$ defines a cone over X if and only if the diagram

$$
\begin{array}{ccc}
 & 1 & \\
{\scriptstyle x}\swarrow & & \searrow{\scriptstyle x} \\
X & \xrightarrow[\;g_*\;]{} & X
\end{array}
$$

commutes for all $g \in G$, i.e., if and only if $g \cdot x = x$ for all g in G. Thus, $\lim X \cong X^G$, the set of G-**fixed points** of X.

Comparing Example 3.2.11 with Examples 3.2.7 and 3.2.9, it looks like the pullback may be constructed as an equalizer of a diagram involving a product. Indeed, given a pair of functions $f\colon B \to A$ and $g\colon C \to A$, the pullback $B \times_A C$ is the equalizer of the diagram

$$
B \times_A C \;\rightarrowtail\; B \times C \mathrel{\substack{(b,c) \mapsto f(b) \\ \longrightarrow \\ \longrightarrow \\ (b,c) \mapsto g(c)}} A.
$$

Similarly, Example 3.2.10 defines the limit of a tower to be a subset of a product. These are special cases of a general theorem, which constructs the limit of any small diagram of sets in an analogous manner.

THEOREM 3.2.13. *Any small limit in* Set *may be expressed as an equalizer of a pair of maps between products. Explicitly, for any small* $F\colon \mathsf{J} \to \mathsf{Set}$, *there is an equalizer diagram*

$$
(3.2.14) \qquad \lim_{\mathsf{J}} F \;\rightarrowtail\; \prod_{j \in \mathrm{ob}\,\mathsf{J}} F j \mathrel{\substack{c \\ \longrightarrow \\ \longrightarrow \\ d}} \prod_{f \in \mathrm{mor}\,\mathsf{J}} F(\mathrm{cod}\,f).
$$

PROOF. By Theorem 3.2.6, elements of the limit of $F\colon \mathsf{J} \to \mathsf{Set}$ correspond to cones with summit 1 over F. The data of such a cone consists of an element λ_j in each set $F j$, indexed by the objects $j \in \mathsf{J}$. The conditions that make the family of elements $(\lambda_j)_{j \in \mathsf{J}}$ into a cone over F are indexed by the (non-identity) morphisms in J: for each morphism f, we demand that the following triangle commutes:

$$
\begin{array}{ccc}
 & 1 & \\
{\scriptstyle \lambda_{\mathrm{dom}\,f}}\swarrow & & \searrow{\scriptstyle \lambda_{\mathrm{cod}\,f}} \\
F(\mathrm{dom}\,f) & \xrightarrow[\;Ff\;]{} & F(\mathrm{cod}\,f)
\end{array}
\qquad \text{i.e.,} \qquad F f(\lambda_{\mathrm{dom}\,f}) = \lambda_{\mathrm{cod}\,f}\,.
$$

The data just described determines the domain of the equalizer diagram (3.2.14): $(\lambda_j)_{j \in \mathsf{J}}$ is an element of the product $\prod_{j \in \mathrm{ob}\,\mathsf{J}} F j$. The conditions determine the codomain and the parallel pair of maps. Note that the equality expressing the cone condition identifies a pair of elements in $\prod_{f \in \mathrm{mor}\,\mathsf{J}} F(\mathrm{cod}\,f)$.

It remains to define the parallel pair of morphisms c and d of (3.2.14). An element $(\lambda_j)_j \in \prod_{j \in \mathrm{ob}\,\mathsf{J}} F j$—thought of as the legs of a cone with summit 1 over F—is sent by c to the element $(\lambda_{\mathrm{cod}\,f})_f \in \prod_{f \in \mathrm{mor}\,\mathsf{J}} F(\mathrm{cod}\,f)$ and is sent by d to the element $(F f(\lambda_{\mathrm{dom}\,f}))_f \in \prod_{f \in \mathrm{mor}\,\mathsf{J}} F(\mathrm{cod}\,f)$. The equalizer is the subset of $\prod_{j \in \mathrm{ob}\,\mathsf{J}} F j$ for which these elements are equal, which is precisely the set of tuples of maps $(\lambda_j\colon 1 \to F j)_{j \in \mathsf{J}}$ that satisfy the cone compatibility conditions. This, together with our explicit descriptions of products and equalizers in Set given in Examples 3.2.7 and 3.2.9, completes the proof that $\lim_{\mathsf{J}} F$ is the equalizer of c and d. $\square$

REMARK 3.2.15. The maps in the equalizer diagram of Theorem 3.2.13 can also be defined categorically, using the universal property of the product $\prod_{f \in \mathrm{mor}\,\mathsf{J}} F(\mathrm{cod}\,f)$. By Remark 3.1.27, to define c and d it is necessary and sufficient to define each component function, by which we mean the composites of c and of d with the projection π_f. The components of c are themselves projections, as displayed in the top triangle:

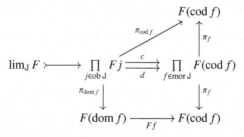

That is, the component of the map c at the indexing element $f \in \operatorname{mor} J$ is the projection from the product $\prod_j Fj$ onto the component indexed by the object $\operatorname{cod} f \in J$.

The component at f of the map d, displayed in the bottom square, is defined by projecting from the product $\prod_j Fj$ onto the component indexed by the object $\operatorname{dom} f \in J$ and then composing with the map $Ff \colon F(\operatorname{dom} f) \to F(\operatorname{cod} f)$. Considering the action of these functions on elements, one can verify that the categorical descriptions agree with the explicit ones described in the proof of Theorem 3.2.13.

Remark 3.2.15 is used to generalize the statement of Theorem 3.2.13 to categories other than Set. By the principle of duality, this generalization will provide a formula for constructing colimits as well; see Theorem 3.4.12.

EXAMPLE 3.2.16. An **idempotent** is an endomorphism $e \colon A \to A$ of some object so that $e \cdot e = e$. The limit of an idempotent in Set is the set of cones with summit 1, i.e., the set of $a \in A$ so that $ea = a$. This is the set of **fixed points** for the idempotent e, often denoted by A^e.

Alternatively, applying Theorem 3.2.13 in the simplified form of Exercise 3.2.ii, the limit A^e is constructed as the equalizer

$$A^e \rightarrowtail\overset{s}{} A \underset{e}{\overset{1}{\rightrightarrows}} A$$

The universal property of the equalizer implies that e factors through s along a unique map r.

$$
\begin{array}{c}
A \\
{\scriptstyle r} \Big| \quad \diagdown\ {\scriptstyle e} \\
\downarrow \quad\ \ \diagdown \\
A^e \rightarrowtail\overset{s}{} A \underset{e}{\overset{1}{\rightrightarrows}} A
\end{array}
$$

The factorization $e = sr$ is said to **split** the idempotent. Now $srs = es = s$ implies that rs and 1_{A^e} both define factorizations of the diagram

$$
\begin{array}{c}
A^e \\
{\scriptstyle rs} \Big|\Big| {\scriptstyle 1_{A^e}} \quad \diagdown\ {\scriptstyle s} \\
\downarrow\downarrow \qquad \diagdown \\
A^e \rightarrowtail\overset{s}{} A \underset{e}{\overset{1}{\rightrightarrows}} A
\end{array}
$$

Uniqueness implies $rs = 1_{A^e}$ so A^e is a retract of A. Conversely, any retract diagram

$$B \rightarrowtail\overset{s}{} A \overset{r}{\twoheadrightarrow} B \qquad\qquad rs = 1_B$$

gives rise to an idempotent sr on A, which is split by B.

Exercises.

EXERCISE 3.2.i. A **small category** can be redefined to be a particular diagram in Set. The data is given by a pair of suggestively-named sets with functions

$$\text{mor } C \underset{\text{cod}}{\overset{\text{dom}}{\underset{\xleftarrow{\text{id}}}{\rightrightarrows}}} \text{ob } C$$

together with a "composition function" yet to be defined. Use a pullback to define the set of composable pairs of morphisms, which serves as the domain for the composition function, and formulate the axioms for a category using commutative diagrams in Set. When Set is replaced by a category E with pullbacks, this defines a **category internal to** E.

EXERCISE 3.2.ii. Show that for any small diagram $F: J \to$ Set, the equalizer diagram (3.2.14) can be modified to yield a slightly smaller equalizer diagram:

$$\lim_J F \rightarrowtail \prod_{j \in \text{ob } J} Fj \underset{d}{\overset{c}{\rightrightarrows}} \prod_{f \in \text{mor } J \backslash \text{ob } J} F(\text{cod } f)$$

in which the second product is indexed only by non-identity morphisms.

EXERCISE 3.2.iii. For any pair of morphisms $f: a \to b, g: c \to d$ in a locally small category C, construct the set of commutative squares

$$\text{Sq}(f, g) := \left\{ \begin{array}{ccc} a & \longrightarrow & c \\ f \downarrow & & \downarrow g \\ b & \longrightarrow & d \end{array} \right\}$$

from f to g as a pullback in Set.

EXERCISE 3.2.iv. Generalize Exercise 3.2.iii to show that for any small category J, any locally small category C, and any parallel pair of functors $F, G: J \rightrightarrows C$, the set $\text{Hom}(F, G)$ of natural transformations from F to G can be defined as a small limit in Set. (Hint: The diagram whose limit is $\text{Hom}(F, G)$ is indexed by a category $J^\S$ whose objects are morphisms in J and which has morphisms $1_x \to f$ and $1_y \to f$ for every $f: x \to y$ in J.)

EXERCISE 3.2.v. Show that for any small category J, any locally small category C, and any parallel pair of functors $F, G: J \rightrightarrows C$, there is an equalizer diagram

$$
\begin{array}{ccc}
C(Fj', Gj') & \xrightarrow{\ \ Ff^*\ \ } & C(Fj, Gj') \\
\pi_{f'} \uparrow & & \uparrow \pi_f \\
\end{array}
$$

$$\text{Hom}(F, G) \rightarrowtail \prod_{j \in \text{ob } J} C(Fj, Gj) \rightrightarrows \prod_{f: j \to j' \in \text{mor } J} C(Fj, Gj')$$

$$
\begin{array}{ccc}
\pi_j \downarrow & & \downarrow \pi_f \\
C(Fj, Gj) & \xrightarrow{\ \ Gf_*\ \ } & C(Fj, Gj')
\end{array}
$$

Note that is not the equalizer diagram obtained by applying Theorem 3.2.13 to the diagram constructed in Exercise 3.2.iv. Rather, this construction gives a second formula for $\text{Hom}(F, G)$ as a limit in Set.

EXERCISE 3.2.vi. Prove that the limit of any small functor $F: C \to$ Set is isomorphic to the set of functors $C \to \int F$ that define a section to the canonical projection $\Pi: \int F \to C$ from the category of elements of F. Using this description of the limit, define the limit cone.

3.3. Preservation, reflection, and creation of limits and colimits

Theorem 3.2.13 would be of somewhat limited interest if it only provided formulae for limits of diagrams of sets. In §3.4, we prove the analogous formulae for limits or colimits of diagrams valued in any category by reducing the proofs of these general results to the case of limits in Set. This strategy succeeds because limits and colimits in a general locally small category are *defined representably* in terms of limits in the category of sets.

The precise meaning of the slogan "limits and colimits are defined representably using limits of sets" is explained in the next section. First, we introduce language to describe a variety of possible ways in which a functor can mediate between the limits and colimits in its domain and codomain categories.

DEFINITION 3.3.1. For any class of diagrams $K\colon \mathsf{J} \to \mathsf{C}$ valued in C, a functor $F\colon \mathsf{C} \to \mathsf{D}$

- **preserves** those limits if for any diagram $K\colon \mathsf{J} \to \mathsf{C}$ and limit cone over K, the image of this cone defines a limit cone over the composite diagram $FK\colon \mathsf{J} \to \mathsf{D}$;
- **reflects** those limits if any cone over a diagram $K\colon \mathsf{J} \to \mathsf{C}$, whose image upon applying F is a limit cone for the diagram $FK\colon \mathsf{J} \to \mathsf{D}$, is a limit cone over K; and
- **creates** those limits if whenever $FK\colon \mathsf{J} \to \mathsf{D}$ has a limit in D, there is some limit cone over FK that can be lifted to a limit cone over K, and moreover F reflects the limits in the class of diagrams.

Definition 3.3.1 dualizes to describe what it means for a functor to preserve, reflect, or create a class of colimits. The terms "preserves," "reflects," and "creates" are used more generally. For instance, Exercise 1.5.iv shows that a fully faithful functor preserves, reflects, and creates isomorphisms.

REMARK 3.3.2. To ground intuition for these terms, the following metaphor may prove helpful: a functor $F\colon \mathsf{C} \to \mathsf{D}$ maps from "upstairs" (the category C) to "downstairs" (the category D). An upstairs diagram (a functor $K\colon \mathsf{J} \to \mathsf{C}$) sits above a downstairs diagram (by composing with F).

Now F **preserves** limits if any upstairs limit cone maps to a downstairs limit cone. F **reflects** limits if any upstairs cone that maps to a downstairs limit cone is also an upstairs limit cone. Finally, F **creates** limits if the mere presence of a downstairs limit cone implies the existence of an upstairs limit cone, and if any cone sitting above a limit cone is a limit cone. The following diagram illustrates this transfer of information in the case of a pullback diagram $c' \to c \leftarrow c''$ in C:

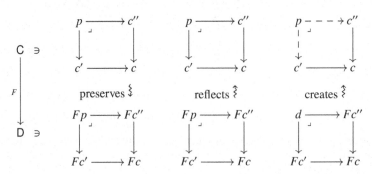

The context in which it is of greatest interest to have a functor $F\colon \mathsf{C} \to \mathsf{D}$ that creates limits or colimits is when the codomain category D is already known to have the (co)limits in question. In such cases, a functor that creates (co)limits both preserves and reflects them:

PROPOSITION 3.3.3. *If $F: C \to D$ creates limits for a particular class of diagrams in C and D has limits of those diagrams, then C admits those limits and F preserves them.*

PROOF. For any diagram $K: J \to C$ in the class, the hypothesis asserts that there is a limit cone $\mu: d \Rightarrow FK$ in D. As F creates these limits, there must be a limit cone $\lambda: c \Rightarrow K$ in C whose image under F is isomorphic to μ (because any two limit cones over a common diagram are isomorphic; see Exercise 3.1.iv). This shows that C admits this class of limits. To see that F preserves them, consider another limit cone $\lambda': c' \Rightarrow K$. By Proposition 3.1.7, the two limit cones in C are isomorphic and by composing isomorphisms we see that the cone $F\lambda': Fc' \Rightarrow FK$ is isomorphic to the limit cone $\mu: d \Rightarrow FK$. This implies that $F\lambda': Fc' \Rightarrow FK$ is again a limit cone, proving that F preserves these limits.   □

Important families of limit-preserving functors appear in the next section and again in the next chapter. For now, let us remark that limit-preservation can be an important hypothesis:

DEFINITION 3.3.4. Let X be a topological space and write $O(X)$ for the poset of open subsets, ordered by inclusion. An I-indexed family of open subsets $U_i \subset U$ is said to **cover** U if the full diagram comprised of the sets U_i and the inclusions of their pairwise intersections $U_i \cap U_j$ has colimit U.[8] A presheaf $F: O(X)^{op} \to Set$ is a **sheaf** if it preserves these colimits, sending them to limits in Set. Applying Theorem 3.2.13, the hypothesis is that for any open cover $\{U_i\}_{i \in I}$ of U, the following is an equalizer diagram:

$$F(U) \overset{F(U_i \hookrightarrow U)}{\rightarrowtail} \prod_{i \in I} F(U_i) \underset{F(U_i \cap U_j \hookrightarrow U_j) \cdot \pi_j}{\overset{F(U_i \cap U_j \hookrightarrow U_i) \cdot \pi_i}{\rightrightarrows}} \prod_{i,j \in I} F(U_i \cap U_j)$$

The following lemmas introduce general classes of examples of functors that reflect or create limits and colimits.

LEMMA 3.3.5. *Any full and faithful functor reflects any limits and colimits that are present in its codomain.*

PROOF. Exercise 3.3.ii.   □

LEMMA 3.3.6. *Any equivalence of categories preserves, reflects, and creates any limits and colimits that are present in its domain or codomain.*

PROOF. Exercise 3.3.iii.   □

With the exception of the equivalences, the functors that one meets in practice that create certain limits and colimits tend to do so *strictly*, in the sense of the following definition.

DEFINITION 3.3.7. A functor $F: C \to D$ **strictly creates** limits for a given class of diagrams if for any diagram $K: J \to C$ in the class and limit cone over $FK: J \to D$,

- there exists a unique lift of that cone to a cone over K, and
- moreover, this lift defines a limit cone in C.

A category J is **connected** if it is non-empty and if, in its underlying directed graph, every pair of objects is connected by a finite zig-zag of morphisms. A diagram is **connected** if it is indexed by a connected category.

PROPOSITION 3.3.8. *For any object $c \in C$, the forgetful functor $\Pi: c/C \to C$ strictly creates*

---

[8]U is also isomorphic to the coproduct $\coprod_{i \in I} U_i$ in $O(X)$, but to state the sheaf condition, it is necessary to describe U as a colimit of the diagram that includes the pairwise intersections $U_j \hookrightarrow U_i \cap U_j \hookrightarrow U_i$.

    (i) *all limits that* C *admits, and*

    (ii) *all connected colimits that* C *admits.*

    A generalization of this result appears as Exercise 3.4.iii.

    PROOF. To define a diagram $(K, \kappa)\colon \mathsf{J} \to c/\mathsf{C}$ is to define a functor $K\colon \mathsf{J} \to \mathsf{C}$ together with a cone $\kappa\colon c \Rightarrow K$ with summit c. We will show that if

$$K\colon \mathsf{J} \xrightarrow{(K,\kappa)} c/\mathsf{C} \xrightarrow{\Pi} \mathsf{C}$$

admits a limit cone in C, then this cone lifts to define a unique cone over (K, κ) in c/C and moreover this cone is a limit cone.

    Given a limit cone $\lambda\colon \ell \Rightarrow K$ in C, there is a unique factorization of the cone $\kappa\colon c \Rightarrow K$ through the limit cone along a map $t\colon c \to \ell$. In the case where J is the parallel pair category

$$a \underset{g}{\overset{f}{\rightrightarrows}} b,$$

this process is illustrated below:

It is straightforward to verify that the object $t\colon c \to \ell$ is a limit for the original diagram in c/C. By construction, $\Pi\colon c/\mathsf{C} \to \mathsf{C}$ preserves this limit.

    Now suppose that J is connected and that K admits a colimit cone $\mu\colon K \Rightarrow p$. We will show that this data lifts to define a unique cone under (K, κ) in c/C that is moreover a colimit cone. The first task is to lift the colimit object p along the functor $\Pi\colon c/\mathsf{C} \to \mathsf{C}$. For this, choose any object $j \in \mathsf{J}$ and define the lifted object to be the composite

$$c \xrightarrow{\kappa_j} Kj \xrightarrow{\mu_j} p$$

of the leg of the cone $\kappa\colon c \Rightarrow K$ and the leg of the colimit cone $\mu\colon K \Rightarrow p$. Because the category J is connected, the composite map $\mu_j \cdot \kappa_j \colon c \to p$ is independent of the choice of $j \in \mathsf{J}$. Now, given a cone under $(K, \kappa)\colon \mathsf{J} \to c/\mathsf{C}$, the underlying cone in C factors uniquely through p, and this factorization lifts to a unique factorization in c/C because the object c is "on the other side of the diagram" K and does not interact with the commutative condition required to define the cone factorization. This proves that $\Pi\colon c/\mathsf{C} \to \mathsf{C}$ strictly creates connected colimits. □

    A final result shows that functor categories inherit limits and colimits, defined "object-wise" in the target category: that is, given a J-indexed diagram in C^A whose objects are functors $F_j\colon \mathsf{A} \to \mathsf{C}$, the value of the limit functor $\lim_{j \in \mathsf{J}} F_j \colon \mathsf{A} \to \mathsf{C}$ at an object $a \in \mathsf{A}$ is the limit of the J-indexed diagram in C whose objects are the objects $F_j(a) \in \mathsf{C}$. This result is analogous to familiar operations on functions: e.g., the sum of a finite collection of abelian group homomorphisms $f_j\colon A \to C$ is the homomorphism defined at $a \in A$ by forming the sum $\sum_j f_j(a)$ in the abelian group C.

    To state this result precisely, we consider the objects of a small indexing category A as a maximal discrete subcategory $\text{ob}\,\mathsf{A} \hookrightarrow \mathsf{A}$.

PROPOSITION 3.3.9. *If* A *is small, then the forgetful functor* $C^A \to C^{ob\,A}$ *strictly creates all limits and colimits that exist in* C. *These limits are defined objectwise, meaning that for each* $a \in A$, *the evaluation functor* $ev_a \colon C^A \to C$ *preserves all limits and colimits existing in* C.

PROOF. The functor category $C^{ob\,A}$ is isomorphic to the ob A-indexed product of the category C with itself, defined in 1.3.12; as the name would suggest, categorical products define products in CAT. By the universal property of products, a diagram of shape J in a product category $\prod_{ob\,A} C$ is an ob A-indexed family of diagrams $J \to C$. It is easy to see that a limit of each of these component diagrams assembles into a limit for the diagram in $C^{ob\,A} \cong \prod_{ob\,A} C$. In particular, $C^{ob\,A}$ has all limits or colimits that C does, and these are preserved by the evaluation functors $ev_a \colon C^{ob\,A} \to C$.

To show that $C^A \to C^{ob\,A}$ strictly creates all limits and colimits, we must show that for any $F \colon J \to C^A$ the ob A-indexed family of objects $\lim_{j\in J} F_j(a)$ extends to a functor on A valued in C. The universal property of the limit construction is used to define the action of this functor on morphisms in A. The uniqueness statement in this universal property then implies that this construction is functorial. The remaining details are left as Exercise 3.3.vi. □

Exercises.

EXERCISE 3.3.i. For any diagram $K \colon J \to C$ and any functor $F \colon C \to D$:

  (i) Define a canonical map colim $FK \to F$ colim K, assuming both colimits exist.
  (ii) Show that the functor F preserves the colimit of K just when this map is an isomorphism.

EXERCISE 3.3.ii. Prove Lemma 3.3.5, that a full and faithful functor reflects both limits and colimits.

EXERCISE 3.3.iii. Prove Lemma 3.3.6, that an equivalence of categories preserves, reflects, and creates any limits and colimits that are present in either its domain or codomain.

EXERCISE 3.3.iv. Prove that $F \colon C \to D$ creates limits for a particular class of diagrams if both of the following hold:

  (i) C has those limits and F preserves them.
  (ii) $F \colon C \to D$ reflects isomorphisms.

EXERCISE 3.3.v. Show that the forgetful functors $U \colon Set_* \to Set$ and $U \colon Top_* \to Top$ fail to preserve coproducts and explain why this result demonstrates that the connectedness hypothesis in Proposition 3.3.8(ii) is necessary.

EXERCISE 3.3.vi. Prove that for any small category A, the functor category C^A again has any limits or colimits that C does, constructed objectwise. That is, given a diagram $F \colon J \to C^A$, with J small, show that whenever the limits of the diagrams

$$ J \xrightarrow{\ F\ } C^A \xrightarrow{\ ev_a\ } C $$

exist in C for all $a \in A$, then these values define the action on objects of $\lim F \in C^A$, a limit of the diagram F. (Hint: See Proposition 3.6.1.)

3.4. The representable nature of limits and colimits

The aim in this section is to explore the meaning of the remark made at the beginning of the preceding one, that all limits and colimits are defined representably in terms of limits in the category of sets.

Fix a small diagram $F\colon J \to C$ in a locally small category C and an object $X \in C$, and consider the composite functor

$$C(X, F-) := J \xrightarrow{\ F\ } C \xrightarrow{\ C(X,-)\ } \text{Set}$$

a diagram of shape J in Set. Theorem 3.2.6 tells us that the limit of $C(X, F-)$ exists and Theorem 3.2.13 tells us how to construct it. An element in the set $\lim_J C(X, F-)$ is an element of the product $\prod_{j \in J} C(X, Fj)$, i.e., a tuple of morphisms $(\lambda_j \colon X \to Fj)_{j \in J}$, subject to some compatibility conditions. There is one condition imposed by each non-identity morphism $f \colon j \to k$ in J, namely that the diagram

$$
\begin{array}{ccc}
 & X & \\
{\scriptstyle \lambda_j}\swarrow & & \searrow{\scriptstyle \lambda_k} \\
Fj & \xrightarrow[\ Ff\ ]{} & Fk
\end{array}
$$

commutes. In summary, an element of $\lim_J C(X, F-)$ is precisely a cone over F with summit X; hence

(3.4.1) $\lim_J C(X, F-) \cong \text{Cone}(X, F).$[9]

Exercise 3.4.i demonstrates that this isomorphism is natural in X. Since the limit of F is defined to be an object that represents the functor $\text{Cone}(-, F)$, we conclude that:

THEOREM 3.4.2. *For any diagram $F\colon J \to C$ whose limit exists there is a natural isomorphism*

(3.4.3) $C(X, \lim_J F) \cong \lim_J C(X, F-).$

The natural isomorphism (3.4.3) expresses the **representable universal property of the limit**: the limit of a J-indexed diagram in a locally small category C is defined representably as the limit of J-indexed diagrams valued in Set.

EXAMPLE 3.4.4. Iterated products of an object $A \in C$ are called **powers** or **cotensors**. For a set I, the I-indexed power of A is denoted $\prod_I A$ or A^I. The representable universal property is expressed by the natural isomorphism

$$C(X, A^I) \cong C(X, A)^I,$$

i.e., a map $h\colon X \to A^I$ is determined by an I-indexed family of maps $h_i\colon X \to A$ defined by composing with each product projection $\epsilon_i\colon A^I \to A$. In the case of $C = \text{Set}$, this notation is consistent with the exponential notation introduced previously: the power A^I is isomorphic to the set of functions from I to A, and the map $\epsilon_i\colon A^I \to A$ evaluates each function at $i \in I$.

There are a number of interpretations of Theorem 3.4.2 that are worth stating explicitly:

PROPOSITION 3.4.5. *Covariant representable functors $C(X, -)\colon C \to \text{Set}$ preserve any limits that exist in a locally small category C, sending them to limits in Set.*

PROOF. Theorem 3.4.2 asserts that the image under $C(X, -)$ of a limit of a diagram F of shape J in C is a limit in Set of the composite diagram

$$J \xrightarrow{\ F\ } C \xrightarrow{\ C(X,-)\ } \text{Set} .$$

---

[9]The size hypotheses are not essential for this isomorphism: for any diagram $F\colon J \to C$, an element of $\lim_J C(X, F-)$ is precisely a cone with summit X over F. This limit is related to the limits considered in Exercises 3.2.iv and 3.2.v with some simplifications arising from the fact that a cone is a natural transformation involving a constant functor.

Moreover, the functor $C(X, -)$ preserves the legs of the limit cone because, by construction, the natural isomorphism of Theorem 3.4.2 commutes with the natural maps to the product:

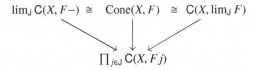

$$\lim_J C(X, F-) \cong \text{Cone}(X, F) \cong C(X, \lim_J F)$$

$$\prod_{j \in J} C(X, Fj)$$

The components of the left-hand diagonal map are the legs of the limit cone for the diagram $C(X, F-)$. The components of the right-hand diagonal are the images of the legs of the limit cone for F under the functor $C(X, -)$.   □

A second interpretation of Theorem 3.4.2 is that the contravariant functor $C(-, \lim F)$ represented by the limit of $F: J \to C$, when considered as an object of the category $\text{Set}^{C^{\text{op}}}$, is the limit of the composite diagram

$$J \xrightarrow{\ F\ } C \xhookrightarrow{\ y\ } \text{Set}^{C^{\text{op}}}$$

whose objects are the representable functors $C(-, Fj)$. Put more concisely, the Yoneda embedding $y: C \hookrightarrow \text{Set}^{C^{\text{op}}}$ preserves all limits that exist in C.

Of course, by the Yoneda lemma, the Yoneda embedding is full and faithful, so Lemma 3.3.5 implies that $y: C \hookrightarrow \text{Set}^{C^{\text{op}}}$ also reflects all limits. In summary:

THEOREM 3.4.6. *Let* C *be any locally small category.*

(i) *Covariant representable functors* $C(X, -)$ *preserve all limits that exist in* C.
(ii) *The covariant Yoneda embedding* $y: C \hookrightarrow \text{Set}^{C^{\text{op}}}$ *both preserves and reflects limits, i.e., a cone over a diagram in* C *is a limit cone if and only if its image defines a limit cone in* $\text{Set}^{C^{\text{op}}}$.

A representable characterization of colimits is obtained by dualizing the preceding discussion. For fixed $F: J \to C$ and $X \in C$, consider the composite functor

$$C(F-, X) := J^{\text{op}} \xrightarrow{\ F\ } C^{\text{op}} \xrightarrow{\ C(-,X)\ } \text{Set}$$

Again, Theorem 3.2.6 tells us that a limit exists and Theorem 3.2.13 provides a construction. An element in the set $\lim_{J^{\text{op}}} C(F-, X)$ is an element of the product $\prod_{j \in J^{\text{op}}} C(Fj, X)$, i.e., a tuple of morphisms $(\lambda_j: Fj \to X)_{j \in J}$, subject to some conditions. There is one condition imposed by each non-identity morphism $f: j \to k$ in J, namely that the diagram

$$Fj \xrightarrow{\ Ff\ } Fk$$
$$\lambda_j \searrow \quad \swarrow \lambda_k$$
$$X$$

commutes, so we see that an element of $\lim_{J^{\text{op}}} C(F-, X)$ is precisely a cone under F with nadir X. The isomorphism

$$\lim_{J^{\text{op}}} C(F-, X) \cong \text{Cone}(F, X)$$

is again natural in X proving:

THEOREM 3.4.7. *For any diagram* $F: J \to C$ *whose colimit exists, there is a natural isomorphism*

$$C(\text{colim}_J F, X) \cong \lim_{J^{\text{op}}} C(F-, X).$$

Theorem 3.4.7 expresses the **representable universal property of the colimit**: the colimit of a J-indexed diagram in a category C is defined representably as the limit of J^{op}-indexed diagrams valued in Set. For instance:

EXAMPLE 3.4.8. For any triple of objects X, Y, Z in a locally small category C, the representable universal property of the coproduct takes the form of a natural isomorphism

$$C(X \sqcup Y, Z) \cong C(X, Z) \times C(Y, Z),$$

in which the set of maps out of a coproduct is expressed as the product of a pair of hom-sets. This answers the puzzle posed in Exercise 2.4.x at the end of Chapter 2.

EXAMPLE 3.4.9. Iterated coproducts of an object $A \in C$ are called **copowers** or **tensors**. For a set I, the I-indexed copower of A is denoted $\coprod_I A$ or $I \cdot A$. The representable universal property is expressed by the natural isomorphism

$$C(\coprod_I A, X) \cong C(A, X)^I,$$

i.e., a map $h: \coprod_I A \to X$ is determined by an I-indexed family of maps $h_i: A \to X$ defined by composing with each coproduct inclusion $\iota_i: A \to \coprod_I A$. In the case of C = Set, the copower $\coprod_I A$ is isomorphic to the set $I \times A$.

EXAMPLE 3.4.10. The representable universal property of a pushout

$$
\begin{array}{ccc}
U & \xrightarrow{f} & V \\
{\scriptstyle g}\downarrow & & \downarrow{\scriptstyle h} \\
W & \xrightarrow{k} & P
\end{array}
$$

asserts that for any $X \in C$, the hom-set $C(P, X)$ is defined by the pullback

$$
\begin{array}{ccc}
C(P, X) & \xrightarrow{C(h,X)} & C(V, X) \\
{\scriptstyle C(k,X)}\downarrow & & \downarrow{\scriptstyle C(f,X)} \\
C(W, X) & \xrightarrow{C(g,X)} & C(U, X)
\end{array}
$$

Theorem 3.4.6 dualizes to:

THEOREM 3.4.11. *Let C be any locally small category.*

(i) *Contravariant representable functors $C(-, X)$ carry colimits in C to limits in Set.*

(ii) *The contravariant Yoneda embedding $y: C^{op} \hookrightarrow Set^C$ both preserves and reflects limits in C^{op}, i.e., a cone under a diagram in C is a colimit cone if and only if its image defines a limit cone in Set^C.*

We now apply Theorems 3.4.6 and 3.4.11 to generalize Theorem 3.2.13, which constructs a generic limit in Set as an equalizer of a pair of maps between products, to arbitrary categories and hence, by the principle of duality, also to colimits.

THEOREM 3.4.12. *The colimit of any small diagram $F: J \to C$ may be expressed as a coequalizer of a pair of maps between coproducts*

$$\coprod_{f \in mor J} F(dom\ f) \underset{c}{\overset{d}{\rightrightarrows}} \coprod_{j \in ob J} Fj \twoheadrightarrow colim_J F$$

In particular, any category with coproducts and coequalizers is cocomplete. Dually, the limit of any small diagram may be expressed as an equalizer (3.2.14) of a pair of maps between products, and any category with products and equalizers is complete.

Our proof will tacitly assume that the category C is locally small, but the argument could be extended to large categories by "passing to a higher set-theoretical universe."

PROOF. Consider a diagram $F\colon \mathsf{J} \to \mathsf{C}$. Dualizing the construction of Remark 3.2.15, define maps d and c

(3.4.13)

$$
\begin{array}{c}
F(\operatorname{dom} f) \\
\end{array}
$$

$$
C \xleftarrow{\quad} \coprod_{j\in\operatorname{ob} \mathsf{J}} Fj \underset{c}{\overset{d}{\rightrightarrows}} \coprod_{f\in\operatorname{mor} \mathsf{J}} F(\operatorname{dom} f)
$$

with the triangle $\iota_{\operatorname{dom} f}$, ι_f at top and $\iota_{\operatorname{cod} f}\uparrow$, $F(\operatorname{cod} f) \xleftarrow{\quad Ff \quad} F(\operatorname{dom} f)$, $\uparrow \iota_f$ below.

between the coproducts indexed by the morphisms and objects of J, respectively: the component of d at $f \in \operatorname{mor} \mathsf{J}$ is defined to be the coproduct inclusion $\iota_{\operatorname{dom} f}$, while the component of c at $f \in \operatorname{mor} \mathsf{J}$ is defined to be the composite of Ff with the coproduct inclusion $\iota_{\operatorname{cod} f}$. By hypothesis, their coequalizer C exists in C. Our aim is to show that C defines a colimit of F.

By Theorem 3.4.11(ii), the Yoneda embedding $y\colon \mathsf{C}^{\operatorname{op}} \to \mathsf{Set}^{\mathsf{C}}$ carries the coequalizer diagram (3.4.13) to an equalizer diagram in $\mathsf{Set}^{\mathsf{C}}$

$$
\mathsf{C}(C, X) \rightarrowtail \mathsf{C}(\coprod_{j\in\operatorname{ob}\mathsf{J}} Fj, X) \underset{\mathsf{C}(c,X)}{\overset{\mathsf{C}(d,X)}{\rightrightarrows}} \mathsf{C}(\coprod_{f\in\operatorname{mor}\mathsf{J}} F(\operatorname{dom} f), X)
$$

with surrounding maps $\mathsf{C}(\iota_{\operatorname{dom} f},X)$, $\mathsf{C}(\iota_f,X)$ to $\mathsf{C}(F(\operatorname{dom} f), X)$, and $\mathsf{C}(\iota_{\operatorname{cod} f},X)$, $\mathsf{C}(\iota_f,X)$ to $\mathsf{C}(F(\operatorname{cod} f), X) \xrightarrow{\mathsf{C}(Ff,X)} \mathsf{C}(F(\operatorname{dom} f), X)$.

comprised of functors and natural transformations in the variable X. Applying Theorem 3.4.11(i) to each object in this equalizer diagram, we obtain an isomorphic diagram

(3.4.14)

$$
\mathsf{C}(C, X) \rightarrowtail \prod_{j\in\operatorname{ob}\mathsf{J}^{\operatorname{op}}} \mathsf{C}(Fj, X) \underset{d}{\overset{c}{\rightrightarrows}} \prod_{f\in\operatorname{mor}\mathsf{J}^{\operatorname{op}}} \mathsf{C}(F(\operatorname{cod} f), X)
$$

with surrounding maps $\pi_{\operatorname{cod} f}$, π_f to $\mathsf{C}(F(\operatorname{cod} f), X)$, and $\pi_{\operatorname{dom} f}$, π_f to $\mathsf{C}(F(\operatorname{dom} f), X) \xrightarrow{\mathsf{C}(Ff,X)} \mathsf{C}(F(\operatorname{cod} f), X)$.

Note that $\mathsf{C}(-, X)$ carries the coproduct inclusion maps to product projection maps. When f is regarded as a morphism in $\mathsf{J}^{\operatorname{op}}$, its domain and codomain are exchanged, so we now write "c" for the map $\mathsf{C}(d, X)$ and "d" for the map $\mathsf{C}(c, X)$.

When X is fixed, applying Proposition 3.3.9 to the evaluation functor $\operatorname{ev}_X\colon \mathsf{Set}^{\mathsf{C}} \to \mathsf{Set}$ tells us that (3.4.14) defines an equalizer diagram in Set. Applying Theorem 3.2.13 to the functor $C(F-, X): J^{op} \to \mathsf{Set}$, we recognize (3.4.14) as the diagram whose equalizer defines the limit $\lim_{J^{op}} C(F-, X)$. Thus, we conclude that

$$\lim_{J^{op}} C(F-, X) \cong C(C, X)$$

for each $X \in C$. These isomorphisms, defined for each $X \in C$, assemble into an isomorphism in $\mathsf{Set}^{ob\,C}$, from which point we apply Proposition 3.3.9 to the forgetful functor $\mathsf{Set}^{C} \to \mathsf{Set}^{ob\,C}$ to conclude that $C(C, -)$ is the limit of the J^{op}-indexed diagram of covariant functors $C(Fj, -)$. Now Theorem 3.4.7 tells us that the coequalizer C is the colimit of $F: J \to C$. □

REMARK 3.4.15 (on generalized elements). A direct proof of Theorem 3.4.12 simply constructs the diagram (3.4.13) in C and checks that the coequalizer of c and d has the universal property that defines $\mathrm{colim}_J F$. There is a sense in which this direct proof is quite similar to our strategy—to first verify the result in Set and then apply representability—by something called the **philosophy of generalized elements**, first articulated by Lawvere [**Law05**].

Elements of a set A stand in bijection with morphisms $a: 1 \to A$ in the category of sets. By composition, morphisms $f: A \to B$ in the category of sets act on elements: the composite fa encodes the element $f(a)$.[10] In any category C, a **generalized element** of an object A is a morphism $a: X \to A$ with codomain A. Morphisms $f: A \to B$ in C act on X-shaped elements by composition: the composite fa encodes an X-shaped generalized element of B. In this terminology, the contravariant represented functor

$$\mathrm{Hom}(-, A): C^{op} \to \mathsf{Set}$$

sends an object X to the set of X-shaped generalized elements of A. Since the Yoneda embedding $C \hookrightarrow \mathsf{Set}^{C^{op}}$ is fully faithful, no information about A is lost by considering instead the functor $\mathrm{Hom}(-, A)$ of generalized elements of A.

In particular, to show that a parallel pair of morphisms $f, g: A \rightrightarrows B$ are equal, it suffices to show that f and g induce the same functions on generalized elements. Or to show that a cone with summit A is a limit cone, it suffices to show, for generalized elements of any shape, that the induced functions define a limit cone in Set: this argument establishes the representable universal property of the limit. See [**Lei08**] for an application of this principle of "doing without diagrams" in more sophisticated contexts.

We conclude this section with an informal illustration of the philosophy of generalized elements by showing that finite products and equalizers can be constructed out of a terminal object 1 and (iterated) pullbacks.

LEMMA 3.4.16. *In any category with a terminal object* 1, *the pullback diagram*

$$
\begin{array}{ccc}
A \times B & \xrightarrow{\pi_B} & B \\
{\scriptstyle \pi_A}\downarrow & \lrcorner & \downarrow{\scriptstyle !} \\
A & \xrightarrow[!]{} & 1
\end{array}
$$

defines the product $A \times B$ *of* A *and* B.

PROOF. It is straightforward to use the universal property of the displayed pullback to verify that the diagram $A \xleftarrow{\pi_A} A \times B \xrightarrow{\pi_B} B$ has the universal property that defines a product. The details are left as Exercise 3.4.iv. □

Hence any category with pullbacks and a terminal object also has all binary, and by iteration finite, products.

---

[10]The standard function notation motivates the usual "composition order."

LEMMA 3.4.17. *In any category with binary products, the pullback diagram*

$$
\begin{array}{ccc}
E & \xrightarrow{\ e\ } & A \\
{\scriptstyle \ell}\downarrow & \lrcorner & \downarrow{\scriptstyle (f,g)} \\
B & \xrightarrow[(1_B,1_B)]{} & B \times B
\end{array}
\qquad \rightsquigarrow \qquad
E \rightarrowtail \xrightarrow{\ e\ } A \underset{g}{\overset{f}{\rightrightarrows}} B
$$

defines an equalizer of the parallel pair $f, g : A \rightrightarrows B$.

PROOF. The representable universal property of the product

$$\mathrm{Hom}(X, B \times B) \cong \mathrm{Hom}(X, B) \times \mathrm{Hom}(X, B)$$

and of the pullback, in the form of Theorem 3.4.6(i), supply a pullback diagram

$$
\begin{array}{ccc}
\mathrm{Hom}(X, E) & \xrightarrow{\quad e_* \quad} & \mathrm{Hom}(X, A) \\
{\scriptstyle \ell_*}\downarrow & \lrcorner & \downarrow{\scriptstyle (f_*,g_*)} \\
\mathrm{Hom}(X, B) & \xrightarrow[(1_*,1_*)]{} & \mathrm{Hom}(X, B) \times \mathrm{Hom}(X, B)
\end{array}
$$

By the description of pullbacks in the category of sets given in Example 3.2.11, composition with e and ℓ defines a bijection between X-shaped generalized elements of E and pairs of generalized elements $(X \xrightarrow{a} A, X \xrightarrow{b} B)$ so that $fa = b = ga$. Thus, composing with e defines an isomorphism between $\mathrm{Hom}(X, E)$ and the subset of $\mathrm{Hom}(X, A)$ consisting of those generalized elements a so that $fa = ga$. By Example 3.2.9, this tells us that

$$\mathrm{Hom}(X, E) \rightarrowtail \xrightarrow{\ e_*\ } \mathrm{Hom}(X, A) \underset{g_*}{\overset{f_*}{\rightrightarrows}} \mathrm{Hom}(X, B)$$

is an equalizer diagram. The conclusion follows from Theorem 3.4.6(ii). □

A diagram is said to be **finite** if its indexing category contains only finitely many morphisms. Lemmas 3.4.16 and 3.4.17 prove:

THEOREM 3.4.18. *Any category with pullbacks and a terminal object has all finite limits. Dually, any category with pushouts and an initial object has all finite colimits.*

Exercises.

EXERCISE 3.4.i. Show that the isomorphism (3.4.1) is natural in X.

EXERCISE 3.4.ii. Explain in your own words why the Yoneda embedding $\mathsf{C} \hookrightarrow \mathsf{Set}^{\mathsf{C}^{\mathrm{op}}}$ preserves and reflects but does not create limits.

EXERCISE 3.4.iii. Generalize Proposition 3.3.8 to show that for any $F \colon \mathsf{C} \to \mathsf{Set}$, the projection functor $\Pi \colon \int F \to \mathsf{C}$:

   (i) strictly creates all limits that C admits and that F preserves, and
   (ii) strictly creates all connected colimits that C admits.

EXERCISE 3.4.iv. Prove Lemma 3.4.16.

3.5. Complete and cocomplete categories

Theorems, such as 3.4.12 and 3.4.18, that reduce the construction of general limits and colimits to a handful of simple ones are particularly useful when it comes to proving that a category is **complete** or **cocomplete**—that is, has all small limits or colimits—or that a functor is **continuous** or **cocontinuous**—that is, preserves all small limits or colimits.

The proofs reduce to checking a few cases. In this section, we describe the constructions of limits and colimits in various categories, in most cases leaving the proofs that these constructions have the claimed universal properties to the reader.

Many of the concrete categories of our acquaintance are complete and cocomplete including Set, Top, Set$_*$, Top$_*$, Cat or CAT, Group, Ab, Ring, Mod$_R$, among others. Here, we restrict consideration to the first six cases. A result from categorical universal algebra specializes to prove that the latter categories are complete and cocomplete (see Theorem 5.6.12).

In the first example, half the work is done already: Theorem 3.2.6 demonstrates that Set is complete.

PROPOSITION 3.5.1. *The category* Set *is cocomplete.*

PROOF. By Theorem 3.4.12, it suffices to show that Set has coproducts and coequalizers. The former are given by disjoint unions, while the coequalizer of a parallel pair of functions $f, g : A \rightrightarrows B$ is the quotient of the set B by the equivalence relation generated by the relations $f(a) \sim g(a)$ for each $a \in A$.   □

PROPOSITION 3.5.2. *The category* Top *of spaces is complete and cocomplete.*

PROOF. The underlying set functor $U : \mathsf{Top} \to \mathsf{Set}$ is represented by the singleton space $*$. Hence, Proposition 3.4.5 implies that if Top has any limits, they must be formed by topologizing the limit of the diagram of underlying sets, as was argued in the case of the cartesian product in Example 3.1.10.

As the name suggests, the product of a family of spaces X_α is formed by giving the cartesian product $\prod_\alpha X_\alpha$ the product topology, which is the coarsest topology so that each of the projection functions $\pi_\beta : \prod_\alpha X_\alpha \to X_\beta$ is continuous. The equalizer of a pair of continuous functions $f, g : X \rightrightarrows Y$ is given by assigning the equalizer $E \subset X$ of the underlying functions the subspace topology, which also may be characterized as the coarsest topology on E so that the inclusion $E \hookrightarrow X$ is continuous.

Dually, we might guess that colimits in Top are formed by topologizing the colimit of the diagram of underlying sets; Theorem 4.5.3 implies that the colimit must be constructed in this manner. The coproduct is formed by assigning the disjoint union $\coprod_\alpha X_\alpha$ the finest topology so that each of the coproduct-inclusions $\iota_\beta : X_\beta \to \coprod_\alpha X_\alpha$ is continuous. The disjoint union of any bases for the spaces X_α defines a basis for $\coprod_\alpha X_\alpha$. The coequalizer of a pair of continuous functions $f, g : X \rightrightarrows Y$ is constructed by first forming the set-theoretic quotient $Y_{/\sim}$ of Y under the equivalence relation generated by the relations $f(x) \sim g(x)$ for each point $x \in X$. The set $Y_{/\sim}$ is then given the quotient topology, under which a subset in $Y_{/\sim}$ is open if and only if its preimage in Y is open; this is the finest topology so that the quotient map $Y \twoheadrightarrow Y_{/\sim}$ is continuous.

Theorem 3.4.12 and its dual now imply that Top has all small limits and colimits. The constructions described here demonstrate moreover that the forgetful functor $U : \mathsf{Top} \to \mathsf{Set}$ preserves them.   □

EXAMPLE 3.5.3. Any limit or colimit in Top can be constructed in the way prescribed in the proof of Proposition 3.5.2: by forming the limit or colimit of the diagram of underlying sets and then topologizing the resulting space with the coarsest (in the case of limits) or finest (in the case of colimits) topologies so that the legs of the limit or colimit cones are continuous.

For instance, consider the diagram $\omega^{\mathrm{op}} \to \mathsf{Top}$ whose objects are circles S^1 and in which each generating map is the "pth power map," the covering map that wraps the domain circle uniformly p times around the codomain circle:

$$\cdots \longrightarrow S^1 \xrightarrow{\ p\ } S^1 \xrightarrow{\ p\ } S^1 \xrightarrow{\ p\ } S^1$$

The inverse limit defines the p-**adic solenoid**.

EXAMPLE 3.5.4. The space of (unordered) configurations of n points in a space X is the topological space whose points are collections of n distinct points in X and whose topology considers two configurations to be "close" if there is a bijection between their points so that each pair of points is "close" to each other.

More formally, the **space of ordered configurations of n points** is a subspace of the product space X^n, namely the complement of the "fat diagonal"

$$\mathrm{PConf}_n(X) := \{(x_1, \ldots, x_n) \in X^n \mid x_i \neq x_j \ \forall i \neq j\}.$$

The symmetric group Σ_n acts on X^n by permuting the coordinates, and this action restricts to the subspace $\mathrm{PConf}_n(X) \subset X^n$. The colimit of the diagram $\mathrm{PConf}_n(X)\colon \mathsf{B}\Sigma_n \to \mathsf{Top}$ defines the **space of configurations of n points**

$$\mathrm{Conf}_n(X) := \mathrm{colim}\left(\mathsf{B}\Sigma_n \xrightarrow{\ \mathrm{PConf}_n(X)\ } \mathsf{Top}\right).$$

Example 3.5.i describes its underlying set.

Completeness and cocompleteness of Set and of Top implies the same for Set_* and Top_* as a special case[11] of a general result:

PROPOSITION 3.5.5. *If C is complete and cocomplete, then so are the slice categories c/C and C/c for any $c \in \mathsf{C}$.*

PROOF. The statements are dual, so it suffices to prove the result for c/C. By Proposition 3.3.8, if C is complete and cocomplete, then c/C has all limits and all connected colimits, created by the forgetful functor $\Pi\colon c/\mathsf{C} \to \mathsf{C}$. Thus, c/C is complete and to prove cocompleteness it remains only to show that c/C has coproducts. By the construction of Lemma 3.4.16, generalized to higher arity coproducts, it suffices to show that c/C has an initial object, which it does: the identity 1_c is initial.

Note that this proof also tells us how the coproduct is constructed: the coproduct of a family of objects $c \to x_\alpha$ is the map from c to the colimit of the diagram defined by these morphisms in C (an "asterisk" with source c and all the morphisms pointing out).[12]   □

Next we consider limits of diagrams of categories. Examples 2.1.5(ix) and (x) reveal that the functors $\mathrm{ob}, \mathrm{mor}\colon \mathsf{Cat} \rightrightarrows \mathsf{Set}$ are represented by the ordinal categories $\mathbb{1}$ and $\mathbb{2}$. By Theorem 3.4.6(i), both functors preserve limits, which means that if the limit of a diagram of categories exists, then its set of objects must be the limit of the underlying diagrams of object-sets and its set of morphisms must be the limit of the underlying diagrams of morphism-sets. Furthermore, domains, codomains, and identities are also preserved, as each of these maps is expressible, by Exercise 2.1.i, as a natural transformation between representable functors. A similar argument involving the category $\mathbb{3}$ proves that composition is preserved as well. These observations suggest candidates for the product of any categories and for the equalizer of any parallel pair of functors, allowing us to prove that:

PROPOSITION 3.5.6. *The categories Cat and CAT are complete.*

---

[11]Exercise 3.3.v explains that Set_* and Top_* are slice categories.

[12]Particularly when C is concrete and c represents the underlying set functor $U\colon \mathsf{C} \to \mathsf{Set}$, so that $c/\mathsf{C} \cong \int U$ is a category of "based objects," the coproduct is typically written with the symbol "$\vee$" in place of the usual "$\sqcup$"; see Example 3.1.24 for instance.

Proof. Higher arity versions of the product, introduced in Definition 1.3.12, define products of categories. The equalizer E of a pair of parallel functors

$$E \rightarrowtail C \overset{F}{\underset{G}{\rightrightarrows}} D$$

is the subcategory of C consisting of those objects $c \in C$ so that $Fc = Gc$ and those morphisms f so that $Ff = Gf$. Applying Theorem 3.4.12, it follows that Cat and CAT admit all small limits.   □

Example 3.5.7. For any functor $F \colon C \to$ Set, the pullback in CAT of

$$C \overset{F}{\longrightarrow} \text{Set} \overset{U}{\longleftarrow} \text{Set}_*$$

is a category whose objects are pairs $c \in C$ and $(X, x) \in$ Set$_*$ so that $Fc = X$. Morphisms are pairs $f \colon c \to c'$ and $g \colon (X, x) \to (X', x')$ so that $Ff = Ug$. Expressing this data more efficiently, we see that the pullback

$$
\begin{array}{ccc}
\int F & \longrightarrow & \text{Set}_* \\
\downarrow & \lrcorner & \downarrow U \\
C & \overset{F}{\longrightarrow} & \text{Set}
\end{array}
$$

is the category $\int F$ of elements of F.[13]

Both Cat and CAT are also cocomplete. Coproducts in each case are straightforward: any category is uniquely expressible as a coproduct of its **connected components**. By Lemma 3.4.17, it suffices to construct pushouts. But these can be rather more complicated than was the case for the corresponding limit notions.

Example 3.5.8. To calculate the pushout,

$$
\begin{array}{ccc}
\mathbb{1} \sqcup \mathbb{1} & \rightarrowtail & 2 \\
\downarrow & & \downarrow \\
\mathbb{1} & \longrightarrow & P
\end{array}
$$

it helps to reflect on its universal property. A functor $P \to C$ corresponds to a cone under the diagram $\mathbb{1} \leftarrow \mathbb{1} \sqcup \mathbb{1} \to 2$ with nadir C. This data defines an endomorphism f in C, a morphism $2 \to C$ so that its domain and codomain define the same object $\mathbb{1} \to C$. In defining P, we must take care not to impose any relations on the composite endomorphisms $f \cdot f \cdots f$. Thus, P must be the free category with one object and one non-identity morphism. Recalling that one-object categories can be identified with monoids, we find that P is quite familiar: it is the monoid $\mathbb{N}$ of natural numbers under addition.

Similarly, the group $\mathbb{Z}$ of integers under addition can be constructed as a pushout

$$
\begin{array}{ccc}
\mathbb{1} \sqcup \mathbb{1} & \rightarrowtail & \mathbb{I} \\
\downarrow & & \downarrow \\
\mathbb{1} & \longrightarrow & B\mathbb{Z}
\end{array}
$$

in which the walking arrow 2 is replaced by the walking isomorphism $\mathbb{I}$, the category with two objects and two non-identity morphisms pointing in opposite directions, which are inverse isomorphisms.

---

[13]The pullback of a contravariant functor $F \colon C^{\mathrm{op}} \to$ Set along $U \colon$ Set$_* \to$ Set defines the opposite of $\int F$.

The proof of cocompleteness of Cat is deferred to Corollary 4.5.16, where it is deduced as a special case of a general result that gives an explicit construction of the colimit of any diagram of categories. The rough idea is to embed the category of categories into a larger category in which colimits are more easily understood and then "reflect" the colimits from the larger category back into Cat.

Before considering a final class of examples, note that completeness and cocompleteness of a category are equivalence-invariant. In particular, a preorder is complete and cocomplete if and only if its skeleton, a poset, is so.

PROPOSITION 3.5.9. *A poset P is complete and cocomplete as a category if and only if it is a **complete lattice**, that is, if and only if every subset $A \subset P$ has both an infimum (greatest lower bound) and a supremum (least upper bound).*

PROOF. In any poset (or preorder), a limit of a diagram is an infimum of its objects, while a colimit of a diagram is a supremum of its objects. Whether or not there are any morphisms in the diagram makes no difference because all diagrams in a preorder commute. In particular, any collection of morphisms with common domain defines a cone over any diagram whose objects are indexed by the codomains.   □

With this work behind us, it is reasonable to ask: What are all these limits and colimits good for? One use of a handful of simple limit constructions is to define a notion of *equivalence relation* internally to a general category. Other significant applications will appear in §4.6, where limits and colimits are used to define functors.

EXAMPLE 3.5.10. In any category with finite limits, the **kernel pair**[14] of a morphism $f\colon X \to Y$ is the pullback of f along itself:

$$
\begin{array}{ccc}
R & \xrightarrow{\ t\ } & X \\
{\scriptstyle s}\big\downarrow & & \big\downarrow{\scriptstyle f} \\
X & \xrightarrow[\ f\ ]{} & Y
\end{array}
$$

These maps define a monomorphism $(s,t)\colon R \rightarrowtail X \times X$, so the object R is always a subobject of the product $X \times X$.[15] In Set, a subset $R \subset X \times X$ defines a **relation** on X. Indeed, subobjects defined by kernel pairs are always **equivalence relations**, in the following categorical sense:

(i) There is a reflexivity map ρ defined by

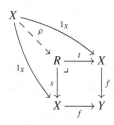

_____

[14]In the absence of a **zero object**—an object that is both initial and terminal, which can be used to define a **zero map** between any pair of objects—kernel pairs are the best analog of kernels. Barry Fawcett acknowledges the pervasiveness of this sort of construction, writing "Category theory tells us: to learn about A, study pairs of maps out of [or into] A" [**Faw86**].

[15]A **subobject** of an object c is a monomorphism with codomain c; see Definition 4.6.8.

that is a section of both s and t, i.e., that defines a factorization of the diagonal

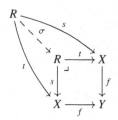

(ii) There is a symmetry map σ defined by

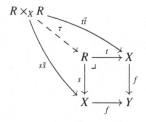

so that $t\sigma = s$ and $s\sigma = t$.

(iii) There is a transitivity map τ whose domain is the pullback of t along s

$$R \times_X R \xrightarrow{\tilde{t}} R \xrightarrow{t} X$$

This diagram defines a cone over the pullback defining R and thus induces a map

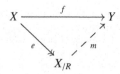

so that $s\tau = s\tilde{s}$ and $t\tau = t\tilde{t}$.

An **equivalence relation** in a category C with finite limits is a subobject $(s,t)\colon R \rightarrowtail X \times X$ equipped with maps ρ, σ, and τ commuting with the morphisms s and t in the manner displayed in the diagrams of (i), (ii), and (iii). When it exists, the coequalizer of the maps $s, t\colon R \rightrightarrows X$ of an equivalence relation defines a quotient object $e\colon X \twoheadrightarrow X_{/R}$. In Set, $X_{/R}$ is the set of R-equivalence classes of elements of X. For equivalence relations arising as kernel pairs, there is a unique factorization

$$X \xrightarrow{f} Y$$

In good situations, such as when C is a *Grothendieck topos* (see §E.4), the map m is a monomorphism and this factorization defines the **image factorization** of the map f: the monomorphism $X_{/R} \rightarrowtail Y$ identifies the **image** $X_{/R}$ as a subobject of Y.

Exercises.

EXERCISE 3.5.i. Let G be a group regarded as 1-object category BG. Describe the colimit of a diagram B$G \to$ Set in group-theoretic terms, as was done for the limit in Example 3.2.12.

EXERCISE 3.5.ii. Prove that the colimit of any small functor $F \colon$ C $\to$ Set is isomorphic to the set $\pi_0(\int F)$ of connected components of the category of elements of F. What is the colimit cone?

EXERCISE 3.5.iii. Prove that the category DirGraph of directed graphs is complete and cocomplete and explain how to construct its limits and colimits. (Hint: Use Proposition 3.3.9.)

EXERCISE 3.5.iv. For a small category J, define a functor $i_0 \colon$ J $\to$ J $\times 2$ so that the pushout

$$
\begin{array}{ccc}
\mathsf{J} & \xrightarrow{\;!\;} & \mathbb{1} \\
{\scriptstyle i_0}\downarrow & & \downarrow{\scriptstyle s} \\
\mathsf{J} \times 2 & \xrightarrow[r]{} & \mathsf{J}^\triangleleft
\end{array}
$$

in Cat defines the **cone** over J, with the functor $s \colon \mathbb{1} \to \mathsf{J}^\triangleleft$ picking out the summit object. Remark 3.1.8 gives an informal description of this category, which is used to index the diagram formed by a cone over a diagram of shape J.

EXERCISE 3.5.v. Describe the limits and colimits in the poset of natural numbers with the order relation $k \le n$ if and only if k divides n.

EXERCISE 3.5.vi. Define a contravariant functor $\mathsf{Fin}^{\mathrm{op}}_{\mathrm{mono}} \to$ Top from the category of finite sets and injections to the category of topological spaces that sends a set with n elements to the space $\mathrm{PConf}_n(X)$ constructed in Example 3.5.4. Explain why this functor does not induce a similar functor sending an n-element set to the space $\mathrm{Conf}_n(X)$.

EXERCISE 3.5.vii. Following [**Gro58**], define a **fiber space** $p \colon E \to B$ to be a morphism in Top. A map of fiber spaces is a commutative square. Thus, the category of fiber spaces is isomorphic to the diagram category Top^2. We are also interested in the non-full subcategory $\mathsf{Top}/B \subset \mathsf{Top}^2$ of fiber spaces over B and maps whose codomain component is the identity. Prove the following:

(i) A map

$$
\begin{array}{ccc}
E' & \xrightarrow{\;g\;} & E \\
{\scriptstyle p'}\downarrow & & \downarrow{\scriptstyle p} \\
B' & \xrightarrow[f]{} & B
\end{array}
$$

of fiber spaces induces a canonical map from the fiber over a point $b \in B'$ to the fiber over its image $f(b) \in B$.

(ii) The fiber of a product of fiber spaces is the product of the fibers.

A projection $B \times F \to B$ defines a **trivial fiber space** over B, a definition that makes sense for any space F.

(iii) Show that the fiber of a trivial fiber space $B \times F \to F$ is isomorphic to F.

(iv) Characterize the isomorphisms in Top/B between two trivial fiber spaces (with a priori distinct fibers) over B.

(v) Prove that the assignment of the set of continuous sections of a fiber space over B defines a functor Sect: Top/B → Set.

(vi) Consider the non-full subcategory Top$^2_{pb}$ of fiber spaces in which the morphisms are the pullback squares. Prove that the assignment of the set of continuous sections to a fiber space defines a functor Sect: (Top$^2_{pb}$)^{op} → Set.

(vii) Describe the compatibility between the actions of the "sections" functors just introduced with respect to the map g of fiber spaces p and q over B and their restrictions along $f: B' \to B$.

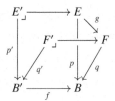

3.6. Functoriality of limits and colimits

Proposition 3.1.7 demonstrates that limits and colimits are each well-defined up to unique isomorphism when they exist. In this section, we establish the functoriality of the construction of limits or colimits of diagrams of a fixed shape, and explore its consequences. An important point is that *functoriality* of the limit is not the same as *canonicity*. In most cases, there are many isomorphic objects that define the limit, and there is seldom a reason to prefer any one to the others. To illustrate this point, we present an example that demonstrates that even when limit objects exist uniquely, as is the case for instance in a skeletal category, it does not necessarily follow that canonical isomorphisms induced by the universal properties of these objects are identities.

PROPOSITION 3.6.1. *If* C *has all* J*-shaped limits, then a choice of a limit for each diagram defines the action on objects of a functor* lim$_J$: C^J → C.

Note that this functor is not canonically defined but rather requires an arbitrary choice of a limit to serve as "the" limit for each diagram.

PROOF. Choose a limit and a limit cone for each diagram $F \in$ C^J. This defines the action of lim$_J$: C^J → C on objects. It remains only to define the action of lim$_J$: C^J → C on morphisms. The vertical composite of a natural transformation $\alpha: F \Rightarrow G$ and the limit cone $\lambda:$ lim$_J F \Rightarrow F$ for F defines a cone

$$\lim_J F \overset{\lambda}{\Longrightarrow} F \overset{\alpha}{\Longrightarrow} G$$

with summit lim$_{JF}$ over G. This cone factors uniquely through the chosen limit cone lim$_J G \Rightarrow G$ for the diagram G via a map between their summits, which we take to be lim$_J \alpha:$ lim$_J F \to$ lim$_J G$. To illustrate, a morphism in C$^{\bullet \to \bullet \leftarrow \bullet}$ is a pair of solid-arrow commutative squares as displayed:

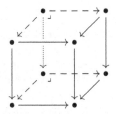

Choosing pullbacks for the top and bottom diagrams, with the legs of the pullback cones displayed in dashes, there is a unique induced dotted arrow morphism between their summits, which defines the action of the functor $\mathsf{C}^{\bullet \to \bullet \leftarrow \bullet} \to \mathsf{C}$ on this morphism. Uniqueness of the universal properties implies that this construction is functorial; see Exercise 3.6.i.   □

EXAMPLE 3.6.2. Recall the functor Path: $\mathsf{Top} \to \mathsf{Set}$ that carries a space X to the set $\mathrm{Path}(X) := \mathsf{Top}(I, X)$, where I is the standard unit interval. Precomposing with the endpoint inclusions $0, 1 \colon * \rightrightarrows I$ defines a functor $P \colon \mathsf{Top} \to \mathsf{Set}^{\bullet \rightrightarrows \bullet}$ that carries a space X to the parallel pair of functions

$$\mathrm{Path}(X) \cong \mathsf{Top}(I, X) \underset{\mathrm{ev}_1}{\overset{\mathrm{ev}_0}{\rightrightarrows}} \mathsf{Top}(*, X) \cong \mathrm{Point}(X)$$

that evaluate a path at its endpoints. Their coequalizer

$$\mathrm{Path}(X) \underset{\mathrm{ev}_1}{\overset{\mathrm{ev}_0}{\rightrightarrows}} \mathrm{Point}(X) \longtwoheadrightarrow \pi_0 X$$

defines the set of **path components** of X, the quotient of the set of points in X by the relation that identifies any pair of points connected by a path. Proposition 3.6.1 tells us that this coequalizer defines a functor, the **path components functor**

$$\pi_0 := \mathsf{Top} \xrightarrow{\ P\ } \mathsf{Set}^{\bullet \rightrightarrows \bullet} \xrightarrow{\ \mathrm{colim}\ } \mathsf{Set}.$$

More generally, the construction of Proposition 3.6.1 implies that a natural transformation between diagrams gives rise to a morphism between their limits or between their colimits, whenever these exist. Whether the codomain category has all limits or colimits of that shape is irrelevant. Moreover, whenever the natural transformation is a natural isomorphism, the induced map between the limits or colimits is an isomorphism.

COROLLARY 3.6.3. *A natural isomorphism between diagrams induces a naturally-defined isomorphism between their limits or colimits, whenever these exist.*

PROOF. The proof of Proposition 3.6.1 defines a limit functor from the full subcategory of C^J spanned by those functors admitting limits in C. This functor, like all functors, preserves isomorphisms (Lemma 1.3.8).   □

The main computational motivation for Eilenberg and Mac Lane's formulation of the concept of naturality was to be able to carefully and correctly formulate this result. While the general theory of limits and colimits had not yet been developed, [**EM45**, 21.2] proves Corollary 3.6.3 in the special case of directed diagrams. Indeed, naturality is a necessary condition in Corollary 3.6.3:

EXAMPLE 3.6.4. For any group G, a pair of G-sets $X, Y \colon \mathrm{B}G \rightrightarrows \mathsf{Set}$ are objectwise isomorphic just when their underlying sets are isomorphic. The functors X and Y are naturally isomorphic just when there exists an *equivariant* isomorphism, that is, a bijection $X \xrightarrow{\cong} Y$

that commutes with the actions by each $g \in G$. This second condition is stronger than the first.

For instance, consider the group $\mathbb{Z}/2$ and the $\mathbb{Z}/2$-sets 2, the two-element set with a trivial $\mathbb{Z}/2$-action, and $2'$, the two-element set where the generator of $\mathbb{Z}/2$ acts by exchanging the two elements. We may regard 2 and $2'$ as functors $B\mathbb{Z}/2 \to \mathsf{Set}$ from which perspective they are objectwise isomorphic (because both sets have two elements) but not naturally isomorphic (because there is no isomorphism that commutes with the $\mathbb{Z}/2$-actions).

The limit of a functor $X \colon BG \to \mathsf{Set}$ is the set of G-fixed points (see Example 3.2.12) and the colimit is the set of G-orbits (by Exercise 3.5.i). In the case of our pair of $\mathbb{Z}/2$-sets, $\lim 2$ has two elements while $\lim 2'$ is empty, and $\mathrm{colim}\, 2$ has two elements while $\mathrm{colim}\, 2'$ is a singleton.

EXAMPLE 3.6.5. To a finite set X, we can associate the set $\mathrm{Sym}(X)$ of its permutations or the set $\mathrm{Ord}(X)$ of its total orderings. Both sets have $|X|!$ elements. These constructions are functorial, not with respect to all maps of finite sets, but with respect to the bijections: the functor $\mathrm{Sym} \colon \mathsf{Fin}_{\mathrm{iso}} \to \mathsf{Fin}$ acts by conjugation and the functor $\mathrm{Ord} \colon \mathsf{Fin}_{\mathrm{iso}} \to \mathsf{Fin}$ acts by translation. However, the objectwise isomorphism $\mathrm{Sym}(X) \cong \mathrm{Ord}(X)$ is not natural, and indeed limits or colimits of restrictions of these diagrams need not be isomorphic.

Combinatorialists refer to a functor $F \colon \mathsf{Fin}_{\mathrm{iso}} \to \mathsf{Fin}$ as a **species**.[16] The image $F(n)$ of the n-element set n is the set of **labeled F-structures** on n. The set of **unlabeled F-structures** on n is defined by restricting $\mathsf{Fin}_{\mathrm{iso}}$ to the full subcategory spanned by the n-element set, i.e., to the symmetric group Σ_n regarded as the group of automorphisms of the object $n \in \mathsf{Fin}_{\mathrm{iso}}$, and forming the colimit of the diagram

$$B\Sigma_n \hookrightarrow \mathsf{Fin}_{\mathrm{iso}} \xrightarrow{F} \mathsf{Fin}.$$

Because Ord and Sym are objectwise isomorphic, the sets of labeled Sym-structures and labeled Ord-structures are isomorphic. However, the set of unlabeled Sym-structures on n is the set of conjugacy classes of permutations of n-elements, while the set of unlabeled Ord-structures on n is trivial: all linear orders on n are isomorphic. See [**Joy81**] for more.

A discussion of the functoriality of chosen limits and colimits should come with a few warnings. Limits, when they exist, are unique up to a unique isomorphism commuting with the maps in the limit cone. But this is not the same thing as saying that limits are unique on the nose: even if there is a unique object satisfying the universal property of the limit, each of its automorphisms gives rise to a distinct limit cone. Moreover, choices of limits of diagrams of fixed shape, as required to define the limit functor of Proposition 3.6.1, can seldom be made compatibly. For instance, Freyd has shown that in any category with pullbacks, it is possible to choose canonical pullbacks so that the "horizontal" composite of canonical pullback squares, in the sense of Exercise 3.1.viii, is the canonical pullback of the composite rectangle; however, it is not possible to also arrange so that the "vertical" composites of canonical pullback squares are canonical pullbacks [**FS90**, §1.4].

Even when the chosen limit objects are equal, the various natural isomorphisms associated to the limit constructions might not be identities. To illustrate this, first observe:

LEMMA 3.6.6. *For any triple of objects X, Y, Z in a category with binary products, there is a unique natural isomorphism $X \times (Y \times Z) \cong (X \times Y) \times Z$ commuting with the projections to X, Y, and Z.*

PROOF. Exercise 3.6.ii.   □

---

[16]In this context, "species" is singular.

Lemma 3.6.6 asserts that the product is naturally associative. It follows that any iteration of binary products can be used to define n-ary products. However, even in a skeletal category, in which the objects $X \times (Y \times Z)$ and $(X \times Y) \times Z$ are necessarily equal, the natural isomorphism may not be the identity. The following example, from [**ML98a**, p. 164], is due to Isbell.

EXAMPLE 3.6.7. Consider sk(Set), a skeletal category of sets. Since sk(Set) is equivalent to a complete category it has all limits and, in particular, has products. Let C denote the countably infinite set. Its product $C \times C = C$ and the product projections $\pi_1, \pi_2 : C \to C$ are both epimorphisms. Suppose the component of the natural isomorphism $C \times (C \times C) \cong (C \times C) \times C$ were the identity. Naturality would then imply that for any triple of maps $f, g, h : C \to C$ that $f \times (g \times h) = (f \times g) \times h$, on account of commutativity of the square

$$
\begin{array}{ccc}
C \times (C \times C) & = & (C \times C) \times C \\
{\scriptstyle f \times (g \times h)} \downarrow & & \downarrow {\scriptstyle (f \times g) \times h} \\
C \times (C \times C) & = & (C \times C) \times C
\end{array}
$$

Now a pair of maps between products are equal if and only if their projections onto components are equal.

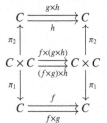

Because π_1 and π_2 are epimorphisms, this implies that $f = f \times g$ and $g \times h = h$. So we conclude that any pair of maps $f, g : C \rightrightarrows C$ must be equal, which is absurd. It follows that the component of the natural associativity isomorphism $C \times (C \times C) \cong (C \times C) \times C$ is not the identity.

Exercises.

EXERCISE 3.6.i. In a category C with pullbacks, prove that the mapping $\lim : C^{\bullet \to \bullet \leftarrow \bullet} \to C$ defined in the proof of Proposition 3.6.1 is functorial.

EXERCISE 3.6.ii. Prove Lemma 3.6.6.

3.7. Size matters

Recall a category is **complete** if it admits a limit for any diagram whose domain is a small category. In this section, we explain the reason for this size restriction. As a warm-up, the following lemma highlights a commonly existing large limit. Somewhat peculiarly, an initial object, which is defined to be the *colimit* of the empty diagram, may also be characterized as the *limit* of the identity functor:

LEMMA 3.7.1. *For any category* C, *the identity functor* $1_C : C \to C$ *admits a limit if and only if* C *has an initial object.*

PROOF. Consider a limit cone $(\lambda_c : \ell \to c)_{c \in C}$. The legs of the limit cone demonstrate that ℓ is a **weakly initial object**, admitting at least one morphism to every object $c \in C$. To show that λ_c is the only morphism from ℓ to c, observe that for any morphism $f : \ell \to c$, commutativity of the limit cone implies that $f\lambda_\ell = \lambda_c$. So to prove that ℓ is initial, it suffices to show that the cone leg $\lambda_\ell \colon \ell \to \ell$ is the identity.

Considering λ_c as a morphism in the diagram 1_C, the cone condition tells us that $\lambda_c\lambda_\ell = \lambda_c$. From this equation, we see that λ_ℓ defines a factorization of the cone $\lambda \colon \ell \Rightarrow 1_C$ through itself. By uniqueness of such factorizations, we conclude that $\lambda_\ell = 1_\ell$, as desired.

The verification that an initial object defines a limit of 1_C is left as Exercise 3.7.i. $\square$

By the size restrictions inherent in the notion of completeness, a complete (large) category need not have an initial object. As it turns out, complete small categories are hard to come by. To explain this, we require a precise definition of the cardinality of a small category.

DEFINITION 3.7.2. The **cardinality** of a small category is the cardinality of the set of its morphisms. A category whose cardinality is less than κ is called κ-**small**.

A κ-**small diagram** is one whose indexing category is κ-small. The following result explains why categories tend only to have limits or colimits of diagrams of smaller size.

PROPOSITION 3.7.3 (Freyd). *Any κ-small category that admits all κ-small limits or all κ-small colimits is a preorder.*

This result remains true under less restrictive completeness hypotheses, which the reader will have no trouble formulating.

PROOF. Let λ be the cardinality of the set of morphisms in a κ-small category C, and suppose there exists a parallel pair $f, g \colon B \rightrightarrows A$ of morphisms with $f \neq g$. By the universal property of the λ-indexed power (see Example 3.4.4), there are 2^λ distinct morphisms $B \to A^\lambda$ whose components are either f or g. By Cantor's diagonalization argument,

$$|C(B, A^\lambda)| \geq 2^\lambda > \lambda = |\operatorname{mor} C|,$$

contradicting the fact that $C(B, A^\lambda)$ is a subset of $\operatorname{mor} C$. If C instead has κ-small copowers, then a dual construction yields a similar contradiction. $\square$

Proposition 3.5.9 reveals that it is possible for a preorder to have all limits or colimits of unrestricted size: a poset P has this property as a category if and only if it is a **complete lattice**.

Exercises.

EXERCISE 3.7.i. Complete the proof of Lemma 3.7.1 by showing that an initial object defines a limit of the identity functor $1_C \colon C \to C$.

3.8. Interactions between limits and colimits

Given a functor of several variables, it is natural to consider limits and colimits formed separately in each variable. Assuming these objects exist, it follows from their universal properties that limits commute with limits and colimits commute with colimits. Because the universal properties of limits and colimits have opposite handedness, in the few cases where limits and colimits commute with each other, this tends not to follow from formal abstract nonsense, but instead most likely has something to do with how these objects are constructed in the particular category in question.

To explore these issues, consider a functor $F \colon I \times J \to C$ of two variables, which, by Exercise 1.7.vii, may also be regarded as a functor $F \colon I \to C^J$ or as $F \colon J \to C^I$. If for each $i \in I$, $\lim_{j \in J} F(i, j)$ exists, then by Proposition 3.3.9 these objects define the values of a diagram $\lim_{j \in J} F(-, j) \colon I \to C$. The following theorem proves that the limit of this I-indexed diagram is isomorphic to the object obtained by exchanging the roles of I and J.

THEOREM 3.8.1. *If the limits* $\lim_{i \in I} \lim_{j \in J} F(i, j)$ *and* $\lim_{j \in J} \lim_{i \in I} F(i, j)$ *associated to a diagram* $F \colon I \times J \to C$ *exist in* C, *they are isomorphic and define the limit* $\lim_{I \times J} F$.

To illustrate, suppose $I = \bullet \rightrightarrows \bullet$ is the category that indexes equalizer diagrams and $J = \bullet \to \bullet \leftarrow \bullet$ is the category that indexes pullbacks. Then Theorem 3.8.1 says that the limit of a diagram indexed by the category

$$I \times J =$$

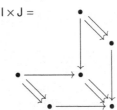

may be formed either by first forming the two pullbacks and then taking the equalizer of the induced maps between them, as displayed below-left

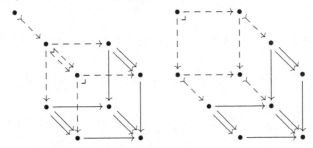

or by taking the equalizers of the three parallel pairs, and then forming the pullback of the induced maps between them, as displayed above-right.

PROOF OF THEOREM 3.8.1. By the Yoneda lemma, it suffices to prove that

$$C(X, \lim_{i \in I} \lim_{j \in J} F(i, j)) \cong C(X, \lim_{(i, j) \in I \times J} F(i, j)) \cong C(X, \lim_{j \in J} \lim_{i \in I} F(i, j)).$$

By Theorem 3.4.6(i), covariant representable functors preserve limits:

$$C(X, \lim_{i \in I} \lim_{j \in J} F(i, j)) \cong \lim_{i \in I} C(X, \lim_{j \in J} F(i, j)) \cong \lim_{i \in I} \lim_{j \in J} C(X, F(i, j)).$$

Arguing similarly for the other limits under consideration, the general statement reduces to the case of proving that for any set-valued functor $H \colon I \times J \to$ Set, in particular for the functors $C(X, F-)$, the limits

$$\lim_{i \in I} \lim_{j \in J} H(i, j) \cong \lim_{(i, j) \in I \times J} H(i, j) \cong \lim_{j \in J} \lim_{i \in I} H(i, j)$$

are isomorphic. On account of the isomorphism of categories $I \times J \cong J \times I$, it suffices to prove the left-hand isomorphism.

By Definition 3.2.3, the set $\lim_{(i, j) \in I \times J} H(i, j)$ is the set of cones with summit 1 over the $I \times J$-indexed diagram H. The set $\lim_{i \in I} \lim_{j \in J} H(i, j)$ is the set of cones with summit 1 over the I-indexed diagram $\lim_{j \in J} H(-, j)$. Such a cone is comprised of legs

$$\left(1 \xrightarrow{\lambda_i} \lim_{j \in J} H(i, j) \right)_{i \in I}$$

that commute with the maps of limits $\lim_{j\in J} H(i, j) \to \lim_{j\in J} H(i', j)$ determined by each morphism $i \to i' \in I$. The map of limits is defined, as in Proposition 3.6.1, by the map of J-indexed diagrams whose components $H(i, j) \to H(i', j)$ are defined by applying the functor $H(-, j)$ to the morphism in I.

By the universal property of the J-indexed limits, each cone leg $\lambda_i \colon 1 \to \lim_{j\in J} H(i, j)$ is itself determined by legs

$$\left(1 \xrightarrow{\lambda_{i,j}} H(i, j)\right)_{j\in J},$$

which must commute with the maps $H(i, j) \to H(i, j')$ induced by each $j \to j'$. Thus, the totality of this data is precisely a cone

$$\left(1 \xrightarrow{\lambda_{i,j}} H(i, j)\right)_{(i,j)\in I\times J}$$

over the $I \times J$-indexed diagram H. Having demonstrated that their elements coincide, we conclude that $\lim_{i\in I} \lim_{j\in J} H(i, j) \cong \lim_{(i,j)\in I\times J} H(i, j)$. $\qquad\square$

By Theorem 3.8.1 and its dual, limits commute with limits, and colimits commute with colimits up to isomorphism. By the uniqueness statements in the universal properties for limits and colimits, it follows that there exist natural isomorphisms

(3.8.2)

$$\begin{array}{ccc} C^{I\times J} & \xrightarrow{\lim_J} & C^I \\ {\scriptstyle \lim_I}\downarrow & \cong & \downarrow{\scriptstyle \lim_I} \\ C^J & \xrightarrow{\lim_J} & C \end{array} \qquad\qquad \begin{array}{ccc} C^{I\times J} & \xrightarrow{\operatorname{colim}_J} & C^I \\ {\scriptstyle \operatorname{colim}_I}\downarrow & \cong & \downarrow{\scriptstyle \operatorname{colim}_I} \\ C^J & \xrightarrow{\operatorname{colim}_J} & C \end{array}$$

for any choices of limit and colimit functors.

The relationship between the construction of limits in one variable and colimits in another is more subtle. These operations seldom commute. However, there is always a canonical comparison morphism:

LEMMA 3.8.3. *For any bifunctor $F \colon I \times J \to C$ so that the displayed limits and colimits exist, there is a canonical map*

$$\kappa \colon \operatorname*{colim}_{i\in I} \lim_{j\in J} F(i, j) \to \lim_{j\in J} \operatorname*{colim}_{i\in I} F(i, j).$$

To remember the direction of the canonical map, recall that colimits are defined via a "mapping out" universal property, while limits are defined via a "mapping in" universal property.

PROOF. By the universal property of the colimit, the map κ may be defined by specifying components

$$\left(\lim_{j\in J} F(i, j) \xrightarrow{\kappa_i} \lim_{j\in J} \operatorname*{colim}_{i'\in I} F(i', j)\right)_{i\in I}$$

that define a cone under the I-indexed diagram $\lim_{j\in J} F(-, j)$. By the universal property of the limit, each component κ_i in turn may be defined by specifying components

$$\left(\lim_{j'\in J} F(i, j') \xrightarrow{\kappa_{i,j}} \operatorname*{colim}_{i'\in I} F(i', j)\right)_{j\in J}$$

that define a cone over the J-indexed diagram $\operatorname{colim}_{i\in I} F(i, -)$. Define $\kappa_{i,j}$ to be the composite

$$\kappa_{i,j} \colon \lim_{j'\in J} F(i, j') \xrightarrow{\pi_{i,j}} F(i, j) \xrightarrow{\iota_{i,j}} \operatorname*{colim}_{i'\in I} F(i', j)$$

of the leg $\pi_{i,j}$ of the limit cone for $F(i, -)$ and the leg $\iota_{i,j}$ of the colimit cone for $F(-, j)$. Because the maps in the diagrams $\lim_{j \in J} F(-, j)$ and $\text{colim}_{i \in I} F(i, -)$ are induced by the maps $F(i, j) \to F(i', j')$ obtained by applying F to morphisms $i \to i' \in I$ and $j \to j' \in J$, the $\kappa_{i,j}$ assemble into a cone, defining κ_i, and the κ_i assemble into a cone defining κ. $\square$

Specializing to the case where C is the poset category $(\mathbb{R}, \leq)$, we have the following immediate corollary:

COROLLARY 3.8.4. *For any pair of sets X and Y and any function $f : X \times Y \to \mathbb{R}$*

$$\sup_{x \in X} \inf_{y \in Y} f(x, y) \leq \inf_{y \in Y} \sup_{x \in X} f(x, y)$$

whenever these infima and suprema exist.

PROOF. The presence of a map $\sup_X \inf_Y f \to \inf_Y \sup_X f$ in the poset category implies the displayed inequality. $\square$

REMARK 3.8.5. A direct proof of Corollary 3.8.4, as might be presented in a real analysis course, is in fact the categorical proof used for Lemma 3.8.3. To show that $\sup_X \inf_Y f$ is bounded above by some quantity it suffices—by what might be called the universal property of the supremum—to show this is the case for each $\inf_Y f(x, -)$. Dually, to show that $\inf_Y \sup_X f$ is bounded below, it suffices to demonstrate the inequality for each $\sup_X f(-, y)$. So it suffices to prove that $\inf_Y f(x, -) \leq \sup_X f(-, y)$, which follows from transitivity, i.e., composition, of the inequalities

$$\inf_Y f(x, -) \leq f(x, y) \leq \sup_X f(-, y).$$

EXAMPLE 3.8.6. Consider the extended real line $\overline{\mathbb{R}} = \mathbb{R} \cup \{-\infty, \infty\}$ with the obvious extended ordering $-\infty \leq x \leq \infty$. As a category, $\overline{\mathbb{R}}$ is complete and cocomplete: the limit of a function $f : X \to \overline{\mathbb{R}}$ is its infimum and the colimit is its supremum. These are ordinary real numbers in the case where the range of f is bounded. The advantage of working in the extended category $\overline{\mathbb{R}}$ is that boundedness hypotheses can be ignored and all limits and colimits exist.

A function $x : \mathbb{N} \to \mathbb{R}$ defines a sequence $(x_n)_{n \in \mathbb{N}}$ of real numbers. The **lim inf** and **lim sup** is defined by regarding the sequence as a bifunctor $\mathbb{N} \times \mathbb{N} \xrightarrow{+} \mathbb{N} \xrightarrow{x} \mathbb{R}$ indexed by the discrete category $\mathbb{N} \times \mathbb{N}$:

$$\lim_{n \to \infty} \inf x_n = \sup_{n \geq 0} \inf_{m \geq n} x_m = \sup_{n \geq 0} \inf_{m \geq 0} x_{n+m} = \underset{n}{\text{colim}} \lim_m x_{n+m}$$

$$\lim_{n \to \infty} \sup x_n = \inf_{n \geq 0} \sup_{m \geq n} x_m = \inf_{n \geq 0} \sup_{m \geq 0} x_{n+m} = \lim_n \underset{m}{\text{colim}} x_{n+m}$$

Having translated these analytic notions into categorical ones, Lemma 3.8.3 applies in the form of an inequality:

$$\lim_{n \to \infty} \inf x_n \leq \lim_{n \to \infty} \sup x_n.$$

The limit of this sequence exists if and only if this inequality is an equality.

In Set and related categories, there are certain classes of limits and colimits for which the map defined in Lemma 3.8.3 is invertible. Theorem 3.8.9 describes the most famous of these results. A comprehensive survey can be found in [**BJLS15**].

DEFINITION 3.8.7. A category J is **filtered** if there is a cone under every finite[17] diagram in J.

---

[17]This definition can be generalized, replacing the smallest infinite cardinal ω by a regular cardinal κ: a category is κ-**filtered** if there is a cone under any diagram with fewer than κ morphisms.

EXAMPLE 3.8.8. The category ω (or, more generally, any ordinal category) and the category with one object and one non-identity idempotent are filtered. The parallel pair category, the category $\bullet \leftarrow \bullet \rightarrow \bullet$, and discrete categories with more than one object are not.

THEOREM 3.8.9. *Filtered colimits commute with finite limits in* Set.

The central observation used in this proof is the following. If $F: \mathsf{J} \to$ Set is a diagram indexed by a small filtered category, then $\mathrm{colim}_{j \in \mathsf{J}} Fj$ is the quotient of the coproduct $\coprod_{j \in \mathsf{J}} Fj$ under the equivalence relation that identifies $x \in Fj$ with $y \in Fk$ if and only if there are maps $f: j \to t$ and $g: k \to t$ in J so that $Ff(x) = Fg(y)$. More generally, any finite collection of elements in $\coprod_{j \in \mathsf{J}} Fj$ that are identified in $\mathrm{colim}_{j \in \mathsf{J}} Fj$ map to a common element in some Ft.

PROOF. Suppose I is a finite category and J is a small filtered category. Our task is to show that for any functor $F: \mathsf{I} \times \mathsf{J} \to$ Set the canonical map
$$\kappa: \mathrm{colim}_{j \in \mathsf{J}} \lim_{i \in \mathsf{I}} F(i, j) \to \lim_{i \in \mathsf{I}} \mathrm{colim}_{j \in \mathsf{J}} F(i, j)$$
is a bijection. By Definition 3.2.3, an element on the right-hand side is a cone λ with summit 1 over $\mathrm{colim}_{j \in \mathsf{J}} F(i, j)$. As I is finite, such a cone is given by a finite collection of elements $\lambda_i \in \mathrm{colim}_{j \in \mathsf{J}} F(i, j)$ satisfying a finite number of compatibility conditions. Since J is filtered, there is some "sufficiently large" $t \in \mathsf{J}$ so that for each $i \in \mathsf{I}$ the λ_i are represented by elements $\lambda_i' \in F(i, t)$ and moreover these elements assemble into a cone
$$\left(1 \xrightarrow{\lambda_i'} F(i, t)\right)_{i \in \mathsf{I}}.$$
This cone defines an element in $\lim_{i \in \mathsf{I}} F(i, t)$ that represents an element on the left-hand side mapping to our chosen element under κ. This proves that κ is surjective.

Now consider a pair of elements in the left-hand set, represented by cones
$$\left(1 \xrightarrow{\alpha_i} F(i, j)\right)_{i \in \mathsf{I}} \quad \text{and} \quad \left(1 \xrightarrow{\beta_i} F(i, k)\right)_{i \in \mathsf{I}}$$
for some $j, k \in \mathsf{J}$. If these elements have a common image under κ, then for each $i \in \mathsf{I}$, there must be some $t_i \in \mathsf{J}$ so that α_i and β_i have a common image in $F(i, t_i)$. As I is finite and J is filtered, there is some $t \in \mathsf{J}$ so that for all i α_i and β_i have a common image in $F(i, t)$. But this says that α and β represent the same element on the left-hand side, proving injectivity of κ. □

The proof of Theorem 3.8.9 is not categorical; the argument relies upon an explicit description of filtered colimits in the category of sets. However, for a large class of categories that will be introduced in Chapter 5, the *categories of models for an algebraic theory*, limits and filtered colimits are created in the category of sets (see Theorem 5.6.5). Hence finite limits and filtered colimits commute in these contexts as well.

Exercises.

EXERCISE 3.8.i. Dualize the construction of Example 3.2.16 to express the splitting of an idempotent as a coequalizer. Explain why these colimits (or limits) are preserved by any functor and conclude that splittings of idempotents commute with both limits and colimits of any shape.

EXERCISE 3.8.ii. Show that if G and H are groups whose orders are coprime, then BG-indexed limits commute with BH-indexed colimits in Set. See [BJLS15] for more examples of this flavor.

CHAPTER 4

Adjunctions

It appears ... that there exists a kind of duality
between the tensor product and the ... functor
Hom ...

Daniel Kan, "Adjoint functors" [**Kan58**]

Vector spaces are sets equipped with a binary operation $v + w$ (addition of vectors) and a unary multiplication $c \cdot v$ for each scalar c in the ground field $\Bbbk$, satisfying certain axioms. The passage from a vector space to its underlying set of "vectors" is functorial, which is to say there is a forgetful functor $U \colon \mathsf{Vect}_{\Bbbk} \to \mathsf{Set}$ that carries a linear transformation to its underlying function.

A more interesting challenge is to reverse the direction of this translation and build a vector space from a set S. Cardinality restrictions might prevent the direct use of S as the set of vectors for a $\Bbbk$-vector space, and even without this obstruction there is no natural way to define vector addition in S. Instead, a more natural way to build a vector space is to let the elements of S serve as a basis: vectors are then finite formal sums[1] $k_1 s_1 + \cdots + k_n s_n$ with $k_i \in \Bbbk$, $s_i \in S$, and $n \geq 0$. This defines a vector space $\Bbbk[S]$ whose dimension is equal to the cardinality of S, called the **free vector space on the set** S. Moreover, as our use of the adjective "natural" would suggest, this construction is functorial, defining a functor $\Bbbk[-] \colon \mathsf{Set} \to \mathsf{Vect}_{\Bbbk}$: a function $f \colon S \to T$ induces a linear map $\Bbbk[f] \colon \Bbbk[S] \to \Bbbk[T]$ defined on the basis elements in the evident way.

There are many instances of free constructions in mathematics. One can define the free (abelian) group on a set, the free ring on a set or on an abelian group, the free module on an abelian group, the free category on a directed graph, and so on. Sometimes there are competing notions of universal constructions: Does the free graph on a set have a single edge between every pair of vertices or none? Does the free topological space on a set have as many open sets as possible or as few? Other times there are none: For instance, there is no "free field." All of these constructions, and many others besides, are clarified by the concept of an *adjunction*, introduced by Kan [**Kan58**].

The universal property of the free vector space functor $\Bbbk[-] \colon \mathsf{Set} \to \mathsf{Vect}_{\Bbbk}$ is expressed by saying that it is *left adjoint* to the underlying set functor $U \colon \mathsf{Vect}_{\Bbbk} \to \mathsf{Set}$: linear maps $\Bbbk[S] \to V$ correspond naturally to functions $S \to U(V)$, which specify the image of the basis vectors $S \subset \Bbbk[S]$. By the Yoneda lemma, this universal property can be used to define the action of the free vector space functor $\Bbbk[-]$ on maps. The forgetful functor $U \colon \mathsf{Vect}_{\Bbbk} \to \mathsf{Set}$ has no *right adjoint*, so in this setting there are no competing notions of free construction.

---

[1]A precise way to define what is meant by this informal phrase is to say that a **finite formal sum** is a finitely supported function $\phi \colon S \to \Bbbk$, corresponding to the expression $\sum_{s \in S} \phi(s)s$. In particular, two finite sums are considered to be identical when they differ only up to reordering of terms, up to consolidating repeated instances of the same term by adding their coefficients, or up to inclusion or deletion of terms whose coefficients are zero.

In this chapter, we explore the general theory of adjoint functors, both in the abstract and aided by a plethora of examples. Two equivalent definitions of an adjunction are presented in §4.1 and §4.2, both of which offer useful perspectives on the adjoint correspondence between certain linear transformations and certain functions just mentioned. §4.3 explores the forms of adjointness that arise between pairs of contravariant functors or between $(n + 1)$-tuples of functors with n-variables. For instance, a two-variable adjunction encodes the duality between the tensor product and hom bifunctors observed by Kan.

The basic calculus of adjunctions is developed in §4.4, with each result proven twice, once using the Yoneda lemma and again via the techniques of "formal category theory," to illustrate two important methods of categorical reasoning. In §4.5, we prove what is perhaps the most frequently applied result in category theory: that right adjoint functors preserve limits while left adjoint functors preserve colimits. This explains why limits tend to be easier to construct than the formally dual colimit notions.

In particular, continuity is a necessary condition for a functor to admit a left adjoint, and cocontinuity is necessary for the existence of a right adjoint. In §4.6, we develop sufficient conditions for a functor to have an adjoint, which can be used to construct the missing adjoint functor in certain contexts.

4.1. Adjoint functors

> Adjoint functors arise everywhere.
>
> ――――――――――――――――――――――――
> Saunders Mac Lane, *Categories for the Working*
> *Mathematician* [**ML98a**]

An adjunction consists of an opposing pair of functors $F\colon \mathsf{C} \rightleftarrows \mathsf{D} \colon G$ that enjoy a special relationship to one another.

DEFINITION 4.1.1 (adjunctions I). An **adjunction** consists of a pair of functors $F\colon \mathsf{C} \to \mathsf{D}$ and $G\colon \mathsf{D} \to \mathsf{C}$ together with an isomorphism

$$(4.1.2) \qquad\qquad \mathsf{D}(Fc, d) \cong \mathsf{C}(c, Gd)$$

for each $c \in \mathsf{C}$ and $d \in \mathsf{D}$ that is natural in both variables. Here F is **left adjoint** to G and G is **right adjoint** to F. The morphisms

$$Fc \xrightarrow{\;f^{\sharp}\;} d \qquad \rightsquigarrow \qquad c \xrightarrow{\;f^{\flat}\;} Gd$$

corresponding under the bijection (4.1.2) are **adjunct** or are **transposes** of each other.[2]

When C and D are locally small, the naturality statement in Definition 4.1.1 asserts that the isomorphisms (4.1.2) assemble into a natural isomorphism between the functors

$$\mathsf{C}^{\mathrm{op}} \times \mathsf{D} \underset{\mathsf{C}(-,G-)}{\overset{\mathsf{D}(F-,-)}{\rightrightarrows}} \Downarrow\cong \mathsf{Set}\,.$$

――――――――――――――

[2]For now, we use "$(-)^{\sharp}$" and "$(-)^{\flat}$" to decorate a pair of adjunct arrows, but in practice it tends to be most convenient to use some symbol, such as "$(-)^{\dagger}$" or "$(\hat{-})$," to signal any adjoint transpose, with no preference as to which of the adjunct pair is decorated in this way.

More explicitly, naturality in D says that for any morphism $k \colon d \to d'$, the left-hand diagram displayed below commutes in Set:

$$
\begin{array}{ccc}
D(Fc, d) & \xrightarrow{\;\cong\;} & C(c, Gd) \\
\scriptstyle{k_*} \downarrow & & \downarrow \scriptstyle{Gk_*} \\
D(Fc, d') & \xrightarrow[\cong]{} & C(c, Gd')
\end{array}
\qquad
\forall Fc \xrightarrow{f^\sharp} d \quad \rightsquigarrow
\qquad
\begin{array}{ccc}
c & \xrightarrow{f^\flat} & Gd \\
& \searrow{\scriptstyle{(k \cdot f^\sharp)^\flat}} & \downarrow \scriptstyle{Gk} \\
& & Gd'
\end{array}
$$

which amounts to the assertion that for any $f^\sharp \colon Fc \to d$ and $k \colon d \to d'$, the transpose of $k \cdot f^\sharp \colon Fc \to d'$ is equal to the composite of $f^\flat \colon c \to Gd$ with $Gk \colon Gd \to Gd'$.

Dually, naturality in C says that for any morphism $h \colon c' \to c$, the left-hand diagram displayed below commutes in Set:

$$
\begin{array}{ccc}
D(Fc, d) & \xrightarrow{\;\cong\;} & C(c, Gd) \\
\scriptstyle{Fh^*} \downarrow & & \downarrow \scriptstyle{h^*} \\
D(Fc', d) & \xrightarrow[\cong]{} & C(c', Gd)
\end{array}
\qquad
\forall Fc \xrightarrow{f^\sharp} d \quad \rightsquigarrow
\qquad
\begin{array}{ccc}
c' & & \\
\scriptstyle{h} \downarrow & \searrow{\scriptstyle{(f^\sharp \cdot Fh)^\flat}} & \\
c & \xrightarrow[f^\flat]{} & Gd
\end{array}
$$

which amounts to the assertion that the transpose of $f^\sharp \cdot Fh \colon Fc' \to d$ is the composite of $h \colon c' \to c$ with $f^\flat \colon c \to Gd$.

The following lemma, whose proof is left as Exercise 4.1.i, provides an equivalent expression of the naturality of a collection of isomorphisms (4.1.2) in a form that tends to be convenient to use in proofs.

LEMMA 4.1.3. *Consider a pair of functors $F \colon C \rightleftarrows D \colon G$ equipped with isomorphisms $D(Fc, d) \cong C(c, Gd)$ for all $c \in C$ and $d \in D$. Naturality of this collection of isomorphisms is equivalent to the assertion that for any morphisms with domains and codomains as displayed below*

(4.1.4)
$$
\begin{array}{ccc}
Fc & \xrightarrow{f^\sharp} & d \\
\scriptstyle{Fh} \downarrow & & \downarrow \scriptstyle{k} \\
Fc' & \xrightarrow[g^\sharp]{} & d'
\end{array}
\qquad \leftrightsquigarrow \qquad
\begin{array}{ccc}
c & \xrightarrow{f^\flat} & Gd \\
\scriptstyle{h} \downarrow & & \downarrow \scriptstyle{Gk} \\
c' & \xrightarrow[g^\flat]{} & Gd'
\end{array}
$$

the left-hand square commutes in D if and only if the right-hand transposed square commutes in C.

NOTATION 4.1.5. A turnstile "⊣" is used to designate that an opposing pair of functors are adjoints: the expressions $F \dashv G$ and $G \vdash F$ and the diagrams

$$
C \underset{G}{\overset{F}{\underset{\perp}{\rightleftarrows}}} D
\qquad
C \underset{F}{\overset{G}{\underset{\top}{\leftrightarrows}}} D
\qquad
D \underset{G}{\overset{F}{\underset{\perp}{\leftrightarrows}}} C
\qquad
D \underset{F}{\overset{G}{\underset{\top}{\rightleftarrows}}} C
$$

all assert that $F \colon C \to D$ is left adjoint to $G \colon D \to C$.

The utility of this versatile notation is most apparent in contexts involving several adjoint functors.

EXAMPLE 4.1.6. The forgetful functor $U \colon \mathsf{Top} \to \mathsf{Set}$ admits both left and right adjoints. The left adjoint constructs a topological space from a set S in such a way that continuous maps from this space to another space T correspond naturally and bijectively to functions $S \to U(T)$. The discrete topology on S has this universal property: writing $D(S)$ for this discrete space, any function $S \to U(T)$ defines a continuous map $D(S) \to T$.

Similarly, the right adjoint constructs a topological space from a set S in such a way that continuous maps from T to this space correspond naturally and bijectively to functions $U(T) \to S$. The indiscrete topology on S has this universal property: writing $I(S)$ for this indiscrete space, any function $U(T) \to S$ defines a continuous function $T \to I(S)$. The natural isomorphisms

$$\mathsf{Top}(D(S), T) \cong \mathsf{Set}(S, U(T)) \quad \text{and} \quad \mathsf{Set}(U(T), S) \cong \mathsf{Top}(T, I(S))$$

assert that the discrete, forgetful, and indiscrete functors define a pair of adjunctions:

An adjunction between preorders is called a **(monotone) Galois connection**. Recall, from Exercise 1.3.ii, that functors $F \colon \mathsf{A} \to \mathsf{B}$ and $G \colon \mathsf{B} \to \mathsf{A}$ between preorders A and B are simply order-preserving functions. An adjunction $F \dashv G$ asserts that

$$Fa \le b \quad \text{if and only if} \quad a \le Gb$$

for all $a \in \mathsf{A}$ and $b \in \mathsf{B}$. In this context, F is the **lower adjoint** and G is the **upper adjoint**.

EXAMPLE 4.1.7. The inclusion of posets $\mathbb{Z} \hookrightarrow \mathbb{R}$, with the usual ordering $\le$, has both left and right adjoints, namely, the ceiling and floor functions. For any integer n and real number r, $n \le r$ if and only if $n \le \lfloor r \rfloor$, where $\lfloor r \rfloor$ denotes the greatest integer less than or equal to r. This function is order-preserving, defining a functor $\lfloor - \rfloor \colon \mathbb{R} \to \mathbb{Z}$ that is right adjoint to the inclusion $\mathbb{Z} \hookrightarrow \mathbb{R}$. Dually, $r \le n$ if and only if $\lceil r \rceil \le n$, and this order-preserving function defines a functor $\lceil - \rceil \colon \mathbb{R} \to \mathbb{Z}$ that is left adjoint to the inclusion.

$$\begin{array}{c} \lceil - \rceil \\ \mathbb{Z} \xleftrightarrow{\quad\quad} \mathbb{R} \,. \\ \lfloor - \rfloor \end{array}$$

EXAMPLE 4.1.8. Consider a function $f \colon A \to B$ between sets. The subsets of A and subsets of B form posets, PA and PB, ordered by inclusion. The map f induces direct image and inverse image functors $f_* \colon PA \to PB$ and $f^{-1} \colon PB \to PA$. The inverse image is right adjoint to the direct image: for $A' \subset A$ and $B' \subset B$, $f(A') \subset B'$ if and only if $A' \subset f^{-1}(B')$.

The inverse image functor has a further right adjoint $f_!$ that carries a subset $A' \subset A$ to the subset of elements of B whose fibers lie entirely in A'. With this definition, $B' \subset f_!(A')$ if and only if $f^{-1}(B') \subset A'$. These functors define a pair of adjunctions:

$$\begin{array}{c} f_* \\ PB \xrightarrow{\;f^{-1}\;} PA \,. \\ f_! \end{array}$$

EXAMPLE 4.1.9. A **propositional function** is a function $P \colon X \to \Omega = \{\bot, \top\}$, which we interpret as declaring, for each $x \in X$, whether $P(x)$ is true or false. The set Ω^X is then the set of propositional functions on X. The set Ω is given the partial order $\bot \le \top$ from which Ω^X inherits a pointwise-defined order: $P \le Q$ if and only if $P(x) \le Q(x)$ for all $x \in X$, which is the case if and only if P implies Q.

The logical operations of universal and existential quantification define functors

$$\forall_X, \exists_X \colon \Omega^X \rightrightarrows \Omega$$

in the expected way: $\forall_X P = \top$ if and only if $P(x) = \top$ for all $x \in X$, and $\exists_X P = \top$ if and only if there exists $x \in X$ with $P(x) = \top$. There is also a constant "dummy variable" functor $\Delta_X \colon \Omega \to \Omega^X$, and one can verify that these functors define a triple of adjoints:

$$\Omega \underset{\underset{\forall_X}{\overset{\bot}{\longleftarrow}}}{\overset{\overset{\exists_X}{\longleftarrow}}{\underset{\Delta_X}{\longrightarrow}}} \Omega^X \, .$$

See [Awo96] for more.

EXAMPLE 4.1.10. There is a large and very important family of "free ⊣ forgetful" adjunctions

$$A \underset{U}{\overset{F}{\underset{\bot}{\rightleftarrows}}} S \, ,$$

with the forgetful functor U defining the right adjoint and the free functor F defining the left adjoint. If one can construct a universal object of type A on an object of type S, and this construction defines a left adjoint to a forgetful functor $U \colon A \to S$, then the term "free" is used in place of "universal" to convey this particular relationship. The reason why left adjoints to forgetful functors are more common than right adjoints has to do with the handedness of the universal properties that one meets in practice. "Cofree" constructions, which define right adjoints to some forgetful functor, are somehow less common.

The following forgetful functors admit left adjoints, defining "free" constructions.

(i) $U \colon \mathsf{Set}_* \to \mathsf{Set}$. The left adjoint carries a set X to the pointed set $X_+ := X \sqcup \{X\}$, with a freely-adjoined basepoint.

(ii) $U \colon \mathsf{Monoid} \to \mathsf{Set}$. The free monoid on a set X is the set $\coprod_{n \geq 0} X^{\times n}$ of finite lists of elements of X, including the empty list.

(iii) $U \colon \mathsf{Ring} \to \mathsf{Ab}$, forgetting the multiplicative structure. The free ring on an abelian group A is $\bigoplus_{n \geq 0} A^{\otimes n}$.

Exercise 5.1.i explores the connection between Examples (ii) and (iii): as observed in Definition 1.6.3, rings are monoids in the category of abelian groups.

(iv) $U \colon \mathsf{Ab} \to \mathsf{Set}$. The free abelian group on a set X is the set $\mathbb{Z}[X] := \bigoplus_X \mathbb{Z}$ of finite formal sums of elements of X with integer coefficients. More generally:

(v) $U \colon \mathsf{Mod}_R \to \mathsf{Set}$. The free R-module on a set X is $R[X] := \bigoplus_X R$. A special case defines the free $\Bbbk$-vector space on a set, defining the left adjoint to $U \colon \mathsf{Vect}_\Bbbk \to \mathsf{Set}$ considered in the introduction to this chapter.

(vi) $U \colon \mathsf{Ring} \to \mathsf{Set}$. By composing the left adjoints to the forgetful functors $\mathsf{Ring} \to \mathsf{Ab} \to \mathsf{Set}$, the free ring on a set X is the free monoid on the free abelian group on the set: $\bigoplus_{n \geq 0} (\bigoplus_X \mathbb{Z})^{\otimes n}$.

(vii) $(-)^\times \colon \mathsf{Ring} \to \mathsf{Group}$, carrying a ring to its group of units. The free ring on a group G is the **group ring** $\mathbb{Z}[G] = \bigoplus_G \mathbb{Z}$, whose elements are finite formal sums of group elements, using the group operation to define the bilinear multiplication law.

(viii) $U \colon \mathsf{Group} \to \mathsf{Set}$. The free group on a set is described in more detail in Examples 4.2.4 and 4.6.6.

(ix) $\mathsf{Ab} \hookrightarrow \mathsf{CMonoid}$, the inclusion of abelian groups into the category of commutative monoids. The left adjoint Gr carries a commutative monoid $(M, +, 0)$ to its **group completion**, also called the **Grothendieck group**: as a set $\mathrm{Gr}(M, +, 0)$ is the quotient of $M \times M$ by the relation $(a, b) \simeq (a', b')$ if and only if there is some $c \in M$ so that $a + b' + c = a' + b + c$.

(x) Group $\hookrightarrow$ Monoid. The left adjoint constructs the "group completion" of a monoid as a quotient of the free group on its underlying set modulo relations arising from the multiplication and unit in the monoid structure.

(xi) The **restriction of scalars** functor ϕ^*: Mod$_S$ → Mod$_R$ associated to a ring homomorphism ϕ: R → S. Its left adjoint, given by $S \otimes_R -$: Mod$_R$ → Mod$_S$, is called **extension of scalars**.

(xii) U: Mod$_R$ → Ab, forgetting the scalar multiplication. This is a special case of the previous example induced from the unique (unital) ring homomorphism $\mathbb{Z}$ → R.

(xiii) ϕ^*: Set^{BG} → Set^{BH}, another restriction of scalars functor induced from a group homomorphism ϕ: H → G. The left adjoint is also called **induction** and can be described analogously to (xi).

Each of these free ⊣ forgetful adjunctions is an instance of a **monadic adjunction**, a special class of adjoint functors that is the subject of §5.5.

EXAMPLE 4.1.11 (Frobenius reciprocity). Example 4.1.10(xiii) can be generalized: Example 6.2.8 proves that for any complete and cocomplete category C and any group homomorphism ϕ: H → G, the restriction functor ϕ^*: C^{BG} → C^{BH} admits both left and right adjoints. These are most commonly considered for subgroup inclusions $H \subset G$, in which case the left adjoint is called **induction** and the right adjoint is called **coinduction**:

Taking C = Vect$_{\Bbbk}$, the adjunction ind$_H^G$ ⊣ res$_H^G$ between the category Vect$_{\Bbbk}^{BH}$ of H-representations and the category of Vect$_{\Bbbk}^{BG}$ of G-representations is referred to in the representation theory literature as **Frobenius reciprocity**.

EXAMPLE 4.1.12. None of the functors

$$\text{Field} \xrightarrow{U} \text{Ring}, \quad \text{Field} \xrightarrow{U} \text{Ab}, \quad \text{Field} \xrightarrow{(-)^{\times}} \text{Ab}, \quad \text{Field} \xrightarrow{U} \text{Set}$$

that forget algebraic structure on fields admit left or right adjoints. To see that no left adjoint is possible, note that there exist maps in the target categories from $\mathbb{Z}$ to fields of any characteristic. As there are no field homomorphisms between fields of differing characteristic, it is not possible to define the value of a hypothetical left adjoint on $\mathbb{Z}$. A similar argument can be used to show that no right adjoints to these functors exist.

EXAMPLE 4.1.13. The forgetful functor U: Cat → DirGraph admits a left adjoint F, defining the **free category** on a directed graph. A **directed graph** G consists of a set V of vertices, a set E of edges, and two functions s, t: $E \rightrightarrows V$ defining the source and target of each directed edge. The free category on G has V as its set of objects. The set of morphisms consists of identities for each vertex together with finite "paths" of edges. Composition is defined by concatenation of paths.

The adjunction supplies a natural bijection between functors $F(G)$ → C and morphisms G → U(C) of directed graphs. A functor $F(G)$ → C, or equally a directed graph morphism G → U(C), defines a diagram in C with no commutativity requirements, since directed graphs do not encode composites. The data of such a diagram is uniquely determined by the images of the vertices and edges in the directed graph G. These edges form **atomic** arrows in the category $F(G)$, admitting no non-trivial factorizations. For instance, as described in

Example 1.1.4(iv), the category ω is free on the directed graph whose vertices are indexed by natural numbers and with edges $n \to n + 1$. On account of the adjunction, an ω-indexed diagram is defined by specifying only the images of the atomic arrows, from an ordinal to its successor.

EXAMPLE 4.1.14. Let $n + 1$ denote the ordinal category, freely generated by the directed graph $0 \to 1 \to 2 \to \cdots \to n$. For each $0 \le i \le n$, there is an injective functor $d^i : n \to n + 1$ where $i \in n + 1$ is the unique object missing from the image. For each $0 \le i < n$, there is also a surjective functor $s^i : n + 1 \to n$ for which $i \in n$ is the unique object with two preimages. These functors define a sequence of $2n + 1$ adjoints:

proving that arbitrarily long finite sequences of adjoint functors exist.

EXAMPLE 4.1.15. The inclusion Groupoid $\hookrightarrow$ Cat admits both left and right adjoints

The right adjoint carries a category to its maximal subgroupoid considered in Lemma 1.1.13: since functors preserve isomorphisms, the image of a functor $G \to C$ whose domain is a groupoid necessarily factors (uniquely) through the maximal subgroupoid of C.

The left adjoint "formally inverts" the morphisms of a category, constructing the corresponding "category of fractions." The category of fractions for C can be constructed as a quotient of the free category on the directed graph obtained by gluing the directed graph underlying C to the directed graph underlying C^{op} along the common discrete subcategory ob C. Objects in the category of fractions are objects of C and morphisms are finite zig-zags of morphisms in C, subject to relations that, for instance, equate the zig-zags

$$a \xrightarrow{\ f\ } b \xleftarrow{\ f\ } a \qquad\qquad b \xleftarrow{\ f\ } a \xrightarrow{\ f\ } b$$

with identities. More details on this construction can be found in [**Bor94a**, §5.2].

Exercises.

EXERCISE 4.1.i. Show that functors $F : C \to D$ and $G : D \to C$ and bijections $D(Fc, d) \cong C(c, Gd)$ for each $c \in C$ and $d \in D$ define an adjunction if and only if these bijections induce a bijection between commutative squares (4.1.4). That is, prove Lemma 4.1.3.

EXERCISE 4.1.ii. Define left and right adjoints to

  (i) ob: Cat $\to$ Set,
 (ii) Vert: Graph $\to$ Set, and
(iii) Vert: DirGraph $\to$ Set.

EXERCISE 4.1.iii. Show that any triple of adjoint functors

$$C \xrightarrow{\quad U \quad} D$$

gives rise to a canonical adjunction $LU \dashv RU$ between the induced endofunctors of C.

EXERCISE 4.1.iv. Prove that the construction of the free R-module on a set described in Example 4.1.10(v) is functorial.[3]

EXERCISE 4.1.v. Use Lemma 4.1.3 to show that if $F\colon C \rightleftarrows D\colon G$ are adjoint functors, with $F \dashv G$, then the comma categories $F \downarrow D$ and $C \downarrow G$, introduced in Exercise 1.3.vi, are isomorphic, via an isomorphism that commutes with the forgetful functors to $C \times D$.

4.2. The unit and counit as universal arrows

Recall that an adjunction is comprised of an opposing pair of functors together with natural isomorphisms

$$(4.2.1) \qquad C \xrightarrow[\ G\ ]{\ F\ } D \qquad\qquad D(Fc, d) \cong C(c, Gd).$$

Fixing $c \in C$, this natural isomorphism asserts that the object $Fc \in D$ represents the functor $C(c, G-)\colon D \to \mathsf{Set}$. By the Yoneda lemma, the natural isomorphism $D(Fc, -) \cong C(c, G-)$ is determined by an element of $C(c, GFc)$, the transpose of 1_{Fc}, denoted by η_c. The naturality of adjoint transposition implies that the maps η_c assemble into the components of a natural transformation $\eta\colon 1_C \Rightarrow GF$:

LEMMA 4.2.2. *Given an adjunction $F \dashv G$, there is a natural transformation $\eta\colon 1_C \Rightarrow GF$, called the **unit** of the adjunction, whose component $\eta_c\colon c \to GFc$ at c is defined to be the transpose of the identity morphism 1_{Fc}.*

PROOF. To prove that η is natural, we must show that the left-hand square commutes for every $f\colon c \to c'$ in C.

$$
\begin{array}{ccc}
c & \xrightarrow{\ \eta_c\ } & GFc \\
{\scriptstyle f}\downarrow & & \downarrow{\scriptstyle GFf} \\
c' & \xrightarrow[\ \eta_{c'}\ ]{} & GFc'
\end{array}
\qquad \rightsquigarrow \qquad
\begin{array}{ccc}
Fc & \xrightarrow{\ 1_{Fc}\ } & Fc \\
{\scriptstyle Ff}\downarrow & & \downarrow{\scriptstyle Ff} \\
Fc' & \xrightarrow[\ 1_{Fc'}\ ]{} & Fc'
\end{array}
$$

This follows from Lemma 4.1.3 and the obvious commutativity of the right-hand transposed square.   □

Dually, fixing $d \in D$ the defining natural isomorphism (4.2.1) of an adjunction $F \dashv G$ says that the object $Gd \in C$ represents the functor $D(F-, d)\colon C^{op} \to \mathsf{Set}$. By the Yoneda lemma, the natural isomorphism $C(-, Gd) \cong D(F-, d)$ is determined by an element of $D(FGd, d)$, the transpose of 1_{Gd}, denoted by ϵ_d. By the dual of Lemma 4.2.2, the maps ϵ_d assemble into the components of a natural transformation $\epsilon\colon FG \Rightarrow 1_D$:

---

[3]This exercise intends to ask for an explicit description of the homomorphism between two formally defined finite formal sums associated to a function between their generating sets. But the reader who is disinclined to think about explicit constructions could instead use the Yoneda lemma to prove functoriality. (Hint: See Proposition 4.3.4.)

LEMMA 4.2.3. *Given an adjunction $F \dashv G$, there is a natural transformation $\epsilon\colon FG \Rightarrow 1_D$, called the* **counit** *of the adjunction, whose component $\epsilon_d\colon FGd \to d$ at d is defined to be the transpose of the identity morphism 1_{Gd}.*

EXAMPLE 4.2.4. The left adjoint F to the forgetful functor $U\colon$ Group $\to$ Set defines the **free group** on a set S. Elements of the group $F(S)$ are finite "words" whose "letters" are elements s of S and formal "inverses" s^{-1}, modulo some evident relations. The empty word serves as the identity element and multiplication is by concatenation; the relations, in particular, identify ss^{-1} and $s^{-1}s$ with the empty word.

The component of the unit of the adjunction at a set S is the function $\eta_S\colon S \to UF(S)$ that sends an element $s \in S$ to the corresponding singleton word. The component of the counit of the adjunction at a group G is the group homomorphism $\epsilon_G\colon FU(G) \to G$ that sends a word, whose letters are elements of the group G and formal inverses, to the product of those symbols, interpreted using the multiplication, inverses, and identity present in the group G.

A similar description, with the unit defining some sort of "singleton" map and the counit defining some form of "evaluation" homomorphism, can be given for many of the other adjunctions listed in Example 4.1.10.

In the terminology of Definition 2.3.3, the components of the unit and the counit define universal elements characterizing the natural isomorphisms

$$\text{Group}(F(S), -) \cong \text{Set}(S, U(-)) \quad \leftrightsquigarrow \quad \eta_S \in \text{Set}(S, UF(S))$$
$$\text{Set}(-, U(G)) \cong \text{Group}(F(-), G) \quad \leftrightsquigarrow \quad \epsilon_G \in \text{Group}(FU(G), G)$$

By Proposition 2.4.8, the components of the unit and the counit enjoy certain dual universal properties to be explored in Exercise 4.2.iv, which will feature prominently in the arguments of §4.6.

Lemmas 4.2.2 and 4.2.3 prove that any adjunction has a unit and a counit. Conversely, if $F\colon C \rightleftarrows D\colon G$ are opposing functors equipped with natural transformations $\eta\colon 1_C \Rightarrow GF$ and $\epsilon\colon FG \Rightarrow 1_D$ satisfying a dual pair of conditions, then this data can be used to define a natural bijection $D(Fc, d) \cong C(c, Gd)$, exhibiting F and G as adjoint functors. That is, the notion of adjunction may be redefined as follows:

DEFINITION 4.2.5 (adjunction II). An **adjunction** consists of an opposing pair of functors $F\colon C \rightleftarrows D\colon G$, together with natural transformations $\eta\colon 1_C \Rightarrow GF$ and $\epsilon\colon FG \Rightarrow 1_D$ that satisfy the **triangle identities**:

$$F \xrightarrow{\;F\eta\;} FGF \qquad\qquad G \xrightarrow{\;\eta G\;} GFG$$
$$\begin{array}{ccc} & \searrow_{1_F} & \Downarrow_{\epsilon F} \\ & & F \end{array} \qquad\qquad \begin{array}{ccc} & \searrow_{1_G} & \Downarrow_{G\epsilon} \\ & & G \end{array}$$

The left-hand triangle asserts that a certain diagram commutes in D^C, while the right-hand triangle asserts that the dual diagram commutes in C^D. The natural transformations $F\eta$, ϵF, ηG, and $G\epsilon$ are defined by whiskering; recall Remark 1.7.6. Their components are defined just as the notation would suggest: $G\epsilon_d$ is $G(\epsilon_d)$, while ϵF_c is ϵ_{Fc}.

Together the triangle identities assert that "the counit is a left inverse of the unit modulo translation." They cannot literally be inverses because the components of the unit $\eta_c\colon c \to GFc$ lie in C while the components of the counit $\epsilon_d\colon FGd \to d$ lie in D, and such morphisms are not composable. But if we apply F to the unit, we obtain a morphism $F\eta_c\colon Fc \to FGFc$ in D whose left inverse is ϵ_{Fc}. And if we apply G to the counit, we obtain a morphism $G\epsilon_d\colon GFGd \to Gd$ whose right inverse is η_{Gd}.

PROPOSITION 4.2.6. *Given a pair of functors $F\colon \mathsf{C} \rightleftarrows \mathsf{D}\colon G$, there exists a natural iso-morphism $\mathsf{D}(Fc, d) \cong \mathsf{C}(c, Gd)$ if and only if there exists a pair of natural transformations $\eta\colon 1_\mathsf{C} \Rightarrow GF$ and $\epsilon\colon FG \Rightarrow 1_\mathsf{D}$ satisfying the triangle identities.*

Proposition 4.2.6 asserts that the definitions of adjunction given in 4.1.1 and in 4.2.5 are equivalent.

PROOF. The dual Lemmas 4.2.2 and 4.2.3 demonstrate that a natural isomorphism $\mathsf{D}(Fc, d) \cong \mathsf{C}(c, Gd)$ gives rise to a pair of natural transformations $\eta\colon 1_\mathsf{C} \Rightarrow GF$ and $\epsilon\colon FG \Rightarrow 1_\mathsf{D}$, whose components are transposes of identity morphisms. It remains to demonstrate the triangle identities. By Lemma 4.1.3, the upper-left-hand and lower-right-hand squares

$$
\begin{array}{ccc}
Fc \xrightarrow{1_{Fc}} Fc & & c \xrightarrow{\eta_c} GFc \\
{\scriptstyle F\eta_c}\downarrow \quad \downarrow{\scriptstyle 1_{Fc}} & \leftrightsquigarrow & {\scriptstyle \eta_c}\downarrow \quad \downarrow{\scriptstyle 1_{GFc}} \\
FGFc \xrightarrow{\epsilon_{Fc}} Fc & & GFc \xrightarrow{1_{GFc}} GFc
\end{array}
$$

$$
\begin{array}{ccc}
FGd \xrightarrow{1_{FGd}} FGd & & Gd \xrightarrow{\eta_{Gd}} GFGd \\
{\scriptstyle 1_{FGd}}\downarrow \quad \downarrow{\scriptstyle \epsilon_d} & \leftrightsquigarrow & {\scriptstyle 1_{Gd}}\downarrow \quad \downarrow{\scriptstyle G\epsilon_d} \\
FGd \xrightarrow{\epsilon_d} d & & Gd \xrightarrow{1_{Gd}} Gd
\end{array}
$$

commute because the transposed squares manifestly do. By Lemma 1.7.1, which defines vertical composition for natural transformations, this is what we wanted to show.

Conversely, the unit and counit can be used to define a natural bijection $\mathsf{D}(Fc, d) \cong \mathsf{C}(c, Gd)$. Given $f^\sharp\colon Fc \to d$ and $g^\flat\colon c \to Gd$, their adjuncts are defined to be the composites:

$$
f^\flat := c \xrightarrow{\eta_c} GFc \xrightarrow{Gf^\sharp} Gd \qquad\qquad g^\sharp := Fc \xrightarrow{Fg^\flat} FGd \xrightarrow{\epsilon_d} d
$$

The triangle identities imply that these operations are inverses. The transpose of the transpose of $f^\sharp\colon Fc \to d$ is equal to the top composite

$$
Fc \xrightarrow{F\eta_c} FGFc \xrightarrow{FGf^\sharp} FGd \xrightarrow{\epsilon_d} d
$$

with ϵ_{Fc}, 1_{Fc}, Fc, $f^\sharp$ labels in the lower triangle.

By naturality of ϵ and one triangle identity, this returns the original $f^\sharp\colon Fc \to d$. The dual diagram chase, left as an exercise for the reader, demonstrates that the transpose of the transpose of $g^\flat\colon c \to Gd$ returns this map. $\square$

REMARK 4.2.7. Proposition 4.2.6 reveals that the data of a **fully-specified adjunction** can be presented in two equivalent forms: it consists of a pair of functors $F\colon \mathsf{C} \rightleftarrows \mathsf{D}\colon G$ together with

(i) a natural family of isomorphisms $\mathsf{D}(Fc, d) \cong \mathsf{C}(c, Gd)$ for all $c \in \mathsf{C}$ and $d \in \mathsf{D}$,

or, equivalently,

(ii) natural transformations $\eta\colon 1_\mathsf{C} \Rightarrow GF$ and $\epsilon\colon FG \Rightarrow 1_\mathsf{D}$ so that $G\epsilon \cdot \eta G = 1_G$ and $\epsilon F \cdot F\eta = 1_F$.

Indeed, either of the unit and the counit alone, satisfying an appropriate universal property, suffices to determine a fully specified adjunction: (i) and (ii) are each equivalent to either of

(iii) a natural transformation $\eta\colon 1_C \Rightarrow GF$ so that the function

$$\mathsf{D}(Fc, d) \xrightarrow{\ G\ } \mathsf{C}(GFc, Gd) \xrightarrow{\ (\eta_c)^*\ } \mathsf{C}(c, Gd)$$

defines an isomorphism for all $c \in \mathsf{C}$ and $d \in \mathsf{D}$,

or, equivalently, and dually,

(iv) a natural transformation $\epsilon\colon FG \Rightarrow 1_D$ so that the function

$$\mathsf{C}(c, Gd) \xrightarrow{\ F\ } \mathsf{D}(Fc, FGd) \xrightarrow{\ (\epsilon_d)_*\ } \mathsf{D}(Fc, d)$$

defines an isomorphism for all $c \in \mathsf{C}$ and $d \in \mathsf{D}$.

In particular, a morphism of adjunctions, introduced in Exercise 4.2.v, is defined to be a morphism of fully-specified adjunctions. On account of the equivalence between (i), (ii), (iii), and (iv), this notion can be presented in several equivalent ways.

The mere presence of unit and counit transformations establishes the following fixed point formulae for a Galois connection between posets.

COROLLARY 4.2.8. *If* A *and* B *are posets and* $F\colon$ A $\to$ B *and* $G\colon$ B $\to$ A *form a Galois connection, with* $F \dashv G$, *then* F *and* G *satisfy the following fixed point formulae*

$$FGF = F \qquad and \qquad GFG = G.$$

PROOF. By the triangle identities $F(a) \leq FGF(a) \leq F(a)$ for all $a \in \mathsf{A}$, whence $F = FGF$. The other formula is dual.   □

For instance, consider the direct-image $\dashv$ inverse-image adjunction

$$PB \underset{f^{-1}}{\overset{f_*}{\underset{\perp}{\rightleftarrows}}} PA$$

induced by a function $f\colon A \to B$ defined in Example 4.1.8. For generic subsets $X \subset A$ and $Y \subset B$, neither of the inclusions

$$X \subset f^{-1}(f(X)) \qquad \text{or} \qquad f(f^{-1}(Y)) \subset Y$$

need be equalities but nonetheless

$$f(X) = f(f^{-1}(f(X))) \qquad \text{and} \qquad f^{-1}(f(f^{-1}(Y))) = f^{-1}(Y).$$

Exercises.

EXERCISE 4.2.i. Prove that any pair of adjoint functors $F\colon$ C $\rightleftarrows$ D$\colon G$ restrict to define an equivalence between the full subcategories spanned by those objects $c \in \mathsf{C}$ and $d \in \mathsf{D}$ for which the components of the unit η_c and of the counit ϵ_d, respectively, are isomorphisms.

EXERCISE 4.2.ii. Explain each step needed to convert the statement of Lemma 4.2.2 into the statement of the dual Lemma 4.2.3.

EXERCISE 4.2.iii. Pick your favorite forgetful functor from Example 4.1.10 and prove that it is a right adjoint by defining its left adjoint, the unit, and the counit, and demonstrating that the triangle identities hold.

EXERCISE 4.2.iv. Each component of the counit of an adjunction is a terminal object in some category. What category?

EXERCISE 4.2.v. A **morphism of adjunctions** from $F \dashv G$ to $F' \dashv G'$ is comprised of a pair of functors

$$
\begin{array}{ccc}
\mathsf{C} & \xrightarrow{\ H\ } & \mathsf{C}' \\
F \downarrow \dashv \uparrow G & & F' \downarrow \dashv \uparrow G' \\
\mathsf{D} & \xrightarrow{\ K\ } & \mathsf{D}'
\end{array}
$$

so that the square with the left adjoints and the square with the right adjoints both commute (i.e., $KF = F'H$ and $HG = G'K$) and satisfying one additional condition, which takes a number of equivalent forms. Prove that the following are equivalent:

(i) $H\eta = \eta' H$, where η and η' denote the respective units of the adjunctions.
(ii) $K\epsilon = \epsilon' K$, where ϵ and ϵ' denote the respective counits of the adjunctions.
(iii) Transposition across the adjunctions commutes with application of the functors H and K, i.e., for every $c \in \mathsf{C}$ and $d \in \mathsf{D}$, the diagram

$$
\begin{array}{ccc}
\mathsf{D}(Fc, d) & \xrightarrow{\ \cong\ } & \mathsf{C}(c, Gd) \\
K\downarrow & & \downarrow H \\
\mathsf{D}'(KFc, Kd) & & \mathsf{C}'(Hc, HGd) \\
\| & & \| \\
\mathsf{D}'(F'Hc, Kd) & \underset{\cong}{\longrightarrow} & \mathsf{C}'(Hc, G'Kd)
\end{array}
$$

commutes.

4.3. Contravariant and multivariable adjoint functors

Kan's original exploration of adjoint functors also considered adjoint relationships involving functors that are contravariant or have more than one variable [**Kan58**]. Indeed, these examples provided the central motivation for Kan's discovery, at least according to an apocryphal account of a question he posed to Eilenberg in a homological algebra seminar at Columbia University [**Mar09**, §4.2]:

> KAN: You have explained how the tensor product can be defined
>   in terms of the hom functor. Can the hom instead be defined
>   in terms of the tensor product?
>
> EILENBERG: No, of course not. That's absurd.

In fact, Kan's intuition was correct. By the Yoneda lemma, any adjoint functor determines its adjoints up to natural isomorphism. Example 4.3.11 explains how to define the hom bifunctor from the tensor product.

Contravariant adjoint functors require little special consideration, arising simply as an application of the principle of duality. Definition 4.1.1 can be dualized in three ways, by replacing C, D, or both C and D by their opposite categories. The latter dualization recovers the original notion of adjunction, with $G\colon \mathsf{D}^{op} \to \mathsf{C}^{op}$ left adjoint to $F\colon \mathsf{C}^{op} \to \mathsf{D}^{op}$ if and only if $F \dashv G$. In particular, any theorem about left adjoints has a dual theorem about right adjoints, which is a very useful duality principle for adjunctions. The other two dualizations lead to new types of adjoint functors.

DEFINITION 4.3.1. A pair of contravariant functors $F\colon \mathsf{C}^{op} \to \mathsf{D}$ and $G\colon \mathsf{D}^{op} \to \mathsf{C}$ are **mutually left adjoint** if there exists a natural isomorphism

$$
\mathsf{D}(Fc, d) \cong \mathsf{C}(Gd, c),
$$

or **mutually right adjoint** if there exists a natural isomorphism

$$D(d, Fc) \cong C(c, Gd).$$

Dualizing Definition 4.2.5, a pair of mutual left adjoints $F: C^{op} \to D$ and $G: D^{op} \to C$ come equipped with a pair of "counit" natural transformations $GF \Rightarrow 1_C$ and $FG \Rightarrow 1_D$, while a pair of mutual right adjoints $F: C^{op} \to D$ and $G: D^{op} \to C$ come equipped with a pair of "unit" natural transformations $1_C \Rightarrow GF$ and $1_D \Rightarrow FG$. The formulation of the triangle identities in each case is left to Exercise 4.3.i.

Mutual right adjoints between preorders form what is sometimes called an **antitone Galois connection**.

EXAMPLE 4.3.2. For a fixed natural number n and algebraically closed field $\Bbbk$, consider the sets of elements in the polynomial ring $\Bbbk[x_1, \ldots, x_n]$ and in the vector space $\Bbbk^n$. There are contravariant functors

$$P(\Bbbk[x_1, \ldots, x_n])^{op} \xrightarrow{V} P(\Bbbk^n) \quad \text{and} \quad P(\Bbbk^n)^{op} \xrightarrow{I} P(\Bbbk[x_1, \ldots, x_n])$$

between the posets of subsets of these sets defined for $S \subset \Bbbk[x_1, \ldots, x_n]$ and $T \subset \Bbbk^n$ by:

$$V(S) := \{(t_1, \ldots, t_n) \in \Bbbk^n \mid f(t_1, \ldots, t_n) = 0, \, \forall f \in S\}$$

$$I(T) := \{f \in \Bbbk[x_1, \ldots, x_n] \mid f(t_1, \ldots, t_n) = 0, \, \forall (t_1, \ldots, t_n) \in T\},$$

and these define mutual right adjoints because

$$T \subset V(S) \quad \text{if and only if} \quad S \subset I(T).$$

This Galois connection is a starting point of modern algebraic geometry.

The unit and counit encode relations

$$T \subset V(I(T)) \quad \text{and} \quad S \subset I(V(S)).$$

A set of points $T \subset \Bbbk^n$ defines a fixed point $T = V(I(T))$ if and only if T is closed in the Zariski topology on $\Bbbk^n = \text{Spec}(\Bbbk[x_1, \ldots, x_n])$, introduced in Example 1.3.7(iv). A set of polynomials $S \subset \Bbbk[x_1, \ldots, x_n]$ defines a fixed point $S = I(V(S))$ if and only if S defines a radical ideal of $\Bbbk[x_1, \ldots, x_n]$; this is **Hilbert's Nullstellensatz**. Exercise 4.2.i now implies that the poset of Zariski closed subsets of $\Bbbk^n$ is isomorphic to the opposite of the poset of radical ideals of $\Bbbk[x_1, \ldots, x_n]$.

EXAMPLE 4.3.3. Let Axiom$_\sigma$ be a set of **axioms**, i.e., sentences in a fixed first-order language whose **signature** σ, specifies a list of function, constant, and relation symbols to be used with the standard logical symbols. Let Struct$_\sigma$ be a set of σ-**structures**, i.e., sets with interpretations of the given constant, relation, and function symbols. For instance, the language of the natural numbers has a constant symbol "0," a binary function symbol "+," and a binary relation symbol "≤." A structure for this language is any set with a specified constant, binary function, and binary relation. Axioms include sentences such as

$$\forall x, y, z, ((x \leq y) \land (y \leq z)) \to (x \leq z) \quad \text{or} \quad \forall x, x + 0 = x,$$

asserting transitivity of the relation "≤" and that the constant "0" serves as a unit for the binary function "+."

Given a set of σ-structures M and a set of axioms A, we write $M \vDash A$ if each of the axioms in A is **satisfied by**—meaning "is true in"—each of the σ-structures in M. For instance, the first displayed sentence is satisfied by a structure if and only if its interpretation of the relation "≤" is transitive.

Form the poset categories $P(\text{Axiom}_\sigma)$ and $P(\text{Struct}_\sigma)$ ordered by inclusion. There are contravariant functors

$$P(\text{Axiom}_\sigma)^{\text{op}} \xrightarrow{\text{True in}} P(\text{Struct}_\sigma) \quad \text{and} \quad P(\text{Struct}_\sigma)^{\text{op}} \xrightarrow{\text{Satisfying}} P(\text{Axiom}_\sigma)$$

which send a set of axioms to the set of σ-structures, called **models**, that satisfy those axioms and send a set of σ-structures to the set of axioms that they satisfy. These are mutual right adjoints, forming what is called the **Galois connection between syntax and semantics** [**Smi**].

Before turning our attention to multivariable functors, we make a useful observation that applies equally to ordinary adjunctions:

PROPOSITION 4.3.4. *Consider a functor $F: A \to B$ so that for each $b \in B$ there exists an object $Gb \in A$ together with an isomorphism*

$$(4.3.5) \qquad B(Fa, b) \cong A(a, Gb), \quad \text{natural in } a \in A.$$

Then there exists a unique way to extend the assignment $G: \text{ob}\, B \to \text{ob}\, A$ to a functor $G: B \to A$ so that the family of isomorphisms (4.3.5) is also natural in $b \in B$.

In other words, if $F: A \to B$ is a functor admitting representations for each functor

$$B(F-, b): A^{\text{op}} \to \text{Set}$$

for each $b \in B$, then this data assembles into a right adjoint $F \dashv G$.

PROOF. Naturality of (4.3.5) in $f: b \to b'$ demands that the function

$$A(a, Gb) \cong B(Fa, b) \xrightarrow{f_*} B(Fa, b') \cong A(a, Gb')$$

equals post-composition by the yet-to-be-defined morphism $Gf: Gb \to Gb'$. This composite function defines a natural transformation $A(-, Gb) \Rightarrow A(-, Gb')$, which by the Yoneda lemma must equal post-composition by a unique morphism $Gf: Gb \to Gb'$ in A. Uniqueness of this definition implies that the assignment $G: \text{mor}\, B \to \text{mor}\, A$ is functorial, as we saw in the proof of Theorem 1.5.9.[4]   □

The argument used to prove Proposition 4.3.4 can be extended to functors of many variables.

PROPOSITION 4.3.6. *Suppose that $F: A \times B \to C$ is a bifunctor so that for each object $a \in A$, the induced functor $F(a, -): B \to C$ admits a right adjoint $G_a: C \to B$. Then:*

(i) *These right adjoints assemble into a unique bifunctor $G: A^{\text{op}} \times C \to B$, defined so that $G(a, c) = G_a(c)$ and so that the isomorphisms*

$$C(F(a, b), c) \cong B(b, G(a, c))$$

are natural in all three variables.

If furthermore for each $b \in B$, the induced functor $F(-, b): A \to C$ admits a right adjoint $H_b: C \to A$, then:

---

[4]More concisely, G is defined to be the unique restriction of the functor

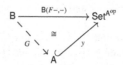

along the Yoneda embedding so that the triangle commutes up to the specified natural isomorphisms (4.3.5).

(ii) *There is a unique bifunctor* $H\colon \mathsf{B}^{\mathrm{op}} \times \mathsf{C} \to \mathsf{A}$ *defined so that* $H(b,c) = H_b(c)$ *and the isomorphisms*

$$\mathsf{C}(F(a,b),c) \cong \mathsf{B}(b,G(a,c)) \cong \mathsf{A}(a,H(b,c))$$

are natural in all three variables.

(iii) *In this case, for each* $c \in \mathsf{C}$, *the functors* $G(-,c)\colon \mathsf{A}^{\mathrm{op}} \to \mathsf{B}$ *and* $H(-,c)\colon \mathsf{B}^{\mathrm{op}} \to \mathsf{A}$ *are mutual right adjoints.*

PROOF. Exercise 4.3.ii. □

DEFINITION 4.3.7. A triple of bifunctors

$$\mathsf{A} \times \mathsf{B} \xrightarrow{F} \mathsf{C}, \quad \mathsf{A}^{\mathrm{op}} \times \mathsf{C} \xrightarrow{G} \mathsf{B}, \quad \mathsf{B}^{\mathrm{op}} \times \mathsf{C} \xrightarrow{H} \mathsf{A}$$

equipped with a natural isomorphism

$$\mathsf{C}(F(a,b),c) \cong \mathsf{B}(b,G(a,c)) \cong \mathsf{A}(a,H(b,c))$$

defines a **two-variable adjunction**.

Particularly, when $F\colon \mathsf{C} \times \mathsf{C} \to \mathsf{C}$ defines some sort of monoidal product, its pointwise-defined right adjoints G and H are called its **left** and **right closures**, respectively. When these are isomorphic, the bifunctor F is called **closed**.

EXAMPLE 4.3.8. The product bifunctor

$$\mathsf{Set} \times \mathsf{Set} \xrightarrow{\times} \mathsf{Set}$$

is closed: the operation called **currying** in computer science defines a family of natural isomorphisms

$$\{A \times B \xrightarrow{f} C\} \cong \{A \xrightarrow{f} C^B\} \cong \{B \xrightarrow{f} C^A\}.$$

Thus, the product and exponential bifunctors

$$\mathsf{Set} \times \mathsf{Set} \xrightarrow{\times} \mathsf{Set}, \quad \mathsf{Set}^{\mathrm{op}} \times \mathsf{Set} \xrightarrow{(-)^{(-)}} \mathsf{Set}, \quad \mathsf{Set}^{\mathrm{op}} \times \mathsf{Set} \xrightarrow{(-)^{(-)}} \mathsf{Set}$$

define a two-variable adjunction.

DEFINITION 4.3.9. A **cartesian closed category** is a category C with finite products in which the product bifunctor

$$\mathsf{C} \times \mathsf{C} \xrightarrow{\times} \mathsf{C}$$

is closed.

In addition to Set, the categories Fin, Cat, Set^{BG}, and, more generally, $\mathsf{Set}^{\mathsf{C}}$ are cartesian closed; this was proven for Cat in Exercise 1.7.vii.

EXAMPLE 4.3.10 (a convenient category of spaces). The category Top is not cartesian closed. However, a famous paper of Steenrod proves that there exists a **convenient category of topological spaces** that is complete, cocomplete, cartesian closed, and sufficiently large so as to contain the CW complexes [**Ste67**]. Steenrod's convenient category is a subcategory of the category Haus of Hausdorff spaces and continuous functions. Previous researchers had argued that a "convenient category of spaces" satisfying Steenrod's list of conditions did not exist, but Steenrod notes:

> The arguments are based on a blind adherence to the customary definitions of the standard operations. These definitions are suitable for the category of Hausdorff spaces, but they need not be for a subcategory. The categorical viewpoint enables us to defrost these definitions and bend them a bit.

What he means is that the "product" on a yet-to-be-defined subcategory cgHaus should be the categorical product of Definition 3.1.9, which, as it turns out, is not preserved by the full inclusion cgHaus $\hookrightarrow$ Haus.

Objects in cgHaus are **compactly generated Hausdorff spaces**: Hausdorff spaces X with the property that any subset $A \subset X$ that intersects each compact subset $K \subset X$ in a closed subset $A \cap K$ is itself closed in X. The inclusion cgHaus $\hookrightarrow$ Haus has a right adjoint $k \colon$ Haus $\to$ cgHaus called k-**ification**: the space $k(X)$ refines the topology on X by adding to the collection of closed sets those subsets $A \subset X$ whose intersections with all compact subsets K are closed.

By Exercise 4.6.ii and the dual of Proposition 4.5.15, the presence of this adjoint implies that cgHaus is complete and cocomplete. The product on cgHaus is the k-ification of the product on Haus; this latter product is preserved by the inclusion Haus $\hookrightarrow$ Top. The pointwise right adjoint is given by the following construction of **function spaces** Map(X, Y): the underlying set of Map(X, Y) is the set of continuous maps $X \to Y$, and the topology is the k-ification of the compact-open topology. See [**Ste67**] for a proof.

Other categories admit a hom bifunctor that is not necessarily right adjoint to the cartesian product. The most famous of these are the "tensor $\dashv$ hom" adjunctions that inspired Kan:

EXAMPLE 4.3.11. Consider the hom bifunctor

$$\mathsf{Ab}^{\mathrm{op}} \times \mathsf{Ab} \xrightarrow{\mathrm{Hom}} \mathsf{Ab},$$

where Hom(A, B) is the group of homomorphisms $A \to B$ with addition defined pointwise in the abelian group B.[5] Fixing the contravariant variable, there is an adjunction

(4.3.12) $\mathsf{Ab} \underset{\mathrm{Hom}(A,-)}{\overset{A \otimes_{\mathbb{Z}} -}{\underset{\perp}{\rightleftarrows}}} \mathsf{Ab}$ $\mathsf{Ab}(A \otimes_{\mathbb{Z}} B, C) \cong \mathsf{Ab}(B, \mathrm{Hom}(A, C))$

defining the tensor product. Once the objects $A \otimes_{\mathbb{Z}} B \in \mathsf{Ab}$ have been defined, Proposition 4.3.4 uses the isomorphisms (4.3.12) to extend this data into a functor $A \otimes_{\mathbb{Z}} - \colon \mathsf{Ab} \to \mathsf{Ab}$. Proposition 4.3.6(i) then extends this data into the **tensor product bifunctor**

$$\mathsf{Ab} \times \mathsf{Ab} \xrightarrow{\otimes_{\mathbb{Z}}} \mathsf{Ab}$$

in such a way that the tensor product and the hom define a two-variable adjunction.

As Kan suspected, this process can be reversed. Given the tensor product bifunctor $\otimes_{\mathbb{Z}} \colon \mathsf{Ab} \times \mathsf{Ab} \to \mathsf{Ab}$, the abelian group Hom$(B, C)$ can be defined as a representation for the functor

(4.3.13) $\mathsf{Ab}^{\mathrm{op}} \xrightarrow{\mathsf{Ab}(-\otimes_{\mathbb{Z}} B, C)} \mathsf{Set}$ $\mathsf{Ab}(A, \mathrm{Hom}(B, C)) \cong \mathsf{Ab}(A \otimes_{\mathbb{Z}} B, C).$

Propositions 4.3.4 and 4.3.6(i) then imply that there is a unique way to extend this construction Hom$\colon$ ob $\mathsf{Ab} \times$ ob $\mathsf{Ab} \to$ ob Ab into a bifunctor Hom$\colon \mathsf{Ab}^{\mathrm{op}} \times \mathsf{Ab} \to \mathsf{Ab}$ in such a way that the isomorphisms (4.3.13) are natural in B and C, as well as A.

EXAMPLE 4.3.14. Temporarily, let Top denote not the ordinary category of topological spaces but the cartesian closed category of spaces defined in Example 4.3.10. In this

---

[5]Recall from §1.1 that this is why the collection of morphisms between a fixed pair of objects in a general category is often denoted "Hom." Here we use "Ab" for the mere set of homomorphisms, which in this instance plays a secondary role.

context, the two-variable adjunction specializes to define an adjunction:

$$\mathsf{Top} \underset{\mathrm{Map}(S^1,-)}{\overset{S^1\times-}{\underset{\perp}{\rightleftarrows}}} \mathsf{Top}$$

where $S^1 \in \mathsf{Top}$ is the unit circle. The space $\mathrm{Map}(S^1, X)$ is the **free loop space** on X; points in $\mathrm{Map}(S^1, X)$ are loops in the space X.

The slice category of Top under the singleton space $*$ defines a convenient category Top_* of based topological spaces. By Exercise 4.3.iv(ii), Top_* admits a two-variable adjunction

$$\mathsf{Top}_* \times \mathsf{Top}_* \overset{\wedge}{\to} \mathsf{Top}_*, \quad \mathsf{Top}_*^{\mathrm{op}} \times \mathsf{Top}_* \overset{\mathrm{Map}_*}{\longrightarrow} \mathsf{Top}_*, \quad \mathsf{Top}_*^{\mathrm{op}} \times \mathsf{Top}_* \overset{\mathrm{Map}_*}{\longrightarrow} \mathsf{Top}_* ,$$

where $\mathrm{Map}_*((X, x), (Y, y))$ denotes the based space of basepoint-preserving continuous functions $(X, x) \to (Y, y)$. The bifunctor $\wedge$ is called the **smash product**.

The space

$$\Omega X := \mathrm{Map}_*(S^1, (X, x))$$

is the **based loop space** on (X, x); the basepoint $x \in X$ implicit in the common notation. The functor Ω has a left adjoint $\Sigma X := S^1 \wedge X$, which constructs the **reduced suspension** of the based space (X, x). This defines the **loops ⊣ suspension adjunction** between based topological spaces:

$$\mathsf{Top}_* \underset{\Omega}{\overset{\Sigma}{\underset{\perp}{\rightleftarrows}}} \mathsf{Top}_* \qquad \mathsf{Top}_*(\Sigma X, Y) \cong \mathsf{Top}_*(X, \Omega Y)$$

Definition 4.3.7 can be generalized to define multivariable adjunctions. Applying Proposition 4.3.6, an n-**variable adjunction** is determined by a functor $F : A_1 \times \cdots \times A_n \to B$ admitting pointwise right adjoints when any $(n - 1)$ of its variables are fixed. See [**CGR14**] for more.

Exercises.

EXERCISE 4.3.i. Dualize Definition 4.2.5 to define mutual left adjoints and mutual right adjoints as a pair of contravariant functors equipped with appropriate natural transformations.

EXERCISE 4.3.ii. Prove Proposition 4.3.6.

EXERCISE 4.3.iii. Show that the contravariant power set functor $P : \mathsf{Set}^{\mathrm{op}} \to \mathsf{Set}$ is mutually right adjoint to itself.

EXERCISE 4.3.iv. Define pointwise adjoints to the following bifunctors, giving rise to examples of two-variable adjunctions.

(i) There is a bifunctor

$$\mathsf{Set}_*^{\mathrm{op}} \times \mathsf{Set}_* \overset{\mathrm{Hom}_*}{\longrightarrow} \mathsf{Set}_* ,$$

where $\mathrm{Hom}_*((X, x), (Y, y))$ is defined to be the set of pointed functions $(X, x) \to (Y, y)$, with the constant function at y serving as the basepoint. Define a two-variable adjunction determined by this bifunctor, the pointwise left adjoints to $\mathrm{Hom}_*((X, x), -)$.

(ii) Describe the left adjoint bifunctor $\mathsf{Set}_* \times \mathsf{Set}_* \overset{\wedge}{\to} \mathsf{Set}_*$ constructed in (i) in a sufficiently categorical way so that Set can be replaced by any cartesian closed category.

(iii) The discussion in Example 4.3.11 extends to any category Mod_R of modules over a *commutative* ring R. If R is not commutative, for a pair of R-modules A and B,

$\mathrm{Hom}(A, B)$ is not necessarily an R-module, but it is still an abelian group. This construction defines a bifunctor

$$\mathsf{Mod}_R^{\mathrm{op}} \times \mathsf{Mod}_R \xrightarrow{\;\mathrm{Hom}\;} \mathsf{Ab}\,.$$

Extend this data to a two-variable adjunction.

(iv) In a similar fashion, define a two-variable adjunction determined by the hom bifunctor

$$\mathsf{Ch}_R^{\mathrm{op}} \times \mathsf{Ch}_R \xrightarrow{\;\mathrm{Hom}\;} \mathsf{Mod}_R\,,$$

where the ring R is commutative and the set $\mathrm{Hom}(A_\bullet, B_\bullet)$ of chain homomorphisms $A_\bullet \to B_\bullet$ inherits an R-module structure with addition and scalar multiplication defined elementwise.

4.4. The calculus of adjunctions

> Perhaps the purpose of categorical algebra is to show that which is formal is formally formal.

J. Peter May, [**May01**]

This section introduces the basic calculus of adjoint functors: proving that adjoints are unique up to unique natural isomorphism, that adjunctions can be composed and whiskered, and that any equivalence can be promoted to an adjoint equivalence in which the inverse equivalences are also adjoints, with either choice of handedness. What further unites these results is that they all admit "formal categorical"[6] proofs, meaning that they can be proven syntactically using objects (categories), morphisms (functors), and morphisms of morphisms (natural transformations), which can be composed horizontally and vertically in the sense described in §1.7. As our interest is as much in these categorical proof techniques as in the results themselves, we give secondary proofs of these results, to illustrate alternate Yoneda-style arguments, and briefly comment on the merits of each approach.

PROPOSITION 4.4.1. *If F and F' are left adjoint to G, then $F \cong F'$, and moreover there is a unique natural isomorphism $\theta: F \cong F'$ commuting with the units and counits of the adjunctions:*

$$
\begin{array}{ccc}
1_C \xRightarrow{\;\eta\;} GF & \qquad & FG \xRightarrow{\;\epsilon\;} 1_D \\[4pt]
\;\;\eta' \searrow \;\;\Downarrow G\theta & & \theta G \Downarrow \;\;\nearrow \epsilon' \\[4pt]
GF' & & F'G
\end{array}
$$

PROOF 1. To define a natural transformation $\theta: F \Rightarrow F'$, it suffices, by Lemma 4.1.3, to define a transposed natural transformation, which we take to be $\eta': 1_C \Rightarrow GF'$

(4.4.2)
$$
\begin{array}{ccc}
Fc \xrightarrow{\;\theta_c\;} F'c & & c \xrightarrow{\;\eta_c'\;} GF'c \\
\;\;\Big\downarrow{Ff} \quad\quad \Big\downarrow{F'f} & \longleftrightarrow\!\!\!\rightsquigarrow & \;\;\Big\downarrow{f} \quad\quad \Big\downarrow{GF'f} \\
Fc' \xrightarrow[\;\theta_{c'}\;]{} F'c' & & c' \xrightarrow[\;\eta_{c'}'\;]{} GF'c'
\end{array}
$$

Remark 4.2.7(iv) provides an explicit formula for θ:

$$\theta := F \xRightarrow{\;F\eta'\;} FGF' \xRightarrow{\;\epsilon F'\;} F'.$$

---

[6]Here, formal category theory is meant in the rather narrow sense of 2-category theory.

Exchanging the roles of F' and F, we also define a natural transformation $\theta' : F' \Rightarrow F$ to be the transpose of $\eta : 1_C \Rightarrow GF$, given by the formula:

$$\theta' := F' \xrightarrow{F'\eta} F'GF \xrightarrow{\epsilon'F} F.$$

The hope is that these natural transformations define inverse isomorphisms. To prove that $\theta' \cdot \theta = 1_F$, it suffices to demonstrate this equality after transposing both $\theta' \cdot \theta$ and 1_F, i.e., to show that $\eta : 1 \Rightarrow GF$ equals the composite

$$1 \xrightarrow{\eta} GF \xrightarrow{GF\eta'} GFGF' \xrightarrow{G\epsilon F'} GF' \xrightarrow{GF'\eta} GF'GF \xrightarrow{G\epsilon' F} GF \ .$$

By naturality of η, this composite equals

$$1 \xrightarrow{\eta'} GF' \xrightarrow{\eta GF'} GFGF' \xrightarrow{G\epsilon F'} GF' \xrightarrow{GF'\eta} GF'GF \xrightarrow{G\epsilon' F} GF \ .$$

By the triangle identity $G\epsilon \cdot \eta G = 1_G$, this reduces to

$$1 \xrightarrow{\eta'} GF' \xrightarrow{GF'\eta} GF'GF \xrightarrow{G\epsilon' F} GF \ .$$

By naturality of η', this equals

$$1 \xrightarrow{\eta} GF \xrightarrow{\eta'GF} GF'GF \xrightarrow{G\epsilon' F} GF \ ,$$

and by the triangle identity $G\epsilon' \cdot \eta'G = 1_G$, this reduces to

$$1 \xrightarrow{\eta} GF$$

as desired. The other diagram chase, proving that $\theta \cdot \theta' = 1_{F'}$, is dual.

From the formula for θ, there are easy diagram chases, left to the reader, that verify that the triangles of natural transformations displayed in the statement commute. The left-hand triangle asserts that the transpose of θ across the adjunction $F \dashv G$ is η', as indeed has been taken as the definition in (4.4.2), thereby proving uniqueness.   □

PROOF 2. By the Yoneda lemma, the composite natural isomorphism

(4.4.3) $D(F'c, d) \cong C(c, Gd) \cong D(Fc, d) \ ,$

defines a natural isomorphism $\theta : F \cong F'$, whose component θ_c is defined to be the image of $1_{F'c}$ under the bijection (4.4.3). The first isomorphism carries $1_{F'c}$ to $\eta'_c : c \to GF'c$. The second isomorphism carries it to its transpose along $F \dashv G$. Thus,

$$\theta_c := Fc \xrightarrow{F\eta'_c} FGF'c \xrightarrow{\epsilon F'_c} F'c \ ,$$

which is the formula used in Proof 1; here we see that it was not necessary to guess the definition of θ. Setting $d = F'c$ in the commutative triangle of natural isomorphisms (4.4.3) proves the compatibility of $\theta : F \cong F'$ with the units of the adjunctions; setting $c = Ud$ proves compatibility with the counits.   □

PROPOSITION 4.4.4. *Given adjunctions $F \dashv G$ and $F' \dashv G'$*

$$C \underset{G}{\overset{F}{\rightleftarrows}} D \underset{G'}{\overset{F'}{\rightleftarrows}} E \qquad \leadsto \qquad C \underset{GG'}{\overset{F'F}{\rightleftarrows}} E$$

the composite $F'F$ is left adjoint to the composite GG'.

Proof 1. There are natural isomorphisms

$$\mathsf{E}(F'Fc, e) \cong \mathsf{D}(Fc, G'e) \cong \mathsf{C}(c, GG'e),$$

the first defined using $F' \dashv G'$ and the second defined using $F \dashv G$.   □

Proof 2. The only reasonable definitions for the unit and counit of $F'F \dashv GG'$ that can be defined from the data of the two adjunctions are:

$$\bar{\eta} := 1_{\mathsf{C}} \xoverset{\eta}{\Longrightarrow} GF \xoverset{G\eta'F}{\Longrightarrow} GG'F'F \qquad\qquad \bar{\epsilon} := F'FGG' \xoverset{F'\epsilon G'}{\Longrightarrow} F'G' \xoverset{\epsilon'}{\Longrightarrow} 1_{\mathsf{E}}$$

which indeed are the natural transformations that result from passing the identities $1_{F'F}$ and $1_{GG'}$ across the natural isomorphisms in Proof 1. The proof of the triangle identities is an entertaining diagram chase left to the reader.   □

Proposition 4.4.5. *Any equivalence*

$$\mathsf{C} \xrightleftharpoons[G]{F} \mathsf{D} \qquad \eta\colon 1_{\mathsf{C}} \cong GF, \quad \epsilon\colon FG \cong 1_{\mathsf{D}}$$

can be promoted to an adjoint equivalence, in which the natural isomorphisms satisfy the triangle identities, by replacing either one of the originally specified natural isomorphisms by a new unit or counit.

Note that the symmetry in the definition implies that an equivalence of categories $F\colon \mathsf{C} \to \mathsf{D}$ may be regarded as either a left or right adjoint; indeed, any chosen equivalence inverse $G\colon \mathsf{D} \to \mathsf{C}$ is both left and right adjoint to F.

Proof 1. Because $\eta\colon 1_{\mathsf{C}} \cong GF$ and $\epsilon\colon FG \cong 1_{\mathsf{D}}$ are isomorphisms, so is the composite

$$\gamma := G \xoverset{\eta G}{\Longrightarrow} GFG \xoverset{G\epsilon}{\Longrightarrow} G$$

but it need not be an identity. Redefining either ϵ or η so as to absorb the isomorphism γ^{-1}—it will not matter which—we will show that the resulting pair of natural isomorphisms define the unit and the counit of an adjunction $F \dashv G$.

Let $\epsilon' := \epsilon \cdot F\gamma^{-1}$. By naturality of η, the diagram

$$G \xoverset{\eta G}{\Longrightarrow} GFG \xoverset{GF\gamma^{-1}}{\Longrightarrow} GFG \xoverset{G\epsilon}{\Longrightarrow} G$$

commutes, proving one triangle identity $G\epsilon' \cdot \eta G = 1_G$. By naturality of η and ϵ', and by this first triangle identity, the diagram

commutes, proving that $\epsilon'_F \cdot F\eta$ is an idempotent. By cancellation, any idempotent isomorphism is necessarily an identity; thus $\epsilon'_F \cdot F\eta = 1_F$, as desired.   □

PROOF 2. If $\eta\colon 1_C \cong GF$ is one of the natural isomorphisms defining an equivalence of categories $F\colon C \rightleftarrows D\colon G$, then the function

$$D(Fc, d) \xrightarrow{\; G \;} C(GFc, Gd) \xrightarrow{\; (\eta_c)^* \;} C(c, Gd)$$

defines a natural isomorphism for all $c \in C$ and $d \in D$: the first map is an isomorphism because G is fully faithful and the second map is an isomorphism because η_c is an isomorphism. By Remark 4.2.7, it follows that F and G define an adjunction with unit η. □

PROPOSITION 4.4.6. *Given an adjunction*

$$C \underset{G}{\overset{F}{\underset{\perp}{\rightleftarrows}}} D$$

post-composition with F and G defines a pair of adjoint functors

$$C^J \underset{G_*}{\overset{F_*}{\underset{\perp}{\rightleftarrows}}} D^J$$

for any small category J, *and pre-composition with F and G also defines an adjunction*

$$E^C \underset{F^*}{\overset{G^*}{\underset{\perp}{\rightleftarrows}}} E^D$$

for any locally small category E. [7]

As in the case of the other results in this section, this can be proven by appealing to either Definition 4.1.1 or Definition 4.2.5. Rather than spell out the details, which are left for the exercises, we instead sketch a third proof.

Definition 4.2.5 reveals that an adjunction can be defined internally to the 2-category CAT introduced in Definition 1.7.8: an adjunction consists of a pair of objects, a pair of 1-morphisms, and a pair of 2-morphisms satisfying certain composition relations. Now a **2-functor** is a morphism between 2-categories: a map on objects, 1-morphisms, and 2-morphisms preserving composition and identities at all levels. There is a 2-dimensional analog of Lemma 1.6.5—that functors preserve commutative diagrams—which says that 2-functors preserve "2-dimensional diagrams" including, in particular, adjunctions. Ignoring size issues, Proposition 4.4.6 is an immediate corollary of the fact that

$$(-)^J\colon \mathsf{CAT} \to \mathsf{CAT} \quad \text{and} \quad E^{(-)}\colon \mathsf{Cat}^{\mathrm{op}} \to \mathsf{CAT}$$

define 2-functors for any small category J and locally small category E.

REMARK 4.4.7. The "syntactic" proofs of Propositions 4.4.1, 4.4.4, 4.4.5, and 4.4.6 (the last of these left to Exercise 4.4.ii) apply immediately to adjunctions defined internally to any 2-category, in place of Cat. This suggests a reason to prefer these proofs to the alternate "Yoneda-style" arguments, which apply only to Cat. However, using analogous arguments to those made in §3.4, these "Yoneda-style" arguments can be bootstrapped from Cat to a general 2-category. As (co)limits in a general category are defined representably in terms of limits in Set, so too—again by the Yoneda lemma—are adjunctions and equivalences in a 2-category are defined representably in terms of adjunctions and equivalences in Cat. This line of reasoning supports the category theorist's joke that "all theorems are Yoneda."

---

[7]Note that these functor categories will not be locally small unless C and D are small. If local smallness of the functor categories is not a concern, then the hypothesis that J is small can be dropped.

Exercises.

EXERCISE 4.4.i. Complete the second proof of Proposition 4.4.4 by verifying the triangle identities.

EXERCISE 4.4.ii. Use the unit and counit associated to an adjunction to prove Proposition 4.4.6.

EXERCISE 4.4.iii. Use the natural bijection between hom-sets to prove the first half of Proposition 4.4.6, that an adjunction $F \dashv G$ induces an adjunction $F_* \dashv G_*$ by post-composition. (Hint: Exercise 3.2.iv might help.)

4.5. Adjunctions, limits, and colimits

In this section, we explore the interaction between adjoint functors and limits and colimits. Our first result provides an adjoint characterization of completeness and cocompleteness of a category C with regard to diagrams of a fixed shape J.

PROPOSITION 4.5.1. *A category* C *admits all limits of diagrams indexed by a small category* J *if and only if the constant diagram functor* $\Delta \colon C \to C^J$ *admits a right adjoint, and admits all colimits of* J-*indexed diagrams if and only if* Δ *admits a left adjoint:*

When these adjoints exist, they define the limit and colimit functors introduced in Proposition 3.6.1. Recall that the axiom of choice is needed to define the action of the limit or colimit functor on objects.

PROOF. These dual statements follow immediately from the defining universal properties of a limit and colimit. For $c \in C$ and $F \in C^J$, the hom-set $C^J(\Delta c, F)$ is the set of natural transformations from the constant J-diagram at c to the diagram F. This is precisely the set of cones over F with summit c, as defined in 3.1.2. There is an object $\lim F \in C$ together with an isomorphism

$$C^J(\Delta c, F) \cong C(c, \lim F)$$

that is natural in $c \in C$ if and only if this limit exists. If such natural isomorphisms exist for all diagrams $F \in C^J$, then Proposition 4.3.4 applies to extend the objects $\lim F$ to a limit functor $\lim \colon C^J \to C$.   □

Proposition 4.5.1 reveals that limits assemble into a right adjoint to the constant diagram functor. The reason that limits appear on the right is because limits are defined as representations for contravariant functors: the universal property of the limit of $F \colon J \to C$ characterizes the functor $C(-, \lim F)$.

Similarly, the value of a right adjoint functor $G \colon D \to C$ on an object $d \in D$ is determined by a characterization of the contravariant representable functor $C(-, Gd)$. Based on these observations, one might expect right adjoints to interact nicely with limits and indeed this is the case:

THEOREM 4.5.2 (RAPL). *Right adjoints preserve limits.*

PROOF. Consider a diagram $K \colon \mathsf{J} \to \mathsf{D}$ admitting a limit cone $\lambda \colon \lim K \Rightarrow K$ in D, illustrated here in the case where the indexing category J is the poset of integers.

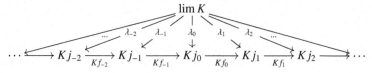

Applying a right adjoint $G \colon \mathsf{D} \to \mathsf{C}$, we obtain a cone $G\lambda \colon G \lim K \Rightarrow GK$ over the diagram GK in C.

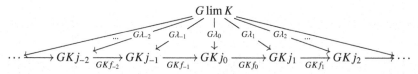

The statement asserts that this is a limit cone for the diagram $GK \colon \mathsf{J} \to \mathsf{C}$. To prove this, consider another cone $\mu \colon c \Rightarrow GK$.

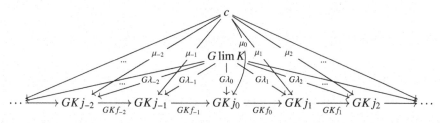

The legs of this cone transpose across the adjunction $F \dashv G$ to define maps in D

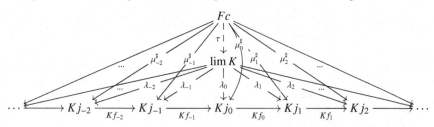

which define a cone $\mu^{\sharp} \colon Fc \Rightarrow K$ over K, by naturality of the transposition relation, as expressed for instance by Lemma 4.1.3. There is a unique factorization $\tau \colon Fc \to \lim K$ of the cone $\mu^{\sharp}$ through the limit cone λ. The map τ transposes to define a map $\tau^{\flat} \colon c \to G \lim K$

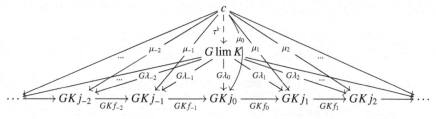

which, by Lemma 4.1.3, defines a factorization of the cone μ through the cone $G\lambda$. This factorization is clearly unique: another such factorization $c \to G \lim K$ would transpose to define a factorization of $\mu^\sharp$ through λ, which, by the universal property of the limit cone λ would necessarily equal τ, and $\tau^\flat$ is the unique transpose of this map. This shows that $G\lambda\colon G\lim K \Rightarrow GK$ has the universal property required to define a limit cone for GK. $\square$

Put more concisely, the argument just presented describes a series of natural isomorphisms:

$$\mathsf{C}^\mathsf{J}(\Delta c, GK) \cong \mathsf{D}^\mathsf{J}(F\Delta c, K) \cong \mathsf{D}^\mathsf{J}(\Delta Fc, K) \cong \mathsf{D}(Fc, \lim_\mathsf{J} K) \cong \mathsf{C}(c, G\lim_\mathsf{J} K),$$

which, by the defining universal property of the limit, says that $G\lim_\mathsf{J} K$ defines a limit for the diagram GK.

Dually, of course:

THEOREM 4.5.3 (LAPC). *Left adjoints preserve colimits.*

Corollaries are ubiquitous. For instance:

COROLLARY 4.5.4. *For any function* $f\colon A \to B$, *the inverse image* $f^{-1}\colon PB \to PA$, *a function between the power sets of A and B, preserves both unions and intersections, while the direct image* $f_*\colon PA \to PB$ *only preserves unions.*

PROOF. Example 4.1.8 describes adjunctions

$$PB \xrightarrow{\;f^{-1}\;} PA\,.$$

Unions are colimits and intersections are limits in the poset category PA. $\square$

COROLLARY 4.5.5. *For any vector spaces* U, V, W,

$$U \otimes (V \oplus W) \cong (U \otimes V) \oplus (U \otimes W).$$

PROOF. The two-variable adjunction of Example 4.3.11 can be defined for the category of modules over any commutative ring, in particular for the category of vector spaces over a field $\Bbbk$. It follows that the functor $U \otimes -\colon \mathsf{Vect}_\Bbbk \to \mathsf{Vect}_\Bbbk$ is left adjoint to $\mathrm{Hom}(U, -)$ and consequently preserves the coproduct $V \oplus W$. This argument also proves that the tensor product distributes over arbitrary direct sums. $\square$

Similarly, on account of the adjunctions $A \times - \dashv (-)^A$ between the product and exponential on Set or its subcategory Fin, Theorems 4.5.2 and 4.5.3 supply proofs of many of the basic operations in arithmetic, first discussed in Example 1.4.9.

COROLLARY 4.5.6. *For any sets* A, B, C, *there are natural isomorphisms*

$$A \times (B + C) \cong (A \times B) + (A \times C) \qquad (B \times C)^A \cong B^A \times C^A \qquad A^{B+C} \cong A^B \times A^C.$$

Consequently, for any cardinals α, β, γ, *cardinal arithmetic satisfies the laws:*

$$\alpha \times (\beta + \gamma) = (\alpha \times \beta) + (\alpha \times \gamma) \qquad (\beta \times \gamma)^\alpha = \beta^\alpha \times \gamma^\alpha \qquad \alpha^{\beta+\gamma} = \alpha^\beta \times \alpha^\gamma.$$

PROOF. The left adjoint $A \times -$ preserves the coproduct $B + C$, the right adjoint $(-)^A$ preserves the product $B \times C$, and the functor $A^-\colon \mathsf{Set}^{\mathrm{op}} \to \mathsf{Set}$, which is mutually right adjoint to itself, carries coproducts in Set to products in Set. The laws of cardinal arithmetic follow by applying the cardinality functor $|-|\colon \mathsf{Set}_{\mathrm{iso}} \to \mathsf{Card}$, which converts these natural isomorphisms to identities in the discrete category of cardinals. $\square$

The forgetful functors of Example 4.1.10 carry any limits that exist in the categories of groups, rings, modules, and so forth to corresponding limits of their underlying sets. Indeed, in §5.6, we show that these forgetful functors create, and not merely preserve, all limits. The dual result, that left adjoints preserve colimits, is less directly useful in such contexts, characterizing only those colimit constructions involving "free" diagrams. For instance:

COROLLARY 4.5.7. *The free group on the set $X \sqcup Y$ is the free product of the free groups on the sets X and Y.*

The free product of not-necessarily free groups will be constructed in §5.6 using language that also describes the construction of the coproducts in the categories of abelian groups, rings, and modules, among many others.

Theorems 4.5.2 and 4.5.3 also have important applications to homological algebra.

COROLLARY 4.5.8. *For any R-S bimodule M, the tensor product $M \otimes_S -$ is right exact.*

Before proving Corollary 4.5.8, we should explain its statement. The term "right exact" comes from homological algebra, which studies functors between *abelian categories*. An abstract definition is given in Definition E.5.1, but by a powerful result, Theorem E.5.2, it suffices to declare that a category is **abelian** if it is a full subcategory of category of modules that contains the zero object and is closed under direct sums, kernels, and cokernels.[8]

The following general definitions of left and right exactness makes sense for any functor, not necessarily between abelian categories.

DEFINITION 4.5.9. A functor is **right exact** if it preserves finite colimits and **left exact** if it preserves finite limits.

The following proposition connects Definition 4.5.9 to the notions of left and right exactness that are used in homological algebra.

PROPOSITION 4.5.10. *A functor $F: \mathsf{A} \to \mathsf{B}$ between abelian categories is left exact if and only if it preserves direct sums and kernels. For such functors, if*

$$0 \to A \to A' \to A''$$

is an exact sequence in A, *then*

$$0 \to FA \to FA' \to FA''$$

is exact in B.

PROOF. Direct sums and kernels are both finite limits, so if F is left exact, these constructions are preserved. Conversely, Theorem 3.4.12 proves that all finite limits can be built from finite products and equalizers. In an abelian category, the equalizer of a parallel pair of maps $f, g: A \rightrightarrows A'$ is the kernel of the map $f - g: A \to A'$. A functor that preserves direct sums also preserves differences of maps, so if F preserves direct sums and kernels it preserves all finite limits.

Now a sequence

$$0 \xrightarrow{\quad i \quad} A \xrightarrow{\quad j \quad} A' \xrightarrow{\quad k \quad} A''$$

is exact in A just when i is the kernel of j (equivalently, when j is a monomorphism), and when j is the kernel of k. A left exact functor preserves these kernels and also the zero object 0. □

---

[8]A **zero object** is an object that is both initial and terminal; see Exercise 1.6.i. Direct sums are defined in Remark 3.1.27. As in Example 3.1.14, the **kernel** of a map $f: A \to A'$ is the equalizer of f with the zero homomorphism, while the **cokernel** is the coequalizer of this pair.

Immediately from Definition 4.5.9 and Theorems 4.5.2 and 4.5.3:

COROLLARY 4.5.11. *For any adjoint functors between abelian categories, the left adjoint is right exact and the right adjoint is left exact. Moreover, both functors are additive.* [9]

Corollary 4.5.8 is a special case:

PROOF OF COROLLARY 4.5.8. For any R-S bimodule M, there is a pair of adjoint functors

$$\mathsf{Mod}_S \underset{\mathrm{Hom}_R(M,-)}{\overset{M \otimes_S -}{\underset{\perp}{\rightleftarrows}}} \mathsf{Mod}_R \,,$$

where the right adjoint carries a left R-module N to the left S-module of R-module homomorphisms $M \to N$. As a left adjoint, Corollary 4.5.11 proves that $M \otimes_S -$ is right exact.   □

For a certain important class of adjoint functors, Theorems 4.5.2 and 4.5.3 take on a stronger form.

DEFINITION 4.5.12. A **reflective subcategory** of a category C is a full subcategory D so that the inclusion admits a left adjoint, called the **reflector** or **localization**:

$$\mathsf{D} \underset{\longrightarrow}{\overset{L}{\underset{\perp}{\leftharpoonup}}} \mathsf{C} \,.$$

Where possible, we identify the full subcategory D with its image in C, declining in particular to introduce notation for the inclusion functor. With this convention in mind, the components of the unit have the form $c \to Lc$; the mnemonic is that "an object looks at its reflection." The following lemma implies that the components of the counit $Ld \cong d$ are isomorphisms. Via the counit, any object in a reflective subcategory $\mathsf{D} \hookrightarrow \mathsf{C}$ is naturally isomorphic to its reflection back into that subcategory.

LEMMA 4.5.13. *Consider an adjunction*

$$\mathsf{C} \underset{G}{\overset{F}{\underset{\perp}{\rightleftarrows}}} \mathsf{D}$$

with counit $\epsilon \colon FG \Rightarrow 1_\mathsf{D}$. Then:

  (i) *G is faithful if and only if each component of ϵ is an epimorphism.*
  (ii) *G is full if and only if each component of ϵ is a split monomorphism.*
  (iii) *G is full and faithful if and only if ϵ is an isomorphism.*

Dually, writing $\eta \colon 1_\mathsf{C} \Rightarrow GF$ for the unit:

  (i) *F is faithful if and only if each component of η is a monomorphism.*
  (ii) *F is full if and only if each component of η is a split epimorphism.*
  (iii) *F is full and faithful if and only if η is an isomorphism.*

PROOF. Exercise 4.5.vi.   □

EXAMPLE 4.5.14. The following define reflective subcategories:

  (i) Compact Hausdorff spaces define a reflective subcategory cHaus $\hookrightarrow$ Top of the category of all topological spaces. The reflector is the functor $\beta \colon$ Top $\to$ cHaus sending a space to its **Stone–Čech compactification**, which is constructed in Example 4.6.12. The universal property of the unit says that any continuous function

---

[9]A functor between abelian categories is **additive** if it preserves direct sums.

$X \to K$ from a space X to a compact Hausdorff space K extends uniquely to the Stone–Čech compactification:

(ii) Abelian groups define a reflective subcategory Ab $\hookrightarrow$ Group of the category of groups. The reflector carries a group G to its **abelianization**, the quotient $G/[G,G]$ by the **commutator subgroup**, the normal subgroup generated by elements $ghg^{-1}h^{-1}$. The quotient maps $G \to G/[G,G]$ define the components of the unit. For any abelian group A, there is an isomorphism $A \cong A/[A,A]$; the commutator subgroup $[A,A]$ is trivial if and only if the group A is abelian. The adjunction asserts that a homomorphism from a group G to an abelian group A necessarily factors through the abelianization of G. A similar construction defines a left adjoint to the inclusion CRing $\hookrightarrow$ Ring of commutative rings into the category of all rings.

(iii) The inclusion $\mathsf{Ab}_{tf} \hookrightarrow \mathsf{Ab}$ of torsion-free abelian groups is reflective. The reflector sends an abelian group A to the quotient A/TA by its torsion subgroup. The quotient maps $A \to A/TA$ define the components of the unit. Any map from A to a torsion-free group factors uniquely through this quotient homomorphism because any torsion elements of A must be contained in its kernel.

(iv) As described in Example 4.1.10(xi), for any ring homomorphism $\phi \colon R \to T$, there exist adjoint functors

$$\mathsf{Mod}_T \underset{\phi^*}{\overset{T \otimes_R -}{\rightleftarrows}} \mathsf{Mod}_R \,,$$

the right adjoint being restriction of scalars and the left adjoint being extension of scalars. The restriction of scalars functor is always faithful because both categories of modules admit faithful forgetful functors to Ab, and is full if and only if $\phi \colon R \to T$ is an epimorphism. For such homomorphisms, restriction of scalars identifies Mod_T as a reflective subcategory of Mod_R.

Epimorphisms in Ring include all surjections but are not limited to the surjections. The localizations define another important class of epimorphisms. Let $S \subset R$ be a monoid under multiplication; that is, S is a multiplicatively closed subset of the ring R. The **localization** $R \to R[S^{-1}]$ is an initial object in the category whose objects are ring homomorphisms $R \to T$ that carry all of the elements of S to units in T. For integral domains, the ring $R[S^{-1}]$ can be constructed as a field of fractions. A similar construction exists for general commutative rings.

(v) Recall that a **presheaf** on a space X is a contravariant functor $O(X)^{op} \to \mathsf{Set}$ from the poset of open subsets of X to sets. A presheaf is a **sheaf** if it preserves certain limits (see Definition 3.3.4). The sheaves define a reflective subcategory Shv_X of the category of presheaves, with the left adjoint called **sheafification**:

$$\mathsf{Shv}_X \underset{\longrightarrow}{\overset{\text{sheafify}}{\longleftarrow}} \mathsf{Set}^{O(X)^{op}}$$

(vi) The category of small categories defines a reflective subcategory of the category $\mathsf{Set}^{\Delta^{op}}$ of **simplicial sets** via an adjunction that is constructed in Exercise 6.5.iv:

$$\mathsf{Cat} \underset{N}{\overset{h}{\rightleftarrows}} \mathsf{Set}^{\Delta^{op}}$$

Here $\Delta \subset$ Cat is the full subcategory whose objects are the finite non-empty ordinals, in this context denoted by $[0], [1], [2], \ldots$, and whose morphisms are all functors, i.e., order-preserving functions, between them. The embedding $N\colon$ Cat $\hookrightarrow$ Set$^{\Delta^{op}}$ carries a small category C to its **nerve**: NC$\colon \Delta^{op} \to$ Set sends $[n] = 0 \to 1 \to \cdots \to n$ to the set of functors NC$_n := $ Cat$([n],$ C$)$. The left adjoint $h\colon$ Set$^{\Delta^{op}} \to$ Cat sends a simplicial set to its **homotopy category**. Restricting to the objects $[0], [1] \in \Delta$, a simplicial set $X\colon \Delta^{op} \to$ Set has an underlying reflexive directed graph

$$X_0 \underset{\longrightarrow}{\overset{\longrightarrow}{\longleftarrow}} X_1 \,.$$

The homotopy category hX is a quotient of the free category generated by this reflexive directed graph modulo relations that arise from elements of the set X_2. In particular, the counit defines an isomorphism hNC $\cong$ C for any category C, proving that the inclusion is full and faithful.

The following result explains our particular interest in reflective subcategories. The presence of a left adjoint to the inclusion of a full subcategory provides complete information about limits and colimits in that subcategory:

PROPOSITION 4.5.15. *If* D $\hookrightarrow$ C *is a reflective subcategory, then:*

   (i) *The inclusion* D $\hookrightarrow$ C *creates all limits that* C *admits.*
   (ii) D *has all colimits that* C *admits, formed by applying the reflector to the colimit in* C.

By Theorems 4.5.2 and 4.5.3, if D has limits or colimits, they must be constructed in the way described in Proposition 4.5.15: limits are preserved by the inclusion D $\hookrightarrow$ C and colimits of diagrams in D, regarded as diagrams in C, are preserved by $L\colon$ C $\to$ D. The real content of this result is that these (co)limits necessarily exist in D. A special case of this completes some unfinished business from §3.5:

COROLLARY 4.5.16. Cat *is complete and cocomplete.*

PROOF. By Proposition 3.3.9, Set$^{\Delta^{op}}$ is complete and cocomplete, with limits and colimits defined objectwise in Set. Proposition 4.5.15 implies that the reflective subcategory Cat inherits these limits, defined objectwise in Set, and also these colimits, defined by applying the homotopy category functor to the colimit in Set$^{\Delta^{op}}$.   □

Proposition 4.5.15(i) appears as Corollary 5.6.6, where this result is deduced as a special case of the more general Theorem 5.6.5.

PROOF OF 4.5.15(ii). For clarity, write $i\colon$ D $\hookrightarrow$ C for the inclusion, the right adjoint to the reflector L. Consider a diagram $F\colon$ J $\to$ D and let $\lambda\colon iF \Rightarrow c$ be a colimit cone for the diagram iF in C. By Theorem 4.5.3, the left adjoint L sends this to a colimit cone $L\lambda\colon LiF \Rightarrow Lc$ in D. Now the counit supplies a natural isomorphism $Li \cong 1_{\mathsf{D}}$. Composing with this natural isomorphism yields a colimit cone $F \cong LiF \Rightarrow Lc$ for the original diagram in D.   □

Exercises.

EXERCISE 4.5.i. When does the unique functor $!\colon$ C $\to$ $\mathbb{1}$ have a left adjoint? When does it have a right adjoint?

EXERCISE 4.5.ii. Suppose the diagonal functor $\Delta\colon$ C $\to$ C^J admits both left and right adjoints. Describe the units and counits of these adjunctions.

EXERCISE 4.5.iii. Use Proposition 4.5.1 to prove that in any complete category limits commute with limits in the sense of the natural isomorphism of limit functors (3.8.2).

EXERCISE 4.5.iv. If $G: \mathsf{D} \to \mathsf{C}$ has a left adjoint F and if D and C admit all limits indexed by a category J, use Propositions 4.5.1, 4.4.4, and 4.4.6 to argue that right adjoints preserve limits by considering the diagram of adjoint functors:

$$
\begin{array}{ccc}
\mathsf{C} & \underset{F}{\overset{G}{\leftrightarrows}} & \mathsf{D} \\
{\scriptstyle\lim}\uparrow\ \vdash\ \downarrow\Delta & & \Delta\downarrow\ \dashv\ \uparrow{\scriptstyle\lim} \\
\mathsf{C}^\mathsf{J} & \underset{G}{\overset{F}{\leftrightarrows}} & \mathsf{D}^\mathsf{J}
\end{array}
$$

The characterization of preservation of limits provided by Exercise 3.3.i might prove useful.

EXERCISE 4.5.v. Show that a morphism $f: x \to y$ in C is a monomorphism if and only if the square

$$
\begin{array}{ccc}
x & \overset{1_x}{\longrightarrow} & x \\
{\scriptstyle 1_x}\downarrow & & \downarrow{\scriptstyle f} \\
x & \underset{f}{\longrightarrow} & y
\end{array}
$$

is a pullback. Conclude that right adjoints preserve monomorphisms, and that left adjoints preserve epimorphisms.

EXERCISE 4.5.vi. Prove Lemma 4.5.13.

EXERCISE 4.5.vii. Consider a reflective subcategory inclusion $\mathsf{D} \hookrightarrow \mathsf{C}$ with reflector $L: \mathsf{C} \to \mathsf{D}$.

   (i) Show that $\eta L = L\eta$, and that these natural transformations are isomorphisms.
   (ii) Show that an object $c \in \mathsf{C}$ is in the **essential image** of the inclusion $\mathsf{D} \hookrightarrow \mathsf{C}$, meaning that it is isomorphic to an object in the subcategory D, if and only if η_c is an isomorphism.
   (iii) Show that the essential image of D consists of those objects c that are **local** for the class of morphisms that are inverted by L. That is, c is in the essential image if and only if the pre-composition functions

$$
\mathsf{C}(b,c) \overset{f^*}{\longrightarrow} \mathsf{C}(a,c)
$$

are isomorphisms for all maps $f: a \to b$ in C for which Lf is an isomorphism in D. This explains why the reflector is also referred to as "localization."

EXERCISE 4.5.viii. Find an example which demonstrates that the inclusion of a reflective subcategory does not create all colimits.

4.6. Existence of adjoint functors

Does the inclusion $\mathsf{Ring} \hookrightarrow \mathsf{Rng}$ of the category of unital rings into the category of possibly non-unital rings have any adjoints? A first strategy to probe a question of this form might be called the "initial and terminal objects test": by Theorems 4.5.2 and 4.5.3, a functor admitting a left adjoint must necessarily preserve all limits while a functor admitting a right adjoint must necessarily preserve all colimits. The ring $\mathbb{Z}$ is initial in Ring but not in Rng, so there can be no right adjoint. The zero ring is terminal in both categories, so a left adjoint to the inclusion might be possible. Indeed, Ring has all limits and the inclusion preserves them.[10]

The search for a left adjoint to Ring $\hookrightarrow$ Rng might be thought of as some sort of formulaic (that is to say *functorial*) optimization problem, whose aim is to adjoin a multiplicative unit to a possibly non-unital ring R in the "most efficient way possible." The Yoneda lemma can be used to make this intuition precise. To define the value of a hypothetical left adjoint on a possibly non-unital ring R, we seek a unital ring R^* together with a natural isomorphism

$$\text{Ring}(R^*, S) \cong \text{Rng}(R, S)$$

for all unital rings S. Note that even if R has a unit, the ring R^* will still differ from R because homomorphisms in Ring must preserve units, while homomorphisms in Rng need not.

To define this natural isomorphism is to define a representation R^* for the functor $\text{Rng}(R, -) \colon \text{Ring} \to \text{Set}$. By Proposition 2.4.8, a representation defines an initial object in the category of elements $\int \text{Rng}(R, -)$. Objects in this category are homomorphisms $R \to S$ whose codomain is a unital ring; morphisms are commutative triangles

$$
\begin{array}{ccc}
 & R & \\
 \swarrow & & \searrow \\
 S & \longrightarrow & S'
\end{array}
$$

whose leg opposite R is a unital ring homomorphism but whose legs with domain R do not necessarily preserve the multiplicative unit, if R happens to have one. This category of elements is isomorphic to the comma category $R \downarrow \text{Ring}$ of non-unital homomorphisms from R to a unital ring. The optimization problem is solved if we can find a unital ring R^* and ring homomorphism $R \to R^*$ that is initial in this category.[11]

The same line of reasoning proves the following general result.

LEMMA 4.6.1. *A functor* $U \colon A \to S$ *admits a left adjoint if and only if for each* $s \in S$ *the comma category* $s \downarrow U$ *has an initial object.*

Comma categories are defined in Exercise 1.3.vi: here, objects of $s \downarrow U$ are pairs $(a \in A, f \colon s \to Ua \in S)$ and a morphism from $f \colon s \to Ua$ to $f' \colon s \to Ua'$ is a map $h \colon a \to a' \in A$ so that the evident triangle commutes in S.

PROOF. The comma category $s \downarrow U$ is isomorphic to the category of elements for the functor $S(s, U-) \colon A \to \text{Set}$. If a left adjoint F exists, then the component of the unit at s defines an initial object $\eta_s \colon s \to UFs$ in this category; see Proposition 2.4.8 or Exercise 4.2.iv.

---

[10]As an application of Theorem 5.6.5, the forgetful functors from Ring or from Rng to Set both create all limits. On account of the commuting triangle

$$
\begin{array}{ccc}
\text{Ring} & \longhookrightarrow & \text{Rng} \\
\scriptstyle U \searrow & & \swarrow \scriptstyle U \\
 & \text{Set} &
\end{array}
$$

it follows that the inclusion preserves all limits.

[11]The optimization problem intuition for the construction of adjoint functors is explained very well on Wikipedia's "adjoint functors" entry (retrieved on April 14, 2015). However, Wikipedia's suggestion that "picking the right category [to express the universal property of the adjoint construction] is something of a knack" is incorrect. A left adjoint to a functor $U \colon A \to S$ at an object $s \in S$ defines an initial object in the category of elements of the functor $S(s, U-) \colon A \to \text{Set}$; dually, a right adjoint defines a terminal object in the category of elements of $S(U-, s) \colon A^{\text{op}} \to \text{Set}$. See Proposition 2.4.8.

Conversely, suppose each $s \downarrow U$ admits an initial object, which we suggestively denote by $\eta_s \colon s \to UFs$. This defines the value of a function $F \colon \mathrm{ob}\,S \to \mathrm{ob}\,A$, which we now extend to a functor; the proof of Proposition 4.3.4 gives an alternate account of this same construction. For each morphism $f \colon s \to s'$ in S, define $Ff \colon Fs \to Fs'$ to be the unique morphism in A making the square

$$
\begin{array}{ccc}
s & \xrightarrow{\;\eta_s\;} & UFs \\
{\scriptstyle f}\Big\downarrow & & \Big\downarrow{\scriptstyle UFf} \\
s' & \xrightarrow{\;\eta_{s'}\;} & UFs'
\end{array}
$$

commute; the fact that η_s is initial in $s \downarrow U$ implies the existence and uniqueness of such a map. The functoriality of F follows from the uniqueness of these choices, so this construction defines a functor $F \colon S \to A$ together with a natural transformation $\eta \colon 1_S \Rightarrow UF$.

This data allows us to define a natural transformation $\phi \colon A(F-, -) \Rightarrow S(-, U-)$ with components

$$\phi_{s,a} \colon A(Fs, a) \to S(s, Ua)$$

defined as in Remark 4.2.7(iii): given a map $g \colon Fs \to a$ in A, define

$$\phi_{s,a}(g) := s \xrightarrow{\;\eta_s\;} UFs \xrightarrow{\;Ug\;} Ua\,.$$

Injectivity and surjectivity of $\phi_{s,a}$ follows immediately from the uniqueness and existence of morphisms from η_s to any particular object $s \to Ua$ in $s \downarrow U$. By Remark 4.2.7, this natural isomorphism proves that $F \dashv U$ with unit η.   □

Recall that a limit-preserving functor is called **continuous**; a colimit-preserving functor is called **cocontinuous**. Lemma 4.6.1 reduces the problem of finding a left adjoint to a continuous functor $U \colon A \to S$ to the problem of finding an initial object in the comma category $s \downarrow U$ defined for each $s \in S$. This comma category, as the category of elements for $S(s, U-) \colon A \to Set$, comes with a canonical forgetful functor $\Pi \colon s \downarrow U \to A$ that carries an object $s \to Ua$ to the object a.

LEMMA 4.6.2. *For any functor* $U \colon A \to S$ *and object* $s \in S$, *the associated forgetful functor* $\Pi \colon s \downarrow U \to A$ *strictly creates the limit of any diagram whose limit exists in* A *and is preserved by* U. *In particular, if* A *is complete and* U *is continuous, then* $s \downarrow U$ *is complete.*

PROOF. Lemma 4.6.2 is a special case of Exercise 3.4.iii, or can be proven directly via a straightforward extension of the argument used to prove Proposition 3.3.8.   □

Lemma 4.6.2 can be used to produce an initial object in $s \downarrow U$. By Lemma 3.7.1, an initial object is equally the *limit* of the identity functor. Applying Lemma 4.6.2, a limit of the identity functor on $s \downarrow U$ exists if and only if the limit of the forgetful functor $\Pi \colon s \downarrow U \to A$ exists in A.

This line of reasoning seems to imply that all continuous functors whose domains are complete should admit left adjoints. This is not the case: the problem is that $s \downarrow U$ is not in general a small category, so even if A admits all small limits it may not admit a limit of the large diagram $\Pi \colon s \downarrow U \to A$. The *adjoint functor theorems*, the two most common of which are discussed here, supply conditions under which this large limit can be reduced to a small one that A possesses.

THEOREM 4.6.3 (General Adjoint Functor Theorem). *Let* $U\colon \mathsf{A} \to \mathsf{S}$ *be a continuous functor whose domain is locally small and complete. Suppose that U satisfies the following* **solution set condition**:

- *For every $s \in \mathsf{S}$ there exists a set of morphisms $\Phi_s = \{f_i\colon s \to Ua_i\}$ so that any $f\colon s \to Ua$ factors through some $f_i \in \Phi_s$ along a morphism $a_i \to a$ in A.*

Then U admits a left adjoint.

The solution set condition says exactly that $\{f_i\colon s \to Ua_i\}$ is a set of jointly weakly initial objects in the category $s \downarrow U$.

DEFINITION 4.6.4.

(i) An object c in a category C is **weakly initial** if for every $x \in \mathsf{C}$ there exists a morphism $c \to x$ (but possibly more than one).
(ii) A collection of objects $\Phi = \{c_i\}$ in a category C is **jointly weakly initial** if for every $x \in \mathsf{C}$ there exists some $c_i \in \Phi$ for which there exists a morphism $c_i \to x$.

In the presence of certain limits, an initial object can be built from a jointly weakly initial set of objects:

LEMMA 4.6.5. *If C is complete, locally small, and has a jointly weakly initial set of objects Φ, then C has an initial object.*

PROOF. Form the limit ℓ of the inclusion into C of the full subcategory spanned by the objects of Φ; because C is locally small, this diagram is small and because C is complete, this limit exists. Using the limit cone $(\kappa_k\colon \ell \to k)_{k\in\Phi}$, we may define a morphism $\lambda_c\colon \ell \to c$ for each $c \in \mathsf{C}$:

- For the objects $k \in \Phi$, define $\lambda_k := \kappa_k$.
- For the remaining objects, choose a morphism $h_c\colon k \to c$, with $k \in \Phi$, and define $\lambda_c := \ell \xrightarrow{\kappa_k} k \xrightarrow{h_c} c$.

The maps λ_c prove that ℓ is weakly initial. To prove that ℓ is initial, we show that the morphisms λ_c define a cone $\lambda\colon \ell \Rightarrow 1_{\mathsf{C}}$ with summit ℓ over the identity functor $1_{\mathsf{C}}\colon \mathsf{C} \to \mathsf{C}$ with the property that the component λ_ℓ is the identity morphism. By the proof of Lemma 3.7.1, this implies that ℓ is initial.

To prove that the maps λ_c define a cone, consider any morphism $f\colon c \to c'$ in C and form the pullback p of $h_{c'}\colon k' \to c'$, the chosen morphism for c' with $k' \in \Phi$, along the composite of f with $h_c\colon k \to c$:

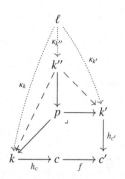

Because Φ is jointly weakly initial, there is some $k'' \to p$ with $k'' \in \Phi$. The dashed composite morphisms live in the full subcategory spanned by the objects in Φ, which means that the top triangles, involving the cone legs κ commute. The commutativity of the rest of the displayed diagram now implies that λ defines a cone.

Finally, the cone condition for λ implies that the triangle

$$
\begin{array}{ccc}
 & \ell & \\
\lambda_\ell \swarrow & & \searrow \lambda_k = \kappa_k \\
\lambda & \xrightarrow[\kappa_k]{} & k
\end{array}
$$

commutes for all $k \in \Phi$. This tells us that λ_ℓ defines a factorization of the limit cone κ through itself; whence $\lambda_\ell = 1_\ell$.   □

The proof of the General Adjoint Functor Theorem proceeds directly by assembling the lemmas in this section.

PROOF OF THEOREM 4.6.3. By Lemma 4.6.1, U admits a left adjoint if and only if for each $s \in S$ the comma category $s \downarrow U$ has an initial object; these initial objects define the value of the left adjoint on objects and the components of the unit of the adjunction. The solution set condition says that $s \downarrow U$ has a jointly weakly initial set of objects. Because A is locally small, $s \downarrow U$ is as well. Because A is complete and U is continuous, Lemma 4.6.2 applies to prove that $s \downarrow U$ is also complete. So we may apply Lemma 4.6.5 to construct an initial object in $s \downarrow U$.   □

EXAMPLE 4.6.6. For example, consider the functor $U : \mathsf{Group} \to \mathsf{Set}$ and a set S. The following construction of the free group on S can be found in [**Lan02**, §I.12]. Let Φ' be the set of representatives for isomorphism classes of groups that can be generated by a set of elements of cardinality at most $|S|$. Let Φ be the set of functions $S \to UG$, with $G \in \Phi'$, whose image generates the group G. This is a jointly weakly initial set of objects in the comma category $S \downarrow U$ and a solution set for $U : \mathsf{Group} \to \mathsf{Set}$.

To construct the initial object, which by Lemma 4.6.1 defines the free group on the set S, form the product, indexed by the set Φ, of the groups appearing as codomains and consider the induced map

$$
S \xrightarrow{\hat\eta} U\left(\prod_{S \to UG \in \Phi} G \right).
$$

The map $\hat\eta$ is the product of the objects of Φ in the category $S \downarrow U$. Let Γ be the subgroup of the product $\prod_{S \to UG \in \Phi} G$ generated by the image of $\hat\eta$. We show that Γ is the free group on S by proving that the codomain restriction $\eta : S \to U\Gamma$ is initial in $S \downarrow U$.

Given a group H and a function $\psi : S \to UH$, let $G \subset H$ be the subgroup generated by the image of S. By construction, ψ factors through its codomain restriction $\phi : S \to UG$, and since G is generated by the image of this map, $\phi \in \Phi$. The map ϕ equals the composite of $\eta : S \to U\Gamma$ with the projection to the component indexed by ϕ, so we have found a way to factor ψ through η via a homomorphism $\Gamma \to H$. The uniqueness of this factorization follows from the fact that Γ is generated by the image of S; as $\psi : S \to UH$ specifies the images of these generators no alternate choices are possible.

Our next adjoint functor theorem requires a few preliminary definitions.

DEFINITION 4.6.7. A **generating set** or **generator** for a category C is a set Φ of objects that can distinguish between distinct parallel morphisms in the following sense: given $f, g : x \rightrightarrows y$, if $f \neq g$ then there exists some $h : c \to x$ with $c \in \Phi$ so that $fh \neq gh$. A **cogenerating set** in C is a generating set in C^{op}.

DEFINITION 4.6.8. A **subobject** of an object $c \in C$ is a monomorphism $c' \rightarrowtail c$ with codomain c. Isomorphic subobjects, that is, subobjects $c' \rightarrowtail c \leftarrowtail c''$ with a commuting isomorphism $c' \cong c''$, are typically identified.

DEFINITION 4.6.9. The **intersection** of a family of subobjects of c is the limit of the diagram of monomorphisms with codomain c.

The induced map from the limit to c is again a monomorphism. By Lemma 1.2.11(ii), the legs of the limit cone are also monomorphisms. Thus, the intersection is the minimal subobject that is contained in every member of the family of subobjects.

THEOREM 4.6.10 (Special Adjoint Functor Theorem). *Let* $U\colon A \to S$ *be a continuous functor whose domain is complete and whose domain and codomain are locally small. Furthermore, if* A *has a small cogenerating set and every collection of subobjects of a fixed object in* A *admits an intersection, then* U *admits a left adjoint.*

In many categories, each object admits only a set's worth of subobjects up to isomorphism, in which case completeness implies that every collection of subobjects admits an intersection. The purpose for these hypotheses is that they are used to construct the initial objects sought for in Lemma 4.6.1.

LEMMA 4.6.11. *Suppose* C *is locally small, complete, has a small cogenerating set* Φ, *and has the property that every collection of subobjects has an intersection. Then* C *has an initial object.*

PROOF. Form the product $p = \prod_{k \in \Phi} k$ of the objects in the cogenerating set and form the intersection $i \hookrightarrow p$ of all of the subobjects of p. We claim that i is initial.

To say that Φ is cogenerating is to say that for every $c \in C$ the canonical map

$$c \rightarrowtail \prod_{k \in \Phi} k^{C(c,k)},$$

whose component at an element $f \in C(c,k)$ is that morphism $f\colon c \to k$, is a monomorphism. There is a map $\prod_{k \in \Phi} k \to \prod_{k \in \Phi} k^{C(c,k)}$, which is the product over $k \in \Phi$ of the maps $\Delta\colon k \to k^{C(c,k)}$ defined to be the identity on each component of the power $k^{C(c,k)}$. The pullback

$$\begin{array}{ccc} p_c & \longrightarrow & c \\ \downarrow & \lrcorner & \downarrow \\ p = \prod_{k \in \Phi} k & \xrightarrow{\;\prod_{k \in \Phi} \Delta\;} & \prod_{k \in \Phi} k^{C(c,k)} \end{array}$$

defines a subobject p_c of p and thus a map $i \to p_c \to c$ from the intersection to c. There cannot be more than one arrow from i to c because the equalizer of two distinct such would define a smaller subobject of p, contradicting minimality of the intersection i. Thus, we conclude that i is an initial object. $\square$

PROOF OF THEOREM 4.6.10. Again, for each $s \in S$, the category $s \downarrow U$ is locally small and complete. Monomorphisms in $s \downarrow U$ are preserved and reflected by $\Pi\colon s \downarrow U \to A$, and so the comma category $s \downarrow U$ has intersections of subobjects created by this functor. If Φ is a cogenerating set for A, then the set

$$\Phi' = \{s \to Ua \mid a \in \Phi\}$$

is cogenerating for $s \downarrow U$; because S is locally small, Φ' is again a set. Applying Lemma 4.6.11, there is an initial object in each comma category $s \downarrow U$, which provides a left adjoint to U by Lemma 4.6.1. $\square$

EXAMPLE 4.6.12. The Special Adjoint Functor Theorem is an abstraction of the construction of the Stone–Čech compactification $\beta\colon$ Top $\to$ cHaus, which defines a left adjoint to the inclusion cHaus $\hookrightarrow$ Top. The unit interval $I = [0, 1] \in \mathbb{R}$ is a cogenerating object in cHaus. To see this, note that if $f \neq g\colon X \rightrightarrows Y$, there must be some $x \in X$ with $f(x) \neq g(x)$. Then Urysohn's lemma can be used to define a continuous function $h\colon Y \to I$ with $hf(x) = 0$ and $hg(x) = 1$, so that in particular $hf \neq hg$. Given a topological space X, Theorem 4.6.10 constructs an initial object in $X \downarrow$ cHaus in the following manner. A cogenerating family in $X \downarrow$ cHaus is given by the set of all maps $X \to I$. By Proposition 3.3.8, the product of these maps, considered as objects in $X \downarrow$ cHaus, is the canonical map

$$X \xrightarrow{\hat{\eta}} \prod_{\mathrm{Hom}(X,I)} I.$$

A subobject of this object is a compact Hausdorff subspace $K \subset \prod_{\mathrm{Hom}(X,I)} I$ containing the image of $\hat{\eta}$. Because $\prod_{\mathrm{Hom}(X,I)} I$ is compact Hausdorff, a subspace is compact Hausdorff if and only if it is closed. Thus, the intersection of all subobjects of $\hat{\eta}\colon X \to \prod_{\mathrm{Hom}(X,I)} I$ is simply the codomain restriction $\eta\colon X \to \beta(X)$, where $\beta(X)$ is the closure of the image of $\hat{\eta}$. This constructs the Stone–Čech compactification.

The adjoint functor theorems have a number of corollaries.

COROLLARY 4.6.13. *Suppose* C *is locally small, complete, has a small cogenerating set, and has the property that every collection of subobjects of a fixed object has an intersection. Then* C *is also cocomplete.*

PROOF. For any small category J, if C is locally small, then C^J is again locally small. The constant diagram functor $\Delta\colon \mathsf{C} \to \mathsf{C}^\mathsf{J}$ preserves limits by Exercise 3.3.vi. Applying Theorem 4.6.10, Δ has a left adjoint, which by Proposition 4.5.1 demonstrates that C has all J-shaped colimits. $\square$

The arguments used to prove the adjoint functor theorems also specialize to give conditions under which a set-valued limit-preserving functor is representable.

COROLLARY 4.6.14. *Suppose* C *is locally small, complete, has a small cogenerating set, and has the property that every collection of subobjects of a fixed object has an intersection. Then any continuous functor* $F\colon \mathsf{C} \to$ Set *is representable.*

PROOF. By Theorem 4.6.10, F has a left adjoint $L\colon$ Set $\to$ C. In particular, there is a natural isomorphism

$$\mathsf{C}(L(*), c) \cong \mathsf{Set}(*, Fc) \cong Fc$$

where $* \in$ Set denotes the singleton set. The object $L(*) \in \mathsf{C}$ represents F. $\square$

THEOREM 4.6.15 (Freyd's Representability Theorem). *Let* $F\colon \mathsf{C} \to$ Set *be a continuous functor and suppose that* C *is complete and locally small. If* F *satisfies the solution set condition:*

- *There exists a set* Φ *of objects of* C *so that for any* $c \in \mathsf{C}$ *and any element* $x \in Fc$, *there exists an* $s \in \Phi$, *an element* $y \in Fs$, *and a morphism* $f\colon s \to c$ *so that* $Ff(y) = x$.

then F *is representable.*

PROOF. The solution set defines a jointly weakly initial set of objects in the comma category $* \downarrow F \cong \int F$, where $* \in$ Set is a singleton set. By Lemma 4.6.2, this category is complete, and so by Lemma 4.6.5 it has an initial object. By Proposition 2.4.8, this defines a representation for F. $\square$

These corollaries aside, the Adjoint Functor Theorems 4.6.3 and 4.6.10 tend to be both useless and widely applicable. Their ineffectualness stems from the fact that solution sets or coseparating sets tend to be hard to find in practice. At the same time, a very simple general characterization of adjoint functors between a large family of categories can be obtained from these theorems.

DEFINITION 4.6.16. Let κ be a regular cardinal.[12] A locally small category C is **locally κ-presentable** if it is cocomplete and if it has a set of objects S so that:

(i) Every object in C can be written as a colimit of a diagram valued in the subcategory spanned by the objects in S.
(ii) For each object $s \in S$, the functor $C(s, -): C \to Set$ preserves κ-filtered colimits.

A functor between locally κ-presentable categories is **accessible** if it preserves κ-filtered colimits.

For example, a large variety of categories whose objects are sets equipped with some sort of "algebraic" structure are locally finitely presentable (meaning locally ω-presentable); see Definition 5.5.5. A locally κ-presentable category is also locally λ-presentable for any $\lambda > \kappa$. Between **locally presentable categories**, meaning categories that are locally κ-presentable for some κ, the adjoint functor theorem takes a particularly appealing form.

THEOREM 4.6.17. *A functor $F: C \to D$ between locally presentable categories*

(i) *admits a right adjoint if and only if it is cocontinuous.*
(ii) *admits a left adjoint if and only if it is continuous and accessible.*

PROOF SKETCH. Theorem 4.6.3 proves (ii): any accessible functor satisfies the solution set condition; see [**AR94**, 1.66].

The dual of Theorem 4.6.10 proves (i): the colimit-generating set of objects S defines a generator for C. A hard theorem proves that any object in a locally presentable category admits at most a set's worth of epimorphic quotients, up to isomorphism [**AR94**, 1.58]. Cocompleteness then implies that cointersections of quotient objects always exist. □

The categories Ring and Rng are locally presentable and the inclusion is accessible, and so Theorem 4.6.17 implies that a left adjoint to the inclusion Ring $\hookrightarrow$ Rng exists. But this general result does not tell us anything useful about how it is constructed. Instead, we sketch a direct construction of the left adjoint that will preview some of the themes from the next chapter.

The inclusion Ring $\hookrightarrow$ Rng commutes with the underlying abelian group functors:

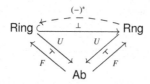

Thus, by Proposition 4.4.4, its left adjoint must commute with the corresponding free functors up to natural isomorphism.

As observed in Example 4.1.10(iii), the free unital ring on an abelian group A is the graded ring $\bigoplus_{n\geq 0} A^{\otimes n}$. Similarly, the free non-unital ring on A is $\bigoplus_{n>0} A^{\otimes n}$. Commutativity of the left adjoints tells us that

$$(\bigoplus_{n>0} A^{\otimes n})^* \cong \bigoplus_{n\geq 0} A^{\otimes n}.$$

That is, the free unital ring on a free non-unital ring is constructed by adjoining a copy of $A^{\otimes 0} := \mathbb{Z}$ and using the componentwise addition and graded multiplication in the graded ring $\bigoplus_{n\geq 0} A^{\otimes n}$ to define the ring structure.

By Proposition 5.4.3, one of the main results to be proven in the next chapter, any non-unital ring R is the coequalizer of a pair of maps between free non-unital rings. By Theorem 4.5.3, R^* is then the coequalizer of the images of those maps between the corresponding free unital rings. From this description we see that $R^* \cong \mathbb{Z} \oplus R$ and can easily guess and verify that the addition is defined componentwise, while the multiplication in R^* is defined by the "graded multiplication" formula:

$$(n, r) \cdot (n', r') := (nn', n'r + nr' + rr').$$

Exercises.

EXERCISE 4.6.i. Give a direct proof of Lemma 4.6.2 along the lines of the argument used in Proposition 3.3.8.

EXERCISE 4.6.ii. Use Theorem 4.6.3 to prove that the inclusion Haus $\hookrightarrow$ Top of the full subcategory of Hausdorff spaces into the category of all spaces has a left adjoint. The left adjoint carries a space to its "largest Hausdorff quotient." Conclude, by applying Proposition 4.5.15, that the category of Hausdorff spaces, as a reflective subcategory of a complete and cocomplete category, is again complete and cocomplete.

EXERCISE 4.6.iii. Suppose C is a locally small category with coproducts. Show that a functor $F : \mathsf{C} \to \mathsf{Set}$ is representable if and only if it admits a left adjoint.

CHAPTER 5

Monads and their Algebras

A monad is just a monoid in the category of
endofunctors, what's the problem?

Philip Walder, fictional attribution by James Iry
[Iry09]

Consider an adjunction

$$C \underset{U}{\overset{F}{\underset{\perp}{\rightleftarrows}}} D \qquad \eta\colon 1_C \Rightarrow UF, \quad \epsilon\colon FU \Rightarrow 1_D$$

from the perspective of the category C ("home"), in ignorance of the category D ("abroad"). From C, only part of the data of the adjunction is visible: the composite endofunctor $UF\colon C \to C$ but neither U nor F individually, and the unit natural transformation but only a whiskered version of the counit. In this chapter, we study the "shadow" cast by an adjunction on the codomain of its right adjoint, the data of which forms a *monad* on C. Monads are interesting in their own right but additionally so because, in unexpectedly common situations, the category D can be recovered from the monad on C underlying the adjunction. When this is the case, objects of D can be represented as *algebras* for the monad on C. In general, a monad can be thought of as a syntactic representation of algebraic structure that is potentially borne by objects in the category on which it acts.[1]

The general definition of a monad is introduced in §5.1 together with a range of examples. As hinted here, any adjunction gives rise to a monad. The converse question, of constructing an adjunction that presents a given monad, is investigated in §5.2, where two universal solutions to this problem are described: the *Kleisli* and *Eilenberg–Moore* categories, which can be constructed for any monad, also arise naturally in several areas of mathematics.

The Eilenberg–Moore category, also called the *category of algebras* for the monad, is our focus for the remainder of the chapter. In §5.3, we pose the question of identifying whether a right adjoint functor $U\colon D \to C$ is equivalent to the forgetful functor from the category of algebras of its monad, in which case the functor is called *monadic*. When D is equivalent to the category of algebras, there are a number of pleasing consequences. The first consequence, proven in §5.4, is that every *algebra* admits a canonical presentation as a quotient of *free algebras*. This construction generalizes the presentation of a group via generators and relations.

The *monadicity theorem*, proven in §5.5, establishes necessary and sufficient conditions for a right adjoint to be monadic. The concluding §5.6 explains our interest in recognizing

---

[1]Dually, the "shadow" cast by an adjunction on the codomain of its left adjoint defines a *comonad*, encoding coalgebraic structure, examples of which are somewhat less common. So that the exposition aligns with the most prominent examples, this dual theory will largely remain unacknowledged.

categories of algebras: this characterization has significant implications for the construction of limits and colimits, which were hinted at in Chapters 3 and 4. For example, monoids, groups, rings, modules over a fixed ring, sets with an action of a fixed group, pointed sets, compact Hausdorff spaces, lattices, and so on all arise as algebras for a monad on the category of sets. Fields do not, which explains why the category of fields shares few of the properties that are common to the categories just described.

5.1. Monads from adjunctions

The abstract definition of a monad is relatively simple to state:

DEFINITION 5.1.1. A **monad** on a category C consists of

- an endofunctor $T : C \to C$,
- a **unit** natural transformation $\eta : 1_C \Rightarrow T$, and
- a **multiplication** natural transformation $\mu : T^2 \Rightarrow T$,

so that the following diagrams commute in C^C:

$$
\begin{array}{ccc}
T^3 & \xrightarrow{T\mu} & T^2 \\
{\scriptstyle \mu T}\big\downarrow & & \big\downarrow{\scriptstyle \mu} \\
T^2 & \xrightarrow[\mu]{} & T
\end{array}
\qquad
\begin{array}{ccc}
T & \overset{\eta T}{\Longrightarrow} T^2 \overset{T\eta}{\Longleftarrow} & T \\
& {\scriptstyle 1_T}\searrow \ \big\downarrow{\scriptstyle \mu}\ \swarrow{\scriptstyle 1_T} & \\
& T &
\end{array}
$$

REMARK 5.1.2. The diagrams in Definition 5.1.1 are reminiscent of the diagrams in Definition 1.6.3. This is no coincidence. Monads, like (topological) monoids, unital rings, and $\Bbbk$-algebras, are all instances of *monoids in a monoidal category*. A **monoidal category** V is a category equipped with a binary functor $V \times V \to V$ and a unit object, together with some additional coherence natural isomorphisms satisfying conditions that are described in §E.2. A monad on C is precisely a monoid in the monoidal category C^C of endofunctors on C, where the binary functor $C^C \times C^C \to C^C$ is composition, and the unit object is the identity endofunctor $1_C \in C^C$. The category of endofunctors is a *strict* monoidal category, in which the coherence natural isomorphisms can be taken to be identities. Nonetheless, the notion of monoid in a monoidal category is rather more complicated than the notion of monad on a category, which is the point of the following joke from Iry's "A Brief, Incomplete, and Mostly Wrong History of Programming Languages" [**Iry09**]:

> 1990—A committee formed by Simon Peyton-Jones, Paul Hudak, Philip Wadler, ... creates Haskell, a pure, non-strict, functional language. Haskell gets some resistance due to the complexity of using monads to control side effects. Wadler tries to appease critics by explaining that "a monad is a monoid in the category of endofunctors, what's the problem?"

One source of intuition is that a monad is the "shadow" cast by an adjunction on the category appearing as the codomain of the right adjoint. Consider the adjunction

$$
\mathsf{Set} \underset{U}{\overset{F}{\rightleftarrows}} \mathsf{Ab} \qquad \eta : 1_{\mathsf{Set}} \Rightarrow UF, \quad \epsilon : FU \Rightarrow 1_{\mathsf{Ab}},
$$

and suppose we have forgotten entirely about the category of abelian groups. What structure remains visible on the category of sets? First, there is an endofunctor, UF, which sends a set to the set of finite formal sums of elements with integer coefficients. There is also the natural transformation η, the "unit map" whose component $\eta_S : S \to UFS$ sends an element of the set S to the corresponding singleton sum. By contrast, the "evaluation map" ϵ, whose component $\epsilon_A \colon FUA \to A$ is the group homomorphism that evaluates a finite formal sum of elements of the abelian group A to its actual sum in A, is not directly visible. There is, however, a related natural transformation $U\epsilon F_S \colon UFUFS \to UFS$ between endofunctors of Set, which is the evaluation map regarded as a function, not a group homomorphism, and considered only in the special case of free groups.

For any adjunction, this triple of data defines a monad:

LEMMA 5.1.3. *Any adjunction*

$$ \mathsf{C} \xrightarrow[\;U\;]{\overset{F}{\underset{\perp}{\longrightarrow}}} \mathsf{D} \qquad\qquad \eta \colon 1_{\mathsf{C}} \Rightarrow UF, \quad \epsilon \colon FU \Rightarrow 1_{\mathsf{D}} $$

gives rise to a monad on the category C *serving as the domain of the left adjoint, with*

- *the endofunctor* T *defined to be* UF,
- *the unit* $\eta \colon 1_{\mathsf{C}} \Rightarrow UF$ *serving as the unit* $\eta \colon 1_{\mathsf{C}} \Rightarrow T$ *for the monad, and*
- *the whiskered counit* $U\epsilon F \colon UFUF \Rightarrow UF$ *serving as the multiplication* $\mu \colon T^2 \Rightarrow T$ *for the monad.*

PROOF. The triangles

$$ UF \overset{\eta UF}{\underset{}{\Longrightarrow}} UFUF \overset{UF\eta}{\underset{}{\Longleftarrow}} UF \qquad\qquad\qquad UFUFUF \overset{UFU\epsilon F}{\Longrightarrow} UFUF $$

commute by the triangle identities for the adjunction. The square commutes by naturality of the vertical natural transformation $U\epsilon \colon UFU \Rightarrow U$.[2]   □

EXAMPLE 5.1.4. The following monads come from familiar adjunctions:

(i) The free ⊣ forgetful adjunction between pointed sets and ordinary sets of Example 4.1.10(i) induces a monad on Set whose endofunctor $(-)_+ \colon$ Set $\to$ Set adds a new disjoint point. The components of the unit are given by the obvious natural inclusions $\eta_A \colon A \to A_+$. The components of the multiplication $\mu_A \colon (A_+)_+ \to A_+$ are defined to be the identity on the subset A and to send the two new points in $(A_+)_+$ to the new point in A_+. By Lemma 5.1.3, or by a direct verification, the diagrams

$$
\begin{array}{ccc}
((A_+)_+)_+ & \overset{(\mu_A)_+}{\longrightarrow} & (A_+)_+ \\
{\scriptstyle \mu_{A_+}}\downarrow & & \downarrow{\scriptstyle \mu_A} \\
(A_+)_+ & \underset{\mu_A}{\longrightarrow} & A_+
\end{array}
\qquad\qquad
\begin{array}{ccc}
A_+ \xrightarrow{\eta_{A_+}} (A_+)_+ \xleftarrow{(\eta_A)_+} A_+ \\
{\scriptstyle 1_{A_+}}\searrow \quad \downarrow{\scriptstyle \mu_A} \quad \swarrow{\scriptstyle 1_{A_+}} \\
A_+
\end{array}
$$

commute. Particularly in computer science, this monad is called the **maybe monad**, for reasons that are explained in Example 5.2.10(i). There is a similar monad on Top, or any category with coproducts, which acts by adjoining a copy of a fixed object (in this case a point).

---

[2] Note that the two composites define the horizontal composite natural transformation (see Lemma 1.7.4):

$$ \mathsf{C} \xrightarrow{F} \mathsf{D} \overset{U \nearrow \overset{\mathsf{C}}{} \searrow F \; \Downarrow\epsilon}{=\!=\!=} \mathsf{D} \overset{U \nearrow \overset{\mathsf{C}}{} \searrow F \; \Downarrow\epsilon}{=\!=\!=} \mathsf{D} \xrightarrow{U} \mathsf{C} $$

(ii) The **free monoid monad** is induced by the free ⊣ forgetful adjunction between monoids and sets of Example 4.1.10(ii). The endofunctor T: Set → Set is defined by

$$TA := \coprod_{n \geq 0} A^n,$$

that is, TA is the set of finite lists of elements in A; in computer science contexts, this monad is often called the **list monad**. The components of the unit $\eta_A : A \to TA$ are defined by the evident coproduct inclusions, sending each element of A to the corresponding singleton list. The components of the multiplication $\mu_A : T^2A \to TA$ are the concatenation functions, sending a list of lists to the composite list. A categorical description of this map is given in Exercise 5.1.i, which demonstrates that the free monoid monad can also be defined in any monoidal category with coproducts that distribute over the monoidal product.

(iii) The free ⊣ forgetful adjunction between sets and the category of R-modules of Example 4.1.10(v) induces the **free R-module monad** $R[-]$: Set → Set. Define $R[A]$ to be the set of finite formal R-linear combinations of elements of A. Formally, a finite R-linear combination is a finitely supported function $\chi : A \to R$, meaning a function for which only finitely many elements of its domain take non-zero values. Such a function might be written as $\sum_{a \in A} \chi(a) \cdot a$. The components $\eta_A : A \to R[A]$ of the unit send an element $a \in A$ to the singleton formal R-linear combination corresponding to the function $\chi_a : A \to R$ that sends a to $1 \in R$ and every other element to zero. The components $\mu_A : R[R[A]] \to R[A]$ of the multiplication are defined by distributing the coefficients in a formal sum of formal sums. Special cases of interest include the **free abelian group monad** and the **free vector space monad**.

(iv) The free ⊣ forgetful adjunction between sets and groups induces the **free group monad** F: Set → Set that sends a set A to the set $F(A)$ of finite words in the letters $a \in A$ together with formal inverses a^{-1}.

(v) The composite adjunction

$$\mathsf{cHaus} \xrightarrow[\underset{\beta}{\quad\perp\quad}]{} \mathsf{Top} \xrightarrow[\underset{U}{\quad\perp\quad}]{D} \mathsf{Set}$$

induces a monad β: Set → Set that sends a set to the underlying set of the Stone–Čech compactification of the discrete space on that set. There is a simpler description: $\beta(A)$ is the set of ultrafilters on A. An **ultrafilter** is a set of subsets of A that is upward closed, closed under finite intersections, and for each subset of A contains either that subset or its complement but not both; see Exercise 5.1.ii.

(vi) The adjunction

$$\mathsf{DirGraph} \xrightarrow[\underset{U}{\quad\perp\quad}]{F} \mathsf{Cat}$$

of Example 4.1.13 induces a monad on DirGraph that carries a directed graph G to the graph with the same vertices but whose edges are finite (directed) paths of edges in G. This is the underlying directed graph of the free category on G. The graph homomorphism $\eta_G : G \to UF(G)$ includes G as the subgraph of "unary" edges.

(vii) By Example 4.3.8, the contravariant power set functor is its own mutual right adjoint:

$$\mathsf{Set} \xrightarrow[\underset{P}{\quad\perp\quad}]{P} \mathsf{Set}^{\mathrm{op}} \qquad\qquad \mathsf{Set}(A, PB) \cong \mathsf{Set}(B, PA)$$

A function from A to the power set of B, or equally a function from B to the power set of A, can be encoded as a function $A \times B \rightarrow \Omega$, i.e., as a subset of $A \times B$. The induced **double power set monad** takes a set A to P^2A. The components of the unit are the "principle ultrafilter" functions $\eta_A : A \rightarrow P^2A$ that send an element a to the set of subsets of A that contain a. The components of the multiplication take a set of sets of sets of subsets to the set of subsets of A with the property that one of the sets of sets of subsets is the set of all sets of subsets of A that include that particular subset as an element.[3] A similar monad can be defined with any other set in place of the two-element set Ω; in computer science contexts, these are called **continuation monads**. This construction can also be generalized to other cartesian closed categories. For instance, there is a similar **double dual monad** on $\mathsf{Vect}_{\Bbbk}$.

EXAMPLE 5.1.5. Monads also arise in nature:

(i) The covariant power set functor $P \colon \mathsf{Set} \rightarrow \mathsf{Set}$ is also a monad. The unit $\eta_A \colon A \rightarrow PA$ sends an element to the singleton subset. The multiplication $\mu_A \colon P^2A \rightarrow PA$ takes the union of a set of subsets. Naturality of the unit with respect to a function $f \colon A \rightarrow B$ makes use of the fact that $f_* \colon PA \rightarrow PB$ is the direct image (rather than inverse image) function. Naturality of the multiplication maps makes use of Corollary 4.5.4; the direct image function, as a left adjoint, preserves unions.

$$
\begin{array}{ccc}
A & \xrightarrow{\;\eta_A\;} & PA \\
{\scriptstyle f}\downarrow & & \downarrow{\scriptstyle f_*} \\
B & \xrightarrow[\;\eta_B\;]{} & PB
\end{array}
\qquad\qquad
\begin{array}{ccc}
P^2A & \xrightarrow{\;\mu_A\;} & PA \\
{\scriptstyle (f_*)_*}\downarrow & & \downarrow{\scriptstyle f_*} \\
P^2B & \xrightarrow[\;\mu_B\;]{} & PB
\end{array}
$$

(ii) A modification of the free monoid monad yields the **free commutative monoid monad** defined by

$$
TA := \coprod_{n \geq 0} (A^{\times n})_{/\Sigma_n} .
$$

Elements are finite unordered lists of elements, or equally finitely supported functions $A \rightarrow \mathbb{N}$.

(iii) There is a monad $- \times \mathbb{N} \colon \mathsf{Set} \rightarrow \mathsf{Set}$. We might think of the second component of an element $(a, n) \in A \times \mathbb{N}$ as a discrete time variable. The unit $A \rightarrow A \times \mathbb{N}$ is defined by $a \mapsto (a, 0)$ and the multiplication $A \times \mathbb{N} \times \mathbb{N} \rightarrow A \times \mathbb{N}$ is defined by $(a, m, n) \mapsto (a, m+n)$. A similar monad exists with $(\mathbb{N}, +, 0)$ replaced by any monoid, and this example can be generalized further from Set to any monoidal category.

(iv) The **Giry monad** acts on the category Meas of measurable spaces and measurable functions. A **measurable space** is a set equipped with a σ-*algebra* of "measurable" subsets, and a function $f \colon A \rightarrow B$ between measurable spaces is **measurable** if the preimage of any measurable subset of B is a measurable subset of A. The Giry monad sends a measurable space A to the measurable space $\mathrm{Prob}(A)$ of *probability measures* on A, equipped with the smallest σ-algebra so that, for each measurable subset $X \subset A$, the evaluation function $\mathrm{ev}_X \colon \mathrm{Prob}(A) \rightarrow I$ is measurable. The unit is the measurable function $\eta_A \colon A \rightarrow \mathrm{Prob}(A)$ that sends each element $a \in A$ to the Dirac measure, which assigns a subset the probability 1 if it contains a and 0 otherwise. The multiplication is defined using integration; see [**Gir82**] or [**Ave16**] for details.

Briefly:

---

[3]This is one of those instances where it is easier to speak mathematics than to speak English: the multiplication is the inverse image function for the map $\eta_{PA} \colon PA \rightarrow P^3A$.

DEFINITION 5.1.6. A **comonad** on C is a monad on C^{op}: explicitly, a comonad consists of an endofunctor $K\colon C \to C$ together with natural transformations $\epsilon\colon K \Rightarrow 1_C$ and $\delta\colon K \Rightarrow K^2$ so that the diagrams dual to Definition 5.1.1 commute in C^C.

A comonad is a comonoid in the category of endofunctors of C. By the dual to Lemma 5.1.3, any adjunction induces a comonad on the domain of its right adjoint.

EXAMPLE 5.1.7. A monad on a preorder $(P, \leq)$ is given by an order-preserving function $T\colon P \to P$ that is so that $p \leq Tp$ and $T^2 p \leq Tp$. If P is a poset, so that isomorphic objects are equal, these two conditions imply that $T^2 p = Tp$. An order-preserving function T so that $p \leq Tp$ and $T^2 p = Tp$ is called a **closure operator**. Dually, a comonad on a poset category $(P, \leq)$ defines a **kernel operator**: an order-preserving function K so that $Kp \leq p$ and $Kp = K^2 p$.

For example, the poset PX of subsets of a topological space X admits a closure operator $TA = \overline{A}$, where $\overline{A}$ is the closure of $A \subset X$, and a kernel operator $KA = A^\circ$, where A° is the interior of $A \subset X$.[4]

Exercises.

EXERCISE 5.1.i. Suppose V is a monoidal category (see §E.2), i.e., a category with a bifunctor $\otimes\colon V \times V \to V$ that is associative up to coherent natural isomorphism[5] and a unit object $* \in V$ with natural isomorphisms $v \otimes * \cong v \cong * \otimes v$. Suppose also that V has finite coproducts and that the bifunctor $\otimes$ preserves them in each variable.[6] Show that $T(X) = \coprod_{n \geq 0} X^{\otimes n}$ defines a monad on V by defining natural transformations $\eta\colon 1_V \Rightarrow T$ and $\mu\colon T^2 \Rightarrow T$ that satisfy the required conditions.

EXERCISE 5.1.ii. Show that the functor $\beta\colon \text{Set} \to \text{Set}$ that carries a set to the set of ultrafilters on that set is a monad by defining unit and multiplication natural transformations that satisfy the unit and associativity laws. (Hint: The ultrafilter monad is a submonad of the double power set monad $P^2\colon \text{Set} \to \text{Set}$.)

EXERCISE 5.1.iii. The adjunction associated to a reflective subcategory of C induces an **idempotent monad** on C. Prove that the following three characterizations of an idempotent monad (T, η, μ) are equivalent:

(i) The multiplication $\mu\colon T^2 \Rightarrow T$ is a natural isomorphism (hence, the appellation "idempotent").
(ii) The natural transformations $\eta T, T\eta\colon T \Rightarrow T^2$ are equal.
(iii) Each component of $\mu\colon T^2 \Rightarrow T$ is a monomorphism.

5.2. Adjunctions from monads

We have seen, in Lemma 5.1.3, that any adjunction presents a monad on the category serving as the domain of its left adjoint. A natural question is whether all monads arise this way. For instance, is there any adjunction that gives rise to the power set monad on Set? Perhaps surprisingly, the answer is yes: any monad T on a category C can be recovered from two (typically distinct) adjunctions that are moreover universal with this property. The initial such adjunction, in a category to be introduced shortly, is between C and the **Kleisli**

---

[4]The **closure** of A is the smallest closed set containing A, equally the intersection of all closed sets containing A; the **interior** is defined dually.

[5]Rather than worry about what this means, feel free to assume that there is a well-defined n-ary functor $V^{\times n} \xrightarrow{\otimes^n} V$ built from the bifunctor $\otimes$.

[6]In particular, $(v \sqcup v') \otimes (w \sqcup w') \cong v \otimes w \sqcup v' \otimes w \sqcup v \otimes w' \sqcup v' \otimes w'$.

category C_T of T. The terminal such adjunction is between C and the **Eilenberg–Moore category** C^T of T, also called the category of T-**algebras**.

The sense in which a monad defines a syntactic encoding of a particular variety of algebraic structure on C is explained by the construction of the category of algebras, the intuition for which we develop in an example. Recall that an affine space, over a fixed field $\Bbbk$, is like a "vector space that has forgotten its origin." An n-dimensional affine space is a set A equipped with a translation automorphism for each n-dimensional vector. This action is simply transitive in the sense that any two points in affine space can be "subtracted" to obtain a unique vector. Choosing an origin $\mathbf{o} \in A$ defines a bijection between points $\mathbf{a} \in A$ in the affine space and vectors $\vec{\mathbf{v}} = \mathbf{a} - \mathbf{o} \in \Bbbk^n$. In summary:

DEFINITION 5.2.1. Given a vector space V over a field $\Bbbk$, an **affine space** is a non-empty set A together with a "translation" function $V \times A \xrightarrow{+} A$ so that

- $\vec{0} + \mathbf{a} = \mathbf{a}$ for all $\mathbf{a} \in A$;
- $(\vec{\mathbf{v}} + \vec{\mathbf{w}}) + \mathbf{a} = \vec{\mathbf{v}} + (\vec{\mathbf{w}} + \mathbf{a})$ for all $\vec{\mathbf{v}}, \vec{\mathbf{w}} \in V$ and $\mathbf{a} \in A$; and
- for any $\mathbf{a} \in A$, the function $- + \mathbf{a} \colon V \to A$ is a bijection.

There is an equivalent definition that allows us to define an affine space without making use of the auxiliary vector space V. If we temporarily fix an origin $\mathbf{o} \in A$, then for any pair of elements $\mathbf{a}, \mathbf{b} \in A$ of the affine space and any scalar $\lambda \in \Bbbk$, we can exploit the bijection $- + \mathbf{o} \colon V \to A$ to see that there is a unique $\mathbf{c} \in A$ so that

$$\mathbf{c} - \mathbf{o} = \lambda(\mathbf{a} - \mathbf{o}) + (1 - \lambda)(\mathbf{b} - \mathbf{o}).$$

This element $\mathbf{c}$ is sensibly denoted by $\lambda\mathbf{a} + (1 - \lambda)\mathbf{b}$ and is independent of the choice of origin. More generally, for any n-tuple $\mathbf{a}_1, \ldots, \mathbf{a}_n \in A$ and scalars $\lambda_1, \ldots, \lambda_n \in \Bbbk$ with $\lambda_1 + \cdots + \lambda_n = 1$, there is a unique element $\lambda_1\mathbf{a}_1 + \cdots + \lambda_n\mathbf{a}_n \in A$ defined analogously as an **affine linear combination** of $\mathbf{a}_i$. This leads to a second equivalent definition of affine space.

DEFINITION 5.2.2. An **affine space** is a non-empty set A in which affine linear combinations can be evaluated.

To make this precise, define $\mathrm{Aff}_\Bbbk(A)$ to be the set of finite formal affine linear combinations $\sum_{i=1}^{n} \lambda_i \mathbf{a}_i$ so that $\sum_{i=1}^{n} \lambda_i = 1$; a sum is **affine** precisely when the coefficients sum to $1 \in \Bbbk$. To say "these can be evaluated" means there is a function $\mathrm{ev}_A \colon \mathrm{Aff}_\Bbbk(A) \to A$. For this map to define a reasonable evaluation function, a few axioms are required:

- If $\eta_A \colon A \to \mathrm{Aff}_\Bbbk(A)$ is the "singleton" function and $\mu_A \colon \mathrm{Aff}_\Bbbk(\mathrm{Aff}_\Bbbk(A)) \to \mathrm{Aff}_\Bbbk(A)$ is the "distributivity" function, then the following diagrams

$$
\begin{array}{ccc}
A \xrightarrow{\;\eta_A\;} \mathrm{Aff}_\Bbbk(A) & \quad & \mathrm{Aff}_\Bbbk(\mathrm{Aff}_\Bbbk(A)) \xrightarrow{\;\mu_A\;} \mathrm{Aff}_\Bbbk(A) \\
{\scriptstyle 1_A}\searrow \quad \downarrow{\scriptstyle \mathrm{ev}_A} & \quad & {\scriptstyle \mathrm{Aff}_\Bbbk(\mathrm{ev}_A)}\downarrow \qquad\qquad \downarrow{\scriptstyle \mathrm{ev}_A} \\
\qquad A & \quad & \mathrm{Aff}_\Bbbk(A) \xrightarrow[\;\mathrm{ev}_A\;]{} A
\end{array}
$$

commute in Set.

The first condition says that the value of a singleton sum $1 \cdot \mathbf{a}$ is the element $\mathbf{a}$. The second condition says that an affine linear combination of affine linear combinations

$$\lambda_1 \cdot (\mu_{11}\mathbf{a}_{11} + \cdots + \mu_{1n_1}\mathbf{a}_{1n_1}) + \cdots + \lambda_k \cdot (\mu_{k1}\mathbf{a}_{k1} + \cdots + \mu_{kn_k}\mathbf{a}_{kn_k})$$

can be evaluated by first distributing—note that $\sum_i \sum_j \lambda_i \mu_{ij} = 1$—and then evaluating or by first evaluating inside each of the k sets of parentheses and then evaluating the resulting affine linear combination. In summary, the precise meaning of Definition 5.2.2 is:

DEFINITION 5.2.3. An **affine space** is an **algebra** for the affine linear combination monad $\mathrm{Aff}_{\Bbbk}(-)\colon \mathsf{Set} \to \mathsf{Set}$.

Let us now introduce the general definition.

DEFINITION 5.2.4. Let C be a category with a monad (T, η, μ). The **Eilenberg–Moore category** for T or the **category of T-algebras** is the category C^T whose:

- objects are pairs $(A \in \mathsf{C}, a\colon TA \to A)$, so that the diagrams

(5.2.5)
$$
\begin{array}{ccc}
A \xrightarrow{\ \eta_A\ } TA & & T^2A \xrightarrow{\ \mu_A\ } TA \\
\searrow_{1_A}\ \ \downarrow a & & {\scriptstyle Ta}\downarrow\downarrow a \\
A & & TA \xrightarrow{\ \ a\ \ } A
\end{array}
$$

commute in C, and
- morphisms $f\colon (A,a) \to (B,b)$ are T-algebra **homomorphisms**: maps $f\colon A \to B$ in C so that the square

$$
\begin{array}{ccc}
TA & \xrightarrow{\ Tf\ } & TB \\
{\scriptstyle a}\downarrow & & \downarrow{\scriptstyle b} \\
A & \xrightarrow{\ f\ } & B
\end{array}
$$

commutes, with composition and identities as in C.

For instance, algebras for the affine space monad $\mathrm{Aff}_{\Bbbk}$ on Set are precisely affine spaces in the sense of Definition 5.2.2. This is a representative example: the abstract definition of 5.2.4, which is a priori a bit strange, precisely captures familiar notions of "algebra," of the variety encoded by the monad (T, η, μ).

EXAMPLE 5.2.6. For instance:

(i) Consider the free pointed set monad of Example 5.1.4(i). An algebra is a set A with a map $a\colon A_+ \to A$ so that the diagrams (5.2.5) commute. The square imposes no additional conditions, but the triangle asserts that the map $a\colon A_+ \to A$ restricts to the identity on the A component of the disjoint union $A_+ = A \sqcup \{*\}$. Thus, the data of an algebra is a set with a specified basepoint $a \in A$, the image of the extra point $*$ under the map $a\colon A_+ \to A$. A morphism $f\colon (A,a) \to (B,b)$ is a map $f\colon A \to B$ so that

$$
\begin{array}{ccc}
A_+ & \xrightarrow{\ f_+\ } & B_+ \\
{\scriptstyle a}\downarrow & & \downarrow{\scriptstyle b} \\
A & \xrightarrow{\ f\ } & B
\end{array}
$$

commutes. The map f_+ carries the extra point in A_+ to the extra point in B_+, so this condition demands that $f(a) = b$. In conclusion, the Eilenberg–Moore category is isomorphic to Set_*, the category of pointed sets.

(ii) For any unital ring R, there is a monad $R \otimes_{\mathbb{Z}} -$ on Ab derived from the adjunction

$$
\mathsf{Ab} \underset{U}{\overset{R\otimes_{\mathbb{Z}}-}{\rightleftarrows}} \mathsf{Mod}_R
$$

of Example 4.1.10(xii). An algebra is an abelian group A equipped with a homomorphism $\cdot\colon R\otimes_{\mathbb{Z}} A \to A$ satisfying a pair of axioms. By the universal property of the tensor product, this homomorphism encodes a $\mathbb{Z}$-bilinear map $(r, a) \mapsto r\cdot a\colon R \times A \to A$, which we call "scalar multiplication." The commutative diagrams

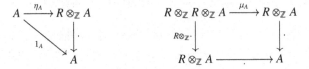

ensure $1 \cdot a = a$ and $r \cdot (r' \cdot a) = (rr') \cdot a$, i.e., that this action is associative and unital. In conclusion, an algebra for the monad $R\otimes_{\mathbb{Z}} -$ on Ab is precisely an R-module.

(iii) An algebra for the free monoid monad of Example 5.1.4(ii) is a set A with a map $\alpha\colon \coprod_{n\geq 0} A^n \to A$, whose component functions $\alpha_n\colon A^n \to A$ specify an n-ary operation on A for each n, satisfying certain conditions. The unit triangle

$$
\begin{array}{ccc}
A & \xrightarrow{\eta_A} & \coprod_{n\geq 0} A^n \\
 & {}_{1_A}\searrow & \downarrow{\scriptstyle\alpha} \\
 & & A
\end{array}
$$

asserts that the unary operation α_1 is the identity. The commutative square

$$
(5.2.7) \qquad
\begin{array}{ccc}
\coprod_{n\geq 0}(\coprod_{m\geq 0} A^m)^n & \xrightarrow{\mu_A} & \coprod_{k\geq 0} A^k \\
{\scriptstyle\coprod_{n\geq 0}\alpha^n}\downarrow & & \downarrow{\scriptstyle\alpha} \\
\coprod_{n\geq 0} A^n & \xrightarrow{\alpha} & A
\end{array}
$$

imposes an associativity condition on the operations. An element of the upper left-hand set is a list

$$((a_{11},\ldots,a_{1m_1}),\ldots,(a_{n1},\ldots,a_{nm_n}))$$

of n varying-length lists of elements of A. Commutativity of (5.2.7) demands that the result of applying the $(m_1 + \cdots + m_n)$-ary operation to the concatenated list is the same as applying the n-ary operation to the results of applying the m_i-ary operations to each sublist.

Recall that a **monoid** is a set A with an associative binary operation $A^2 \to A$ and a unit element $e \in A$. A monoid defines an algebra for the free monoid monad: the nullary operation $1 \to A$ picks out the unit element e, the unary operation $A \to A$ is necessarily the identity, the binary operation $A^2 \to A$ is given, and the operations of higher arity are determined by iterating the binary operation; associativity implies that this is well-defined.

Conversely, any algebra for the free monoid monad defines a monoid, whose unit and binary operations are defined to be the components $\alpha_0\colon 1 \to A$ and $\alpha_2\colon A^2 \to A$ of the map $\alpha\colon \coprod_{n\geq 0} A^n \to A$. The associativity condition (5.2.7) implies in particular that the ternary product of three elements $a_1, a_2, a_3 \in A$ is equal to both of the iterated binary products $(a_1 \cdot a_2) \cdot a_3$ and $a_1 \cdot (a_2 \cdot a_3)$:

$$\alpha_2(\alpha_2(a_1, a_2), a_3) = \alpha_3(a_1, a_2, a_3) = \alpha_2(a_1, \alpha_2(a_2, a_3)).$$

The square (5.2.7) also implies that the composite of the binary operation with the nullary operation in one of its inputs is equal to the identity unary operation.

By Exercise 5.2.i, algebra morphisms correspond precisely to monoid homomorphisms. Thus, we conclude that the category of algebras for the free monoid monad is isomorphic to the category Monoid of associative unital monoids.

(iv) An algebra for the closure closure operator[7] on the poset of subsets of a topological space X is exactly a closed subset of X. Dually, a coalgebra[8] for the interior kernel operator is exactly an open subset.

Various notations are common for the Eilenberg–Moore category, many involving the string "adj" to emphasize the interpretation of its objects as "algebras" of some sort. The notation used here, while less evocative, has the virtue of being concise.

LEMMA 5.2.8. *For any monad (T, η, μ) acting on a category* C, *there is an adjunction*

$$C \underset{U^T}{\overset{F^T}{\rightleftarrows}} C^T$$

between C *and the Eilenberg–Moore category whose induced monad is* (T, η, μ).

PROOF. The functor $U^T : C^T \to C$ is the evident forgetful functor. The functor $F^T : C \to C^T$ carries an object $A \in C$ to the **free T-algebra**

$$F^T A := (TA, \mu_A : T^2 A \to TA)$$

and carries a morphism $f : A \to B$ to the **free T-algebra morphism**

$$F^T f := (TA, \mu_A) \overset{Tf}{\longrightarrow} (TB, \mu_B).$$

Note that $U^T F^T = T$.

The unit of the adjunction $F^T \dashv U^T$ is given by the natural transformation $\eta : 1_C \Rightarrow T$. The components of the counit $\epsilon : F^T U^T \Rightarrow 1_{C^T}$ are defined as follows:

$$\epsilon_{(A,a)} := (TA, \mu_A) \overset{a}{\to} (A, a)$$

$$\begin{array}{ccc} T^2 A & \overset{Ta}{\longrightarrow} & TA \\ \mu_A \downarrow & & \downarrow a \\ TA & \underset{a}{\longrightarrow} & A \end{array}$$

That is, the component of the counit at an algebra $(A, a) \in C^T$ is given by the algebra structure map $a : TA \to A$; the commutative square demonstrates that this map defines a T-algebra homomorphism $a : (TA, \mu_A) \to (A, a)$. Note, in particular, that $U^T \epsilon F^T_A = \mu_A$, so that the monad underlying the adjunction $F^T \dashv U^T$ is (T, η, μ). The straightforward verifications that the unit and counit satisfy the triangle identities are left to Exercise 5.2.ii. □

A second solution to the problem of finding an adjunction that induces a given monad is given by the Kleisli category construction.

DEFINITION 5.2.9. Let C be a category with a monad (T, η, μ). The **Kleisli category** C_T is defined so that

- its objects are the objects of C, and
- a morphism from A to B in C_T, depicted as $A \rightsquigarrow B$, is a morphism $A \to TB$ in C.

Identities and composition are defined using the monad structure:

---

[7]This is not a typo: Example 5.1.7 describes a closure operator on the poset of subsets of a topological space that sends a subset to its closure.

[8]Interpreting Definition 5.2.4 for a monad (T, η, μ) on C^{op} defines the category of **coalgebras** for the comonad on C; see Exercise 5.2.iii.

- The unit $\eta_A \colon A \to TA$ defines the identity morphism $A \rightsquigarrow A \in C_T$.
- The composite of a morphism $f \colon A \to TB$ from A to B with a morphism $g \colon B \to TC$ from B to C is defined to be

$$A \xrightarrow{\ f\ } TB \xrightarrow{\ Tg\ } T^2C \xrightarrow{\ \mu_C\ } TC.$$

The verification that these operations are associative and unital is left as Exercise 5.2.iv.

EXAMPLE 5.2.10. For instance:

(i) Objects in the Kleisli category for the maybe monad on Set are sets. A map $A \rightsquigarrow B$ in the Kleisli category is a function $A \to B_+$, which may be thought of as a partially-defined function from A to B: the elements of A that are sent to the free basepoint have "undefined" output. The composite of two partial functions is the maximal partially-defined function. Thus, the Kleisli category is the category Set^∂ of sets and partially-defined functions.

(ii) Objects in the Kleisli category for the free monoid monad on Set are sets and a map from A to B is a function $A \to \coprod_{n \geq 0} B^n$. For each input element $a \in A$, the output is an element $(b_1, \ldots, b_k) \in \coprod_{n \geq 0} B^n$, i.e., a list of elements of B.

(iii) For a fixed set S of "states," the adjunction $S \times - \dashv (-)^S$ induces a monad $(S \times -)^S$ on Set. A map $A \rightsquigarrow B$ in the Kleisli category is, by adjunction, a function $f \colon A \times S \to B \times S$ that takes an input $a \in A$ together with the current state $s \in S$ and returns an output $f_s(a)$, dependent on both the input and the state, together with a new "updated" state $s' \in S$. Monads of this form make an appearance in certain functional programming languages such as Haskell, where the updated state is interpreted as a "side effect" of the computation f.

(iv) The Kleisli category for the Giry monad of Example 5.1.5(iv) is the category of measurable spaces and **Markov kernels**. For instance, a finite set defines a measurable space with discrete σ-algebra. An endomorphism of this object in the Kleisli category is a **discrete time Markov chain**.

LEMMA 5.2.11. *For any monad (T, η, μ) acting on a category C, there is an adjunction*

$$C \underset{U_T}{\overset{F_T}{\underset{\perp}{\rightleftarrows}}} C_T$$

between C and the Kleisli category whose induced monad is (T, η, μ).

PROOF. The functor F_T is the identity on objects and carries a morphism $f \colon A \to B$ in C to the morphism $A \rightsquigarrow B$ in C_T defined by

$$F_t f := A \xrightarrow{f} B \xrightarrow{\eta_B} TB.$$

The functor U_T sends an object $A \in C_T$ to $TA \in C$ and sends a morphism $A \rightsquigarrow B$ represented by $g \colon A \to TB$ to

$$U_T g := TA \xrightarrow{Tg} T^2 B \xrightarrow{\mu_B} TB.$$

Elementary diagram chases, left as Exercise 5.2.v, demonstrate that both mappings are functorial. Note in particular that $U_T F_T = T$. From the definition of the hom-sets in C_T, there are isomorphisms

$$C_T(F_T A, B) \cong C(A, TB) \cong C(A, U_T B),$$

which are natural in both variables. This establishes the adjunction $F_T \dashv U_T$. $\square$

For any monad (T, η, μ) on C, there is a category Adj_T whose objects are fully-specified adjunctions

$$\mathsf{C} \underset{U}{\overset{F}{\rightleftarrows}} \mathsf{D} \qquad\qquad \eta: 1_\mathsf{C} \Rightarrow UF, \quad \epsilon: FU \Rightarrow 1_\mathsf{D}$$

inducing the monad (T, η, μ) on C. A morphism

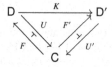

from $F \dashv U$ to $F' \dashv U'$ is a functor $K: \mathsf{D} \to \mathsf{D}'$ commuting with both the left and right adjoints, i.e., so that $KF = F'$ and $U'K = U$. Because the units of $F \dashv U$ and $F' \dashv U'$ are the same, by Exercise 4.2.v the commutativity of K with the right and left adjoints implies further that

(i) the whiskered composites $K\epsilon = \epsilon'K$ of K with the counits of $F \dashv U$ and $F' \dashv U'$ coincide, and

(ii) K carries the transpose in D of a morphism $c \to Ud = U'Kd$ to the transpose of this morphism in D'.

PROPOSITION 5.2.12. *The Kleisli category C_T is initial in Adj_T and the Eilenberg–Moore category C^T is terminal. That is, for any adjunction $F \dashv U$ inducing the monad (T, η, μ) on C, there exist unique functors*

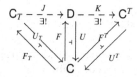

commuting with the left and right adjoints.

Because the functors J and K define adjunction morphisms, they must commute with the counits of the adjunctions and preserve transposition, facts that can be used to deduce the correct definitions of these functors.

PROOF. We argue that there exist unique adjunction morphisms from the Kleisli adjunction to $F \dashv U$ and from $F \dashv U$ to the Eilenberg–Moore adjunction. Write $\epsilon: FU \Rightarrow 1_\mathsf{D}$ for the counit of $F \dashv U$. The hypothesis that this adjunction induces the monad (T, η, μ) on C implies that $U\epsilon F = \mu$.

On objects $c \in \mathsf{C}_T$, we are forced to define $Jc = Fc$ so that $JF_T = F$. A morphism $f: c \rightsquigarrow c'$ in C_T is the transpose, under $F_t \dashv U_T$, of the representing morphism $f: c \to Tc' = U_Tc'$ in C, so the requirement that J commutes with transposition forces us to define Jf to be the transpose

$$Jf := Fc \xrightarrow{Ff} FUFc' \xrightarrow{\epsilon F_{c'}} Fc'$$

of f under $F \dashv U$. The proof that these definitions are functorial, left to Exercise 5.2.vi, makes use of the fact that $U\epsilon F = \mu$.

To define the functor K on an object $d \in \mathsf{D}$ so that $U^T K = U$, we seek a suitable T-algebra structure for the object $Ud \in \mathsf{C}$. For any algebra $(c, \gamma: Tc \to c) \in \mathsf{C}^T$, the algebra structure map γ appears as the component of the counit of $F^T \dashv U^T$ at the object (c, γ); see the proof of Lemma 5.2.8. Thus, γ can be recognized as the morphism $\gamma: (Tc, \mu_c) \to (c, \gamma)$

from the free T-algebra on c to the given T-algebra that is the transpose of the identity on $c = U^T(c, \gamma)$. The fact that K preserves transposes now tells us that we should define $Kd = (Ud, U\epsilon_d)$; the proof that $U\epsilon_d \colon UFUd \to Ud$ is an algebra structure map makes use of the fact that $U\epsilon F = \mu$. On morphisms, the condition $U^T K = U$ forces us to define the image of $f \colon d \to d'$ to be $Uf \colon (Ud, U\epsilon_d) \to (Ud', U\epsilon_{d'})$, and indeed this is a morphism of algebras. Functoriality in this case is obvious. ☐

Proposition 5.2.12 implies, in particular, that there is a unique functor from the Kleisli category for any monad to the Eilenberg–Moore category that commutes with the free and forgetful functors from and to the underlying category. The following result characterizes its image.

LEMMA 5.2.13. *Let (T, η, μ) be a monad on C. The canonical functor $K \colon \mathsf{C}_T \to \mathsf{C}^T$ from the Kleisli category to the Eilenberg–Moore category is full and faithful and its image consists of the free T-algebras.*

For ease of future reference, recall:

DEFINITION 5.2.14. A **free T-algebra** is an object

$$F^T c := (Tc, \mu_c \colon T^2 c \to Tc)$$

in the category of algebras that is in the image of the free functor $F^T \colon \mathsf{C} \to \mathsf{C}^T$.

PROOF. The proof of Proposition 5.2.12 supplies the definition of the functor on objects: $Kc := (Tc, \mu_c)$, Tc being the object $U_T c$ and μ_c being U_T of the component of the counit of the Kleisli adjunction at the object $c \in \mathsf{C}_T$.

Applying Exercise 4.2.v, for each pair $c, c' \in \mathsf{C}_T$, the action of the functor K on the hom-set from c to c'

$$\mathsf{C}_T(c, c') \xrightarrow{\ K\ } \mathsf{C}^T(Kc, Kc') = \mathsf{C}^T((Tc, \mu_c), (Tc', \mu_{c'}))$$

commutes with the transposition natural isomorphisms identifying both of these hom-sets with $\mathsf{C}(c, Tc')$. In particular, this map must also be an isomorphism, demonstrating that the functor $K \colon \mathsf{C}_T \to \mathsf{C}^T$ is full and faithful. ☐

The upshot of Lemma 5.2.13 is that the Kleisli category for a monad embeds as the full subcategory of free T-algebras and all maps between such. Lemma 5.2.13 also tells us precisely when the Kleisli and Eilenberg–Moore categories are equivalent: this is the case when all algebras are free. For instance, all vector spaces are free modules over any chosen basis, so the Kleisli and Eilenberg–Moore categories for the free vector space monad are equivalent. Another equivalence of this form, for the maybe monad on Set, appears in disguise in Example 1.5.6.

Exercises.

EXERCISE 5.2.i. Pick your favorite monad on Set induced from a free ⊣ forgetful adjunction involving some variety of algebraic structure borne by sets, identify its algebras, and show that algebra morphisms are precisely the homomorphisms, in the appropriate sense.

EXERCISE 5.2.ii. Fill in the remaining details in the proof of Lemma 5.2.8 to show that the free and forgetful functors relating a category C with a monad T to the category C^T of T-algebras are adjoints, inducing the given monad T.

EXERCISE 5.2.iii. Dualize Definition 5.2.4 and Lemma 5.2.8 to define the category of coalgebras for a comonad together with its associated forgetful–cofree adjunction.

EXERCISE 5.2.iv. Verify that the Kleisli category is a category by checking that the composition operation of Definition 5.2.9 is associative and unital.

EXERCISE 5.2.v. Verify that the assignments F_T and U_T defined in the proof of Lemma 5.2.11 are functorial.

EXERCISE 5.2.vi. Fill in the remaining details in the proof of Proposition 5.2.12: verify the functoriality of J and K and show that the whiskered composites of the counits of these three adjunctions with K and J agree.

EXERCISE 5.2.vii. From the canonical functor $K\colon \mathsf{C}_T \to \mathsf{C}^T$ of Lemma 5.2.13 that embeds the Kleisli category into the Eilenberg–Moore category for a monad T on C, one can define a "restricted Yoneda" functor

$$\mathsf{C}^T \xrightarrow{\ \mathsf{C}^T(K-,-)\ } \mathsf{Set}^{\mathsf{C}_T^{\mathrm{op}}}$$

$$(A,\alpha) \quad \mapsto \quad \mathsf{C}^T(K-,(A,a))$$

that constructs a presheaf on the Kleisli category for each T-algebra (A,a). Show that the presheaves in the image of this functor restrict along $F_T\colon \mathsf{C} \to \mathsf{C}_T$ to representable functors in $\mathsf{Set}^{\mathsf{C}^{\mathrm{op}}}$. In fact, this functor defines an isomorphism between the Eilenberg–Moore category and the full subcategory of presheaves on the Kleisli category that restrict along F_T to representable functors.

5.3. Monadic functors

The free $\dashv$ forgetful adjunction

$$\mathsf{Set} \underset{U}{\overset{F}{\underset{\perp}{\rightleftarrows}}} \mathsf{Ab}$$

between sets and abelian groups described in Example 4.1.10(iv) induces a monad $\mathbb{Z}[-]$ on Set. Given a set S, $\mathbb{Z}[S]$ is defined to be the set of finite formal $\mathbb{Z}$-linear combinations of elements of S, and the map $\mathbb{Z}[f]\colon \mathbb{Z}[S] \to \mathbb{Z}[T]$ associated to a function $f\colon S \to T$ carries a formal sum $n_1 s_1 + \cdots + n_k s_k$, with $n_i \in \mathbb{Z}$ and $s_i \in S$, to the formal sum $n_1 f(s_1) + \cdots + n_k f(s_k)$; see Exercise 4.1.iv. By Proposition 5.2.12, there is a unique functor

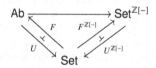

from the category of abelian groups to the category of $\mathbb{Z}[-]$-algebras that commutes with the left and right adjoints. The image of an abelian group A is the algebra consisting of the underlying set of A together with the "evaluation" map $\epsilon_A\colon \mathbb{Z}[A] \to A$ that interprets a finite formal sum as a finite sum, a single element of A.

In fact, this functor defines an isomorphism of categories: to equip a set A with an "evaluation" function $\alpha\colon \mathbb{Z}[A] \to A$, so that diagrams

$$
\begin{array}{ccc}
A & \xrightarrow{\ \eta_A\ } & \mathbb{Z}[A] \\
& \searrow{\scriptstyle 1_A} & \downarrow{\scriptstyle \alpha} \\
& & A
\end{array}
\qquad\qquad
\begin{array}{ccc}
\mathbb{Z}[\mathbb{Z}[A]] & \xrightarrow{\ \mu_A\ } & \mathbb{Z}[A] \\
{\scriptstyle \mathbb{Z}[\alpha]}\downarrow & & \downarrow{\scriptstyle \alpha} \\
\mathbb{Z}[A] & \xrightarrow[\ \alpha\ ]{} & A
\end{array}
$$

commute is precisely to give A the structure of an abelian group. The commutative triangle is a "sanity check" for the evaluation function, asserting that it sends a singleton sum $a \in \mathbb{Z}[A]$ to the element a. The remaining group axioms are imposed by the commutative rectangle: in particular, the values assigned to the sums of sums $(a_1 + a_2) + (a_3)$ and $(a_1) + (a_2 + a_3)$ are equal.

Moreover, an algebra homomorphism

$$
\begin{array}{ccc}
\mathbb{Z}[A] & \xrightarrow{\ \mathbb{Z}[f]\ } & \mathbb{Z}[B] \\
{\scriptstyle \alpha}\downarrow & & \downarrow{\scriptstyle \beta} \\
A & \xrightarrow[\ f\ ]{} & B
\end{array}
$$

is a function $f: A \to B$ that preserves the evaluation of formal sums, i.e., a group homomorphism.

The observations just made, that the canonical comparison functor defines an isomorphism from the category of abelian groups to the category of algebras for the free abelian group monad, are expressed by saying that the category of abelian groups is *monadic* over the category of sets.

DEFINITION 5.3.1.

- An adjunction $\mathsf{C} \underset{U}{\overset{F}{\underset{\perp}{\rightleftarrows}}} \mathsf{D}$ is **monadic** if the canonical comparison functor from D to the category of algebras for the induced monad on C defines an equivalence of categories.
- A functor $U: \mathsf{D} \to \mathsf{C}$ is **monadic** if it admits a left adjoint that defines a monadic adjunction.

EXAMPLE 5.3.2. The Kleisli adjunction for the maybe monad is identified in Example 5.2.10(i) as the adjunction between the category of sets and the category of sets and partially-defined functions:

$$
\mathsf{Set} \underset{}{\overset{}{\underset{\perp}{\rightleftarrows}}} \mathsf{Set}^{\partial} .
$$

This adjunction is monadic. The associated category of algebras is Set_+ and the canonical comparison functor $\mathsf{Set}^{\partial} \to \mathsf{Set}_+$ is part of the equivalence of categories described in Example 1.5.6.

Many of the monadic functors that one meets in practice, such as the example $U: \mathsf{Ab} \to \mathsf{Set}$ considered above, are in fact **strictly monadic**, meaning that the canonical comparison functor defines an isomorphism of categories. Whether this comparison functor is an equivalence or an isomorphism will not make a substantial difference to the theory.

A particularly simple family of examples of monadic functors is given by inclusions of reflective subcategories (see Definition 4.5.12). Consider a reflective subcategory $\mathsf{D} \hookrightarrow \mathsf{C}$ with reflector L. The induced endofunctor $L: \mathsf{C} \to \mathsf{C}$ defines a monad on C with unit $\eta_C: C \to LC$ and multiplication a natural isomorphism $L^2 C \cong LC$. A monad whose multiplication natural transformation is invertible is called an **idempotent monad**; see Exercises 4.5.vii and 5.1.iii.

PROPOSITION 5.3.3. *The inclusion* $D \hookrightarrow C$ *of a reflective subcategory is monadic: that is,
the functor*

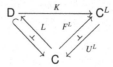

from D *to the category of algebras* C^L *for the induced idempotent monad L is an equivalence
of categories.*

PROOF. An L-algebra is an object $C \in C$ together with a map $c: LC \to C$ that is
a retraction of the unit component η_C. In fact, η_C and c are inverse isomorphisms. By
naturality of η, $\eta_C \cdot c = Lc \cdot \eta_{LC}$, but $\eta_{LC} = L\eta_C$, as both maps are left inverse to the
isomorphism μ_C, and so $Lc \cdot \eta_{LC} = Lc \cdot L\eta_C = 1_{LC}$. So if $C \in C$ admits the structure of an
L-algebra, then its unit component must be invertible. In fact, this necessary condition also
suffices: the map $\eta_C^{-1}: LC \to C$ automatically satisfies the associativity condition required
to define an L-algebra structure on the object C. In conclusion, there is no additional
structure provided by a lift of an object $C \in C^L$ to an L-algebra; rather being an L-algebra
is a *condition* on the object C, namely whether the unit component η_C is invertible.

Exercise 4.5.vii reveals that an object C is in the essential image of $D \hookrightarrow C$ if and only
if η_C is an isomorphism. This proves that $K: D \to C^L$ is essentially surjective. Naturality
of the unit η implies that this functor is fully faithful, so we conclude that K defines an
equivalence of categories.   □

In §5.5, we prove the monadicity theorem, which allows one to easily detect when the
comparison functor defines an equivalence or isomorphism. In particular, the reader need
not remain concerned for very long about the remaining details in the sketched proof of the
monadicity of abelian groups.

Exercises.

EXERCISE 5.3.i. Reading between the lines in the proof of Proposition 5.3.3, prove that the
category of algebras for an idempotent monad on C defines a reflective subcategory of C.
Show further that the Kleisli category for an idempotent monad is also reflective.

5.4. Canonical presentations via free algebras

The reason for our particular interest in monadic functors, such as $U: Ab \to Set$, is
that the forgetful functor $U^T: C^T \to C$ associated to the category of algebras for a monad T
on C has certain special properties, which are enumerated in Propositions 5.4.3 and 5.4.9,
Lemma 5.6.1, Theorem 5.6.5, and Proposition 5.6.11. These properties are transferred
along the equivalence to any monadic functor. The first of these, discussed in this section
and used in the next to prove the monadicity theorem, demonstrate that any algebra for a
monad admits a canonical presentation as a quotient of free algebras.

On account of the isomorphism of categories $Ab \cong Set^{\mathbb{Z}[-]}$, the adjunction

(5.4.1)
$$Set \underset{U}{\overset{\mathbb{Z}[-]}{\rightleftarrows}} Ab$$

is monadic, and so we build intuition by first introducing these canonical presentations in
the special case of abelian groups. To avoid inelegant pedantry, we use the same notation
for the monad $\mathbb{Z}[-]$ on Set and for the free abelian group functor $\mathbb{Z}[-]: Set \to Ab$.

The context—sets and functions or groups and homomorphisms—will always indicate the appropriate target category.

Any abelian group A has a presentation that can be defined in terms of the free and forgetful functors (5.4.1). If G is any set of elements of an abelian group A, there is a canonical homomorphism $\mathbb{Z}[G] \to A$ that sends a finite $\mathbb{Z}$-linear combination of these elements to the element of A that is their sum. The set G is a set of **generators** for A precisely when this map is surjective. **Relations** involving these generators, meaning the terms that are to be set equal to zero, are elements of the group $\mathbb{Z}[G]$, so again there is a canonical "evaluation" homomorphism $\mathbb{Z}[R] \to \mathbb{Z}[G]$ from the free group on a set R of relations to the free group on the generators. The set $G \subset A$ of generators and $R \subset \mathbb{Z}[G]$ of relations gives a **presentation** of A if the quotient map $\mathbb{Z}[G] \twoheadrightarrow A$ is a coequalizer of the evaluation map and the zero homomorphism

$$(5.4.2) \qquad \mathbb{Z}[R] \underset{\text{evaluation}}{\overset{0}{\rightrightarrows}} \mathbb{Z}[G] \longrightarrow\!\!\!\!\!\rightarrow A,$$

in which case one often writes $A = \langle G \mid R \rangle$.

Ad hoc presentations as described by (5.4.2) can be a useful way to define abelian groups, but they are unlikely to be functorial: a homomorphism $\varphi \colon A \to A'$ is unlikely to carry the chosen presentation for A to sets of generators and relations for A'. There is, however, a canonical—by which we mean *functorial*—presentation of any abelian group. Rather than choose a proper subset of generators, we take all of the elements of A to be generators; the canonical evaluation homomorphism $\alpha \colon \mathbb{Z}[A] \twoheadrightarrow A$ is certainly surjective. Similarly, rather than choose any particular set of relations in $\mathbb{Z}[A]$, we take all of the elements of $\mathbb{Z}[A]$ to be "relations." Here we do not intend to send every formal sum of elements of A to zero; the result would be the trivial group. Instead, we generalize the meaning of presentation, making use of the fact that coequalizers are more flexible than cokernels. Proposition 5.4.3 will demonstrate that the diagram

$$\mathbb{Z}[\mathbb{Z}[A]] \underset{\mu_A}{\overset{\mathbb{Z}[\alpha]}{\rightrightarrows}} \mathbb{Z}[A] \overset{\alpha}{\longrightarrow\!\!\!\!\!\rightarrow} A$$

is always a coequalizer. That is, any abelian group is the quotient of the free abelian group on its underlying set modulo the relation that identifies a formal sum with the element to which it is evaluated.

In general:

PROPOSITION 5.4.3. *Let (T, η, μ) be a monad on* C *and let* $(A, TA \overset{\alpha}{\to} A)$ *be a T-algebra. Then*

$$(5.4.4) \qquad (T^2A, \mu_{TA}) \underset{\mu_A}{\overset{T\alpha}{\rightrightarrows}} (TA, \mu_A) \overset{\alpha}{\longrightarrow\!\!\!\!\!\rightarrow} (A, \alpha)$$

is a coequalizer diagram in C^T.

Proposition 5.4.3 can be proven directly by a diagram chase, but we prefer to deduce it as a special case of a more general result, the statement of which requires some new definitions.

DEFINITION 5.4.5. A **split coequalizer** diagram consists of maps

$$x \underset{g}{\overset{f}{\rightrightarrows}} y \underset{s}{\overset{h}{\rightleftarrows}} z$$

so that $hf = hg$, $hs = 1_z$, $gt = 1_y$, and $ft = sh$.

The condition $hf = hg$ says that this triple of maps defines a **fork**, i.e., h is a cone under the parallel pair $f, g\colon x \rightrightarrows y$.

LEMMA 5.4.6. *The underlying fork of a split coequalizer diagram is a coequalizer. Moreover, it is an **absolute colimit**: any functor preserves this coequalizer.*

PROOF. Given a map $k\colon y \to w$ so that $kf = kg$, we must show that k factors through h; uniqueness of a hypothetical factorization follows because h is a split epimorphism. The factorization is given by the map $ks\colon z \to w$, as demonstrated by the following easy diagram chase:

$$ksh = kft = kgt = k.$$

Now clearly split coequalizers, which are just diagrams of a particular shape, are preserved by any functor, so the equationally-witnessed universal property of the underlying fork is also preserved by any functor.   □

EXAMPLE 5.4.7. For any algebra $(A, \alpha\colon TA \to A)$ for a monad (T, η, μ) on C, the diagram

$$T^2 A \underset{\eta_{TA}}{\overset{T\alpha}{\underset{\mu_A}{\rightrightarrows}}} TA \underset{\eta_A}{\overset{\alpha}{\rightrightarrows}} A$$

defines a split coequalizer diagram in C.

Note that while the fork of Example 5.4.7 lifts along $U^T\colon \mathsf{C}^T \to \mathsf{C}$ to a diagram (5.4.4) of algebras, the splittings η_A and η_{TA} do not. This situation is captured by the following general definition.

DEFINITION 5.4.8. Given a functor $U\colon \mathsf{D} \to \mathsf{C}$:

- A U-**split coequalizer** is a parallel pair $f, g\colon x \rightrightarrows y$ in D together with an extension of the pair $Uf, Ug\colon Ux \rightrightarrows Uy$ to a split coequalizer diagram

$$Ux \underset{t}{\overset{Uf}{\underset{Ug}{\rightrightarrows}}} Uy \underset{s}{\overset{h}{\rightrightarrows}} z$$

  in C.

- U **creates coequalizers of** U-**split pairs** if any U-split coequalizer admits a coequalizer in D whose image under U is isomorphic to the fork underlying the given U-split coequalizer diagram in C, and if any such fork in D is a coequalizer.

- U **strictly creates coequalizers of** U-**split pairs** if any U-split coequalizer admits a unique lift to a coequalizer in D for the given parallel pair.

PROPOSITION 5.4.9. *For any monad (T, η, μ) acting on a category C, the monadic forgetful functor $U^T\colon \mathsf{C}^T \to \mathsf{C}$ strictly creates coequalizers of U^T-split pairs.*

PROOF. Suppose given a parallel pair $f, g\colon (A, \alpha) \rightrightarrows (B, \beta)$ in C^T that admits a U^T-splitting:

$$A \underset{t}{\overset{f}{\underset{g}{\rightrightarrows}}} B \underset{s}{\overset{h}{\rightrightarrows}} C.$$

We must show that C lifts to an algebra (C, γ) and h lifts to an algebra map that is a coequalizer of f and g in C^T, and that moreover these lifts are unique with this property. To define the algebra structure map γ, note that the functor $T\colon \mathsf{C} \to \mathsf{C}$ preserves the split coequalizer diagram; in particular, by Lemma 5.4.6, Th is the coequalizer of Tf and Tg. The algebra structure maps α and β define a diagram

$$
\begin{array}{ccccc}
TA & \overset{Tf}{\underset{Tg}{\rightrightarrows}} & TB & \overset{Th}{\twoheadrightarrow} & TC \\
\alpha \downarrow & & \beta \downarrow & & \exists! \mid \gamma \\
A & \underset{g}{\overset{f}{\rightrightarrows}} & B & \overset{h}{\longrightarrow} & C
\end{array}
$$

in which the square defined by f and Tf and the square defined by g and Tg both commute. Thus,

$$h \cdot \beta \cdot Tf = h \cdot f \cdot \alpha = h \cdot g \cdot \alpha = h \cdot \beta \cdot Tg,$$

which says that $h \cdot \beta$ defines a cone under the pair $Tf, Tg\colon TA \rightrightarrows TB$. By the universal property of their coequalizer Th, there is a unique map $\gamma\colon TC \to C$ so that the right-hand square above commutes. Once we show that the pair (C, γ) is a T-algebra, this commutative square will demonstrate that $h\colon (B, \beta) \to (C, \gamma)$ is a T-algebra homomorphism.

It remains to check that the diagrams

$$
\begin{array}{ccc}
C & \overset{\eta_C}{\longrightarrow} & TC \\
& 1_C \searrow & \downarrow \gamma \\
& & C
\end{array}
\qquad\qquad
\begin{array}{ccc}
T^2 C & \overset{\mu_C}{\longrightarrow} & TC \\
T\gamma \downarrow & & \downarrow \gamma \\
TC & \underset{\gamma}{\longrightarrow} & C
\end{array}
$$

commute. This will follow from the corresponding conditions for the T-algebra (B, β) and the fact that the coequalizer maps are epimorphisms. Specifically, we have diagrams

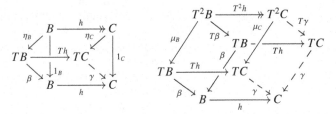

in which all but the right-most face of each prism is known to commute. It follows that $\gamma \cdot \eta_C \cdot h = 1_C \cdot h$ and $\gamma \cdot \mu_C \cdot T^2 h = \gamma \cdot T\gamma \cdot T^2 h$, and we conclude that the right-most faces commute by canceling the epimorphisms h and $T^2 h$. Thus, (C, γ) is a T-algebra.

Finally, we show that $h\colon (B, \beta) \to (C, \gamma)$ is a coequalizer in C^T. Given a cone $k\colon (B, \beta) \to (D, \delta)$ so that $kf = kg$, there is a unique factorization

$$
\begin{array}{ccccc}
A & \overset{f}{\underset{g}{\rightrightarrows}} & B & \overset{h}{\longrightarrow} & C \\
& & & k \searrow & \exists! \mid j \\
& & & & D
\end{array}
$$

using the universal property of the coequalizer in C. To check that j lifts to a map of T-algebras, it again suffices to verify that $j \cdot \gamma = \delta \cdot Tj$ after precomposing with the epimorphism Th. The result follows from an easy diagram chase, using the fact that h and k are algebra maps:

$$j \cdot \gamma \cdot Th = j \cdot h \cdot \beta = k \cdot \beta = \delta \cdot Tk = \delta \cdot Tj \cdot Th. \qquad \square$$

PROOF OF PROPOSITION 5.4.3. Example 5.4.7 shows that the fork (5.4.4) is part of a U^T-split coequalizer. In particular, $\alpha\colon TA \to A$ is an absolute coequalizer of the pair $T\alpha, \mu_A\colon T^2A \rightrightarrows TA$ in C. The proof of Proposition 5.4.9 demonstrates that this coequalizer lifts to a coequalizer in C^T.   □

The results of Propositions 5.4.3 and 5.4.9 extend to general monadic functors with one caveat: as monadic adjunctions are only required to be equivalent to and not isomorphic to the adjunction involving the category of algebras, a monadic functor $U\colon D \to C$ might only create, rather than strictly create, U-split coequalizers.

COROLLARY 5.4.10. *If* $C \underset{U}{\overset{F}{\rightleftarrows}} D$ *is a monadic adjunction, then*

(i) $U\colon D \to C$ *creates coequalizers of U-split pairs.*
(ii) *For any $D \in D$, there is a coequalizer diagram*

$$(5.4.11) \qquad FUFUD \underset{\epsilon_{FUD}}{\overset{FU\epsilon_D}{\rightrightarrows}} FUD \overset{\epsilon_D}{\longrightarrow} D$$

involving the counit $\epsilon\colon FU \Rightarrow 1_D$ *of the adjunction.*

PROOF. To say that $U\colon D \to C$ is monadic is to say that there is an equivalence of categories

$$D \overset{K}{\underset{\simeq}{\longrightarrow}} C^T$$
$$U \searrow \quad \nearrow U^T$$
$$C$$

where T is the monad UF.

For (i), if $f, g\colon A \rightrightarrows B$ is a U-split pair in D, then commutativity $U = U^T K$ implies that $Kf, Kg\colon KA \rightrightarrows KB$ is a U^T-split pair in C^T. Proposition 5.4.9 lifts the fork of the U-split coequalizer diagram in C to a coequalizer diagram in C^T. Any inverse equivalence $L\colon C^T \overset{\simeq}{\to} D$ to the functor K can be used to map this data to a coequalizer for the pair $f, g\colon A \rightrightarrows B$. As $KL \cong 1_{C^T}$, this coequalizer maps via $U\colon D \to C$ to a fork that is isomorphic to the fork of the given U-split coequalizer.[9] Note also that U preserves and reflects all coequalizer diagrams that are U-split because U^T has this property and K preserves and reflects all coequalizers (see Lemma 3.3.6). This completes the proof that U creates U-split coequalizers.

For (ii), the fork (5.4.11) is U-split by the diagram

$$UFUFUD \underset{U\epsilon_{FUD}}{\overset{UFU\epsilon_D}{\rightrightarrows}} UFUD \overset{U\epsilon_D}{\longrightarrow} UD$$
$$\overset{\eta_{UFUD}}{\longleftarrow} \qquad \overset{\eta_{UD}}{\longleftarrow}$$

in C. By (i), the functor $U\colon D \to C$ reflects U-split coequalizers, so the fork (5.4.11) defines a coequalizer diagram in D.   □

Exercises.

EXERCISE 5.4.i. The coequalizer of a parallel pair of morphisms f and g in the category Ab is equally the cokernel of the map $f - g$. Explain how the canonical presentation of an abelian group described in Proposition 5.4.3 defines a presentation of that group, in the usual sense.

---

[9]If $KL = 1_{C^T}$, as is the case if these functors define an isomorphism of categories, then this coequalizer in D is a lift of the U-split coequalizer in C and $U\colon D \to C$ strictly creates U-split coequalizers.

5.5. Recognizing categories of algebras

There are many versions of the monadicity theorem, which characterizes those adjunctions for which objects in the domain of the right adjoint can be regarded as algebras for the monad defined on the codomain. The one presented here is due to Beck [**Bec67**].

THEOREM 5.5.1 (monadicity theorem). *A right adjoint functor* $U \colon \mathsf{D} \to \mathsf{C}$ *is monadic if and only if it creates coequalizers of U-split pairs.*

Recalling Definition 5.4.8, the meaning of this result is the following. Given an adjunction $F \dashv U$ inducing a monad T on C, Proposition 5.2.12 constructs the canonical comparison functor:

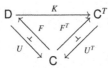

By Theorem 5.5.1, the following are equivalent:
  (i) K is an equivalence (respectively, isomorphism) of categories.
  (ii) U creates (respectively, strictly creates) coequalizers of U-split pairs.

We leave the proof of the strict version of the monadicity theorem, that D is isomorphic to the category of algebras if and only if $U \colon \mathsf{D} \to \mathsf{C}$ strictly creates coequalizers of U-split pairs, to Exercise 5.5.i and focus our attention on the equivalence-invariant statement.

PROOF. Corollary 5.4.10 proves the implication (i)$\Rightarrow$(ii), so we assume (ii) and use it to construct an inverse equivalence $L \colon \mathsf{C}^T \to \mathsf{D}$ to the functor K. We have $U^T K = U$ and $KF = F^T$, so if L is to define an inverse equivalence to K, we must have $U^T \cong UL$ and $F \cong LF^T$, conditions that we use to guide the definition of L.

In fact, we define L so that the second of these conditions holds strictly, namely, for each $A \in \mathsf{C}$, define

$$L(TA, \mu_A) := FA$$

and define L to carry a free map $Tf \colon (TA, \mu_A) \to (TB, \mu_B)$ between free algebras to the map $Ff \colon FA \to FB$. It remains to extend the definition of L to non-free algebras and maps.

An equivalence of categories preserves all limits and colimits. With this in mind, for any algebra $(A, \alpha) \in \mathsf{C}^T$ define $L(A, \alpha)$ to be the coequalizer

$$(5.5.2) \qquad FUFA \underset{\epsilon_{FA}}{\overset{F\alpha}{\rightrightarrows}} FA \dashrightarrow L(A, \alpha),$$

in D; this construction is arranged so that the coequalizer (5.4.4) is preserved by L. By Example 5.4.7, the parallel pair $F\alpha, \epsilon_{FA} \colon FUFA \rightrightarrows FA$ is U-split, so the hypothesis that U creates U-split coequalizers implies that the coequalizer (5.5.2) exists in D. The action of $L \colon \mathsf{C}^T \to \mathsf{D}$ on a morphism $f \colon (A, \alpha) \to (B, \beta)$ is defined to be the unique map between coequalizers induced by a commutative diagram of parallel pairs:

$$
\begin{array}{ccccc}
FUFA & \overset{F\alpha}{\underset{\epsilon_{FA}}{\rightrightarrows}} & FA & \twoheadrightarrow & L(A, \alpha) \\
{\scriptstyle FUFf}\downarrow & & \downarrow{\scriptstyle Ff} & & \downarrow{\scriptstyle \exists! \, | \, Lf} \\
FUFB & \underset{\epsilon_{FB}}{\overset{F\beta}{\rightrightarrows}} & FB & \twoheadrightarrow & L(B, \beta)
\end{array}
$$

Uniqueness of factorizations through colimit cones implies that this definition is functorial; see Proposition 3.6.1.

The functor K carries the parallel pair (5.5.2) to the parallel pair (5.4.4). As U^T strictly creates coequalizers of U^T-split pairs, we see that $KL \cong 1_{\mathsf{C}^T}$. It remains only to prove that $LK \cong 1_{\mathsf{D}}$. Given $D \in \mathsf{D}$, the object LKD is defined to be the coequalizer of

$$FUFUD \overset{FU\epsilon_D}{\underset{\epsilon_{FUD}}{\rightrightarrows}} FUD$$

By Corollary 5.4.10(ii), LKD must therefore be isomorphic to the coequalizer (5.4.11). The isomorphism induced between two coequalizers for the same diagram is unique, and thus defines a natural isomorphism $LK \cong 1_{\mathsf{D}}$. $\square$

COROLLARY 5.5.3. *The following categories are monadic over* Set *via the free $\dashv$ forgetful adjunctions described in Example 4.1.10.*

   (i) Monoid, Group, Ab, Ring, *and other variants, such as commutative rings or monoids, or non-unital versions of the preceding*
   (ii) Mod$_R$, Vect$_{\Bbbk}$, Aff$_{\Bbbk}$, Set^{BG}
   (iii) Lattice *or the categories of meet or join semilattices* [10]
   (iv) Set$_*$

PROOF. We prove that monoids are monadic over sets and leave the proofs of the other cases, which are similar, to Exercise 5.5.ii.

Recall that a **monoid** is a set M together with maps $\eta\colon 1 \to M$ and $\mu\colon M \times M \to M$ so that the diagrams displayed in Definition 1.6.2 commute in Set. A monoid homomorphism $f\colon (M,\eta,\mu) \to (M',\eta',\mu')$ is a function $f\colon M \to M'$ so that the diagrams

commute in Set.

To apply Theorem 5.5.1, it suffices to prove that $U\colon$ Monoid $\to$ Set strictly creates the coequalizer of any pair of monoid homomorphisms $f, g\colon M \rightrightarrows M'$ whose underlying functions extend to a split coequalizer diagram

in Set. That is, we must show that there is a unique way to turn the set M'' into a monoid so that $h\colon M' \to M''$ is a monoid homomorphism, and moreover this fork defines a coequalizer in the category of monoids. The proof will closely resemble the argument used to prove Proposition 5.4.9.

For h to define a homomorphism, the unit for M'' must be defined to be the composite $\eta'' := 1 \overset{\eta'}{\to} M' \overset{h}{\to} M''$. To define the multiplication for M'', we use the fact that the split

---

[10]A **lattice** is a poset with all finite **meets** (limits) and **joins** (colimits), and a morphism of lattices is an order-preserving function that preserves these. A **meet semilattice** only has meets, while a **join semilattice** only has joins.

coequalizer diagram is preserved by any functor, in particular by $(-)^2 \colon$ Set $\to$ Set. This yields a parallel pair of coequalizers

$$
\begin{array}{ccccc}
M \times M & \underset{g\times g}{\overset{f\times f}{\rightrightarrows}} & M' \times M' & \overset{h\times h}{\twoheadrightarrow} & M'' \times M'' \\
\downarrow{\mu} & & \downarrow{\mu'} & & \downarrow{\exists!\,\mu''} \\
M & \underset{g}{\overset{f}{\rightrightarrows}} & M' & \underset{h}{\twoheadrightarrow} & M''
\end{array}
$$

the induced map between which we define to be the multiplication for μ''. Note our definition is arranged so that h defines a monoid homomorphism, and moreover, by the uniqueness of factorizations through the coequalizer $h\times h$, no other choice for μ'' is possible.

The proofs that (M'', η'', μ'') is associative and unital, and that this monoid defines the coequalizer of the monoid homomorphisms $f, g \colon M \rightrightarrows M'$, proceed via the same arguments used in the proof of Proposition 5.4.9: these properties are deduced from the analogous properties of the monoid (M', η', μ') after canceling a quotienting epimorphism. $\square$

There is a unifying term that encompasses each of the categories listed in Corollary 5.5.3.

DEFINITION 5.5.4. A functor is **finitary** if it preserves filtered colimits.

In particular, a monad $T \colon \mathsf{C} \to \mathsf{C}$ is **finitary** if it preserves filtered colimits in C. If a right adjoint is finitary, then so is its monad because its left adjoint preserves all colimits.

DEFINITION 5.5.5. A category A is a **category of models for an algebraic theory** if there is a finitary monadic functor $U \colon \mathsf{A} \to$ Set.

Any category of models for an algebraic theory is locally finitely presentable, in the sense of Definition 4.6.16 [**AR94**, 2.78]. Another familiar category is monadic over sets, though it is not a model of an algebraic theory because its monad is not finitary.

COROLLARY 5.5.6. *The underlying set functor* $U \colon$ cHaus $\to$ Set *from the category of compact Hausdorff spaces is monadic.*

The following proof is due to Paré [**Par71**]. As in the proof of Corollary 5.5.3, a key point is that split coequalizers are **absolute** coequalizers, preserved by any functor.

PROOF. A topological space may be defined to be a set X equipped with a **closure operator** $\overline{(-)} \colon PX \to PX$ on its set of subsets satisfying four axioms for all $A, B \in PX$:

(i) $\overline{\emptyset} = \emptyset$,   (ii) $A \subset \overline{A}$,   (iii) $\overline{\overline{A}} = \overline{A}$,   (iv) $\overline{A \cup B} = \overline{A} \cup \overline{B}$.

A function $f \colon X \to Y$ between topological spaces is **continuous** if and only if $f(\overline{A}) \subset \overline{f(A)}$ for all $A \subset X$ and **closed**[11] if and only if $f(\overline{A}) = \overline{f(A)}$. All continuous functions between compact Hausdorff spaces are closed; that is, a function $f \colon X \to Y$ between compact Hausdorff spaces lifts to a map in cHaus if and only if it preserves the closure operator.

Consider a pair of continuous functions $f, g \colon X \rightrightarrows Y$ in cHaus so that there exists an absolute coequalizer diagram

$$
X \underset{g}{\overset{f}{\rightrightarrows}} Y \overset{h}{\twoheadrightarrow} Z
$$

---

[11]A continuous function is **closed** if it preserves closed subsets.

in **Set**. The covariant power set functor $P: \mathsf{Set} \to \mathsf{Set}$ preserves this, giving rise to a coequalizer diagram of direct image functions:

$$
\begin{array}{ccccc}
PX & \underset{g_*}{\overset{f_*}{\rightrightarrows}} & PY & \overset{h_*}{\twoheadrightarrow} & PZ \\
\overline{(-)}\Big\downarrow & & \overline{(-)}\Big\downarrow & & \exists!\,|\,\overline{(-)}\Big\downarrow \\
PX & \underset{g_*}{\overset{f_*}{\rightrightarrows}} & PY & \overset{h_*}{\twoheadrightarrow} & PZ
\end{array}
$$

Because f and g are maps of compact Hausdorff spaces, the direct image functions commute with the closure operators, and so the commutative diagram of parallel pairs induces a function that we optimistically denote by $\overline{(-)}: PZ \to PZ$.

It is straightforward to verify that this function defines a closure operator on Z so that Z is a topological space. Because the function $h: Y \to Z$ is surjective, continuous, and closed and Y is compact Hausdorff, so is Z.[12] By our characterization of maps in cHaus, it is clear that we have constructed the unique lift of the absolute coequalizer diagram along $U: \mathsf{cHaus} \to \mathsf{Set}$.

To see that $h: Y \to Z$ has the universal property of a coequalizer in cHaus, consider a continuous function $k: Y \to W$ so that $kf = kg$. On underlying sets, there is a unique factorization $\ell: Z \to W$ of k through h, so it remains only to show that ℓ is continuous and closed. That is, we must prove that the right-hand square commutes

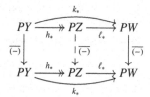

which follows from the fact that the left-hand square and exterior rectangle commute and the fact that h_* is an epimorphism. $\square$

EXAMPLE 5.5.7. Other examples of monadic functors include:

(i) Example 5.2.6(ii) reveals that the forgetful functor $U: \mathsf{Mod}_R \to \mathsf{Ab}$ is monadic. The induced monad on Ab is given by $R \otimes_{\mathbb{Z}} -: \mathsf{Ab} \to \mathsf{Ab}$.

(ii) The proof of Corollary 5.5.3 can be adapted to show that the forgetful functor $U: \mathsf{Ring} \to \mathsf{Ab}$ is monadic. The induced monad on Ab is the free monoid monad $TA := \bigoplus_{n\geq 0} A^{\otimes n}$.

(iii) If C has coproducts (in fact, copowers suffice) and J is small, then Exercise 5.5.v demonstrates that the restriction functor $\mathsf{C}^\mathsf{J} \to \mathsf{C}^{\mathrm{ob}\,\mathsf{J}}$, which carries a J-indexed diagram in C to the ob J-indexed family of objects in its image, is monadic. Special cases include, for instance, the functor $\mathsf{Vect}_{\mathbb{k}}^{BG} \to \mathsf{Vect}_{\mathbb{k}}$ that carries a G-representation to its underlying vector space or analogous functors for G-objects in any category.

(iv) $U: \mathsf{Cat} \to \mathsf{DirGraph}$ is also monadic. The monad forms (the underlying graph of) the free category on a directed graph, where identities and composites of all

---

[12]Compactness of Y and surjectivity of the continuous function h imply that Z is compact. Because Y is Hausdorff, its points are closed, and since h is closed, points are also closed in Z. To show that Z is Hausdorff, note that distinct points $z \neq z' \in Z$ have disjoint closed preimages. Because the compact Hausdorff space Y is normal, the preimages $h^{-1}(z)$ and $h^{-1}(z')$ have disjoint open neighborhoods. The images of their complements are closed subsets of Z, one containing z and the other containing z', whose complements are the desired disjoint open neighborhoods of z and z'.

finite directed paths are formally added. This functor factors through a second monadic functor U: Cat → rDirGraph, whose target is the category whose objects are **reflexive directed graphs,** with specified "identity" endoarrows of each vertex.

The construction of the inverse equivalence to the canonical functor K: D → C^T given in the proof of Theorem 5.5.1 can be adapted to various other hypotheses on the given right adjoint U: D → C. In the categorical literature, these variant theorems are traditionally assigned three-letter acronyms. Theorem 5.5.1 is sometimes called PTT, "precise tripleability[13] theorem," "precise" because the characterization involves a condition that is necessary, as well as sufficient. Statements of the "crude tripleability theorem" (CTT) and "vulgar tripleability theorem" (VTT) can be found in the exercises to [**ML98a,** §VI.7]. Another variant is the "reflexive tripleability theorem":

PROPOSITION 5.5.8 (RTT). *If* U: D → C *has a left adjoint and if*

 (i) D *has coequalizers of reflexive[14] pairs,*
 (ii) U *preserves coequalizers of reflexive pairs, and*
 (iii) U *reflects isomorphisms,*

then U *is monadic.*

PROOF. Exercise 5.5.iii.   □

Proposition 5.5.8 has a neat application:

THEOREM 5.5.9 (Paré). *The contravariant power set functor* P: Set^{op} → Set *is monadic.*

The argument presented here, due to Paré [**Par74**], also holds when Set is replaced by any *elementary topos* (a further generalization of the Grothendieck toposes introduced in §E.4). It makes use of the following lemma, whose conclusion is sometimes called the **Beck–Chevalley condition,** that asserts that for certain commutative squares, the induced direct and inverse image functions on power sets commute:

LEMMA 5.5.10. *For any pullback diagram of monomorphisms, as displayed on the left,*

$$
\begin{array}{ccc}
A' & \xrightarrow{\ g\ } & B' \\
{\scriptstyle a}\Big\uparrow & \lrcorner & \Big\uparrow{\scriptstyle b} \\
A & \xrightarrow[\ f\ ]{} & B
\end{array}
\qquad\qquad
\begin{array}{ccc}
PB' & \xrightarrow{\ g^{-1}\ } & PA' \\
{\scriptstyle b_*}\Big\downarrow & & \Big\downarrow{\scriptstyle a_*} \\
PB & \xrightarrow[\ f^{-1}\ ]{} & PA
\end{array}
$$

the right hand square commutes.

PROOF. For any $X \in PB'$, consider a commutative rectangle

$$
\begin{array}{ccc}
Y & \longrightarrow & X \\
\Big\downarrow & & \Big\downarrow \\
A' & \xrightarrow{\ g\ } & B' \\
{\scriptstyle a}\Big\uparrow & \lrcorner & \Big\uparrow{\scriptstyle b} \\
A & \xrightarrow[\ f\ ]{} & B
\end{array}
$$

---

[13]"Triple" is an antiquated synonym for "monad."
[14]A parallel pair f, g: $A \rightrightarrows B$ is **reflexive** if both maps admit a common section s: $B \to A$.

To say that $a_*g^{-1}(X) = Y$ is to say that the top square is a pullback. To say that $f^{-1}b_*(X) = Y$ is to say that the composite rectangle is the pullback. By Exercise 3.1.viii, these conditions are equivalent. □

By Exercise 4.5.v, any monomorphism $f\colon A \hookrightarrow B$ defines a pullback square

$$
\begin{array}{ccc}
A & =\!=\!= & A \\
\| & & \downarrow{\scriptstyle f} \\
A & \rightarrowtail & B \\
& {\scriptstyle f} &
\end{array}
$$

so a special case of Lemma 5.5.10 tells us that for any monomorphism, the composite

$$PA \xrightarrow{\ f_*\ } PB \xrightarrow{\ f^{-1}\ } PA$$

is the identity.

PROOF OF THEOREM 5.5.9. We apply Proposition 5.5.8. On account of the cartesian closed structure of Set:

$$\mathsf{Set}(A, PB) \cong \mathsf{Set}(A \times B, \Omega) \cong \mathsf{Set}(B, PA)\,,$$

the functor $P\colon \mathsf{Set}^{\mathrm{op}} \to \mathsf{Set}$ has a left adjoint (namely, itself). By Theorem 3.2.6, Set has finite limits, in particular, equalizers of coreflexive pairs proving (i). Given an equalizer of a pair of maps with a common retraction:

$$
E \overset{e}{\rightarrowtail} A \underset{g}{\overset{f}{\rightrightarrows}} B
\qquad\qquad
A \underset{g}{\overset{f}{\rightrightarrows}} B \xrightarrow{\ s\ } A
$$

the square

$$
\begin{array}{ccc}
E & \overset{e}{\rightarrowtail} & A \\
{\scriptstyle e}\big\downarrow & & \big\downarrow{\scriptstyle f} \\
A & \rightarrowtail & B \\
& {\scriptstyle g} &
\end{array}
$$

is a pullback of monomorphisms: the retraction s implies that f and g are split monic, which implies that the legs of any cone over this pullback diagram must be equal, proving that the displayed square is a pullback. Applying Lemma 5.5.10, it follows that

$$
PB \underset{g^{-1}}{\overset{f^{-1}}{\rightrightarrows}} PA \underset{e_*}{\overset{e^{-1}}{\leftrightarrows}} PE
$$

is a split coequalizer diagram, proving (ii).

To show (iii), note that P is faithful: given a parallel pair $f, g\colon A \rightrightarrows B$, the composites

$$
B \xrightarrow{\ \eta\ } PB \underset{g^{-1}}{\overset{f^{-1}}{\rightrightarrows}} PA
$$

with the singleton map $\eta\colon B \to PB$ transpose to define functions $A \times B \rightrightarrows \Omega$ that classify the subsets $(1, f)\colon A \rightarrowtail A \times B$ and $(1, g)\colon A \rightarrowtail A \times B$. It follows from the universal property of the subobject classifier Ω, described in Example 2.1.6(i), that if $f^{-1} = g^{-1}$, then $f = g$. By Exercise 1.6.iii, any faithful functor reflects both monomorphisms and epimorphisms.

In Set, or any elementary topos, the isomorphisms are precisely those maps that are both monic and epic, so it follows that P reflects isomorphisms. Proposition 5.5.8 now implies that the contravariant power set functor is monadic.   □

Exercises.

EXERCISE 5.5.i. Prove the strict version of the monadicity theorem: that the domain D of a right adjoint $U: \mathsf{D} \to \mathsf{C}$ is isomorphic to the category of algebras if and only if U strictly creates coequalizers of U-split pairs.

EXERCISE 5.5.ii. Use the monadicity theorem to prove that another of the categories listed in Corollary 5.5.3 is monadic over sets.

EXERCISE 5.5.iii. Prove Proposition 5.5.8.

EXERCISE 5.5.iv. For any group G, the forgetful functor $\mathsf{Set}^{BG} \to \mathsf{Set}$ admits a left adjoint that sends a set X to the G-set $G \times X$, with G acting on the left. Prove that this adjunction is monadic by appealing to the monadicity theorem.

EXERCISE 5.5.v. Generalizing Exercise 5.5.iv, for any small category J and any cocomplete category C the forgetful functor $\mathsf{C}^{\mathsf{J}} \to \mathsf{C}^{\mathrm{ob}\,\mathsf{J}}$ admits a left adjoint Lan: $\mathsf{C}^{\mathrm{ob}\,\mathsf{J}} \to \mathsf{C}^{\mathsf{J}}$ that sends a functor $F \in \mathsf{C}^{\mathrm{ob}\,\mathsf{J}}$ to the functor $\mathrm{Lan}F \in \mathsf{C}^{\mathsf{J}}$ defined by

$$\mathrm{Lan}F(j) = \coprod_{x \in \mathsf{J}} \coprod_{\mathsf{J}(x,j)} Fx .$$

(i) Define $\mathrm{Lan}F$ on morphisms in J.
(ii) Define Lan on morphisms in $\mathsf{C}^{\mathrm{ob}\,\mathsf{J}}$.
(iii) Use the Yoneda lemma to show that Lan is left adjoint to the forgetful (restriction) functor $\mathsf{C}^{\mathsf{J}} \to \mathsf{C}^{\mathrm{ob}\,\mathsf{J}}$.
(iv) Prove that this adjunction is monadic by appealing to the monadicity theorem.

EXERCISE 5.5.vi. Describe a more general class of functors $K: \mathsf{I} \to \mathsf{J}$ between small categories so that for any cocomplete C the restriction functor $\mathrm{res}_K: \mathsf{C}^{\mathsf{J}} \to \mathsf{C}^{\mathsf{I}}$ strictly creates colimits of res_K-split parallel pairs. All such functors admit left adjoints and are therefore monadic. Challenge: describe the left adjoint (or see Theorem 6.2.1).

EXERCISE 5.5.vii. Consider the Kleisli category Set_T for a monad T acting on Set and choose a skeleton $\mathsf{N} \xrightarrow{\simeq} \mathsf{Fin} \hookrightarrow \mathsf{Set}$ for the full subcategory of finite sets. Let L be the opposite of the full subcategory of the Kleisli category spanned by the objects $0, 1, 2, \ldots$ in N, so that there is an identity-on-objects functor:

$$
\begin{array}{ccc}
\mathsf{N}^{\mathrm{op}} & \overset{I}{\dashrightarrow} & \mathsf{L} \\
{\scriptstyle\wr|}\downarrow & & \downarrow \\
\mathsf{Fin}^{\mathrm{op}} \rightarrowtail & \mathsf{Set}^{\mathrm{op}} \xrightarrow{F_T} & \mathsf{Set}_T^{\mathrm{op}}
\end{array}
$$

(i) Show that the categories N^{op} and L have strictly associative finite products that are preserved by the functor $I: \mathsf{N}^{\mathrm{op}} \to \mathsf{L}$.

When the monad T is finitary, the functor $I: \mathsf{N}^{\mathrm{op}} \to \mathsf{L}$ defines its associated **Lawvere theory**. Because the objects in the category L are all iterated finite products of the object 1, the essential data in the Lawvere theory is the set $\mathsf{L}(n, 1)$ of "n-ary operations." It is possible to recover the category of T-algebras, that is, the category of models for this algebraic theory, from this data **[Lin66]**. A **model** of the Lawvere theory in Set is a finite product preserving functor $\mathsf{L} \to \mathsf{Set}$. A morphism between models is a natural transformation.

(ii) Define a functor from the category of T-algebras to the category of models for the Lawvere theory $I\colon \mathsf{N}^{\mathrm{op}} \to \mathsf{L}$. (Hint: See Exercise 5.2.vii.)

5.6. Limits and colimits in categories of algebras

Recall that a category A is **monadic over** C if there is an adjunction

$$\mathsf{C} \underset{U}{\overset{F}{\underset{\perp}{\rightleftarrows}}} \mathsf{A}$$

that is equivalent to the adjunction between C and the category of algebras for the monad UF. By Theorem 5.5.1, this is the case if and only if the functor $U\colon \mathsf{A} \to \mathsf{C}$ creates coequalizers of U-split pairs, a somewhat strange technical condition that can be relatively practical to check. Our aim in this section is to present a few results from categorical universal algebra, which describe some of the common features of categories that are equivalent to categories of algebras. In particular, these properties hold for Set$_*$, Monoid, Group, Ab, Ring, CRing, Mod$_R$, Aff$_\Bbbk$, Set^{BG}, Lattice, and cHaus, all of which are monadic over Set,[15] and sometimes fail for Poset, Top, and Field, which are not.[16]

A basic property common to all monadic functors is:

LEMMA 5.6.1. *If $U\colon \mathsf{A} \to \mathsf{C}$ is monadic, then U reflects isomorphisms: for any morphism $f\colon a \to a'$ in A, if Uf is an isomorphism in C, then f is an isomorphism in A.*

A functor is **conservative** if it reflects isomorphisms.

PROOF. A monadic functor $U\colon \mathsf{A} \to \mathsf{C}$ is equivalent to the forgetful functor from the category of algebras for the induced monad on C, and so it suffices to demonstrate that $U^T\colon \mathsf{C}^T \to \mathsf{C}$ has this property. Recall that a morphism $f\colon (A,\alpha) \to (A',\alpha')$ in C^T is a map $f\colon A \to A'$ in C so that the left-hand diagram

$$
\begin{array}{ccc}
TA & \xrightarrow{\;Tf\;} & TA' \\
{\scriptstyle\alpha}\downarrow & & \downarrow{\scriptstyle\alpha'} \\
A & \xrightarrow{\;\;f\;\;} & A'
\end{array}
\qquad\qquad
\begin{array}{ccc}
TA' & \xrightarrow{\;Tf^{-1}\;} & TA \\
{\scriptstyle\alpha'}\downarrow & & \downarrow{\scriptstyle\alpha} \\
A' & \xrightarrow{\;\;f^{-1}\;\;} & A
\end{array}
$$

commutes in C, whence the right-hand diagram also commutes whenever the inverse $f^{-1}\colon A' \to A$ exists in C.   □

Lemma 5.6.1 answers the question motivated by Example 1.1.10.

COROLLARY 5.6.2. *Any bijective continuous function between compact Hausdorff spaces is a homeomorphism.*[17]

REMARK 5.6.3. Unlike the other corollaries listed in the introduction, this is not the most efficient proof that bijective continuous functions between compact Hausdorff spaces are

---

[15]For ease of exposition, these applications are focused on categories that are monadic over sets, but the categorical results apply to monads acting on any category, and more sophisticated applications frequently take advantage of this.

[16]The functor $U\colon$ Field $\to$ Set cannot be monadic: Example 4.1.12 observes that it fails to admit any adjoints. Proofs that the forgetful functors $U\colon$ Poset $\to$ Set and $U\colon$ Top $\to$ Set are not monadic will appear shortly.

[17]More generally, any continuous bijection from a compact space to a Hausdorff one is a homeomorphism, but this statement is no true generalization: compactness of the domain and surjectivity of the map imply that the codomain is also compact, while Hausdorffness of the codomain and injectivity of the map imply that the domain is also Hausdorff.

homeomorphisms,[18] but we submit that this roundabout proof of Corollary 5.6.2 is nonetheless interesting. By monadicity of the functor $U\colon$ cHaus $\to$ Set and Proposition 5.4.3, any compact Hausdorff space is canonically a quotient of the Stone–Čech compactification of its underlying set. A continuous bijection induces a natural isomorphism between the Stone–Čech compactifications, and thus a homeomorphism between the canonical quotients.

COROLLARY 5.6.4. *Any bijective homomorphism in a category of models for an algebraic theory is an isomorphism.*

For instance, the inverse of a bijective homomorphism between groups is also a homomorphism. This, of course, is not difficult to prove. Corollary 5.6.4 eliminates the redundancy of proving the same result over and over again in many similar contexts. The contrapositive to Lemma 5.6.1 demonstrates that the categories Top or Poset are not monadic over sets: there exist non-invertible maps that act as the identity on underlying sets.

We now turn our focus to limits and colimits in a category of algebras.

THEOREM 5.6.5. *A monadic functor* $U\colon$ A $\to$ C *creates*

   (i) any limits that C *has, and*
   (ii) any colimits that C *has and the monad and its square preserve.*

PROOF. By Lemma 3.3.6, an equivalence A $\simeq$ C^T creates all limits and colimits present in either category, and so it suffices to prove this result for the forgetful functor $U^T\colon C^T \to C$ from the category of algebras for the induced monad.

To prove (i), consider a diagram $D\colon$ J $\to C^T$, spanning objects $(Dj, \gamma_j) \in C^T$, so that the underlying diagram $U^T D\colon$ J $\to$ C admits a limit cone $(\mu_j\colon L \to Dj)_{j \in J}$ in C. We wish to lift the summit L and the legs of the limit cone to a limit cone for the diagram D in C^T. The fact that D is a diagram of algebras implies that the algebra structure maps assemble into the components of a natural transformation $\gamma\colon TD \Rightarrow D \in C^J$ from the underlying C-valued diagram TD to the underlying C-valued diagram D. The composite natural transformation

$$TL \overset{T\mu}{\Longrightarrow} TD \overset{\gamma}{\Longrightarrow} D$$

defines a cone with summit TL over the diagram $D\colon$ J $\to$ C, which factors through the limit cone to define a unique map $\lambda\colon TL \to L$ in C so that the diagrams

$$
\begin{array}{ccc}
TL & \overset{T\mu_j}{\longrightarrow} & TDj \\
{\scriptstyle \lambda}\big\downarrow & & \big\downarrow{\scriptstyle \gamma_j} \\
L & \underset{\mu_j}{\longrightarrow} & Dj
\end{array}
$$

---

[18]To wit, a closed subset of a compact space is compact and so is its continuous image. If the target space is Hausdorff, then this compact subspace is closed. A continuous closed bijection is a homeomorphism.

commute for each $j \in J$. To verify that (L, λ) is a T-algebra, we must show that the left-most faces in the prisms

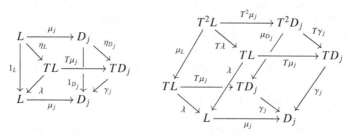

commute in C, knowing already that the remaining faces do. Appealing to the universal property of the limit cone $(\mu_j : L \to D_j)_{j \in J}$, it suffices to prove that these faces commute after composing with μ_j for each $j \in J$, and this is what the commutativity of the remaining faces of each prism demonstrates.[19] The verification that this is a limit cone proceeds similarly and is left to the reader.

A similar argument works for (ii) with the additional hypothesis that T and T^2 preserve the colimit cone in C under consideration. If $L \in C$ is the nadir of a colimit cone under a diagram $U^T D : J \to C$, then its algebra structure map $\lambda : TL \to L$ is induced by the universal property of the colimit cone under $TU^T D$ with nadir TL.[20] The details are left as Exercise 5.6.ii.   □

An immediate corollary completes the unfinished business of Proposition 4.5.15.

COROLLARY 5.6.6. *The inclusion of a reflective subcategory creates all limits. In particular, a reflective subcategory of a complete category is complete.*

PROOF. Proposition 5.3.3 proves that the inclusion of a reflective subcategory is monadic. By Theorem 5.6.5(i), it follows that this inclusion creates all limits that exist in the codomain.   □

Another corollary is the completeness result promised in §3.5.

COROLLARY 5.6.7. *Any category that is monadic over* Set *is complete, with limits created by the monadic forgetful functor.*

PROOF. Theorem 3.2.6 proves that Set is complete. Theorem 5.6.5 demonstrates that any category that is monadic over Set is also complete.   □

EXAMPLE 5.6.8. For instance, the *p*-**adic integers** are defined to be the limit of an ω^{op}-indexed diagram of rings

$$\mathbb{Z}_p := \lim\nolimits_n \mathbb{Z}/p^n \longrightarrow \cdots \longrightarrow \mathbb{Z}/p^3 \longrightarrow \mathbb{Z}/p^2 \longrightarrow \mathbb{Z}/p$$

By Example 3.2.10, because $U : \mathrm{Ring} \to \mathrm{Set}$ preserves limits, as a set

$$\mathbb{Z}_p = \left\{ (a_1 \in \mathbb{Z}/p, a_2 \in \mathbb{Z}/p^2, a_3 \in \mathbb{Z}/p^3, \ldots) \,\middle|\, a_n \equiv a_m \bmod p^{\min(n,m)} \right\}.$$

That is, a *p*-adic integer is a sequence of elements $a_n \in \mathbb{Z}/p^n$ that are compatible modulo congruence.

The underlying set functor also creates the limit cone, which tells us that the ring structure on this set of elements must be defined in such a way so that the projection functions

---

[19]This is the dual of the argument given in the proof of Proposition 5.4.9.

[20]The fact that the monad T preserves split coequalizers was used in this way in the proof of Proposition 5.4.9.

$\mathbb{Z}_p \to \mathbb{Z}/p^n$ are ring homomorphisms. This tells us that addition and multiplication of elements is "componentwise."

COROLLARY 5.6.9. Set *is cocomplete.*

PROOF. The contravariant power set functor P: $\mathsf{Set}^{\mathrm{op}} \to \mathsf{Set}$ is monadic, so the colimit of a diagram is created from the limit of the corresponding diagram on power sets.   □

COROLLARY 5.6.10. $\mathsf{Mod}_R \to \mathsf{Ab}$ *creates all colimits that* Ab *admits.*

PROOF. By Example 4.3.11, for any pair of abelian groups A and A', there is a natural isomorphism

$$\mathsf{Ab}(R \otimes_{\mathbb{Z}} A, A') \cong \mathsf{Ab}(A, \mathrm{Hom}_{\mathbb{Z}}(R, A'))$$

involving the tensor product of abelian groups and the abelian group $\mathrm{Hom}_{\mathbb{Z}}(R, A')$ of group homomorphisms $R \to A'$. In particular, the monad $R \otimes_{\mathbb{Z}} -$: $\mathsf{Ab} \to \mathsf{Ab}$ has a right adjoint $\mathrm{Hom}_{\mathbb{Z}}(R, -)$. By Theorem 4.5.3, $R \otimes_{\mathbb{Z}} -$ preserves all colimits and so Theorem 5.6.5(ii) applies to all diagrams in Ab.   □

In fact, Ab and any category of models for an algebraic theory is cocomplete. However, the construction of those colimits that are not preserved by the monad is more complicated. To explore this problem, consider the monadic forgetful functor U: $\mathsf{Group} \to \mathsf{Set}$. Both Set and Group admit coproducts; in Set these are simply disjoint unions, while in Group they are given by the free product. The **free product** of groups G and H is the group $G * H$ of words whose letters are drawn from the elements of G and H together with formal inverses, modulo relations defined using the group operations in each group. Note, in particular, that U: $\mathsf{Group} \to \mathsf{Set}$ does not preserve (and so, in particular, does not create) coproducts. However, there is a more precise description of the free product $G * H$ that can be stated entirely in terms of the free ⊣ forgetful adjunction

$$\mathsf{Set} \underset{U}{\overset{F}{\underset{\perp}{\rightleftarrows}}} \mathsf{Group},$$

which generalizes to define coproducts relative to any monadic adjunction.

A first approximation to the free product $G * H$ is given by $F(UG \sqcup UH)$, the free group on the disjoint union of the underlying sets of G and H. Elements of this group are words whose letters are elements of G or H, but it remains to impose the relations generated by words in G and words in H. These words in G and words in H are elements of the set $UFUG \sqcup UFUH$, defined to be the disjoint union of the underlying sets of the free groups on the underlying sets of G and H. The free group on this set $F(UFUG \sqcup UFUH)$ is the group of words of words in the letters G and H so that each subword is exclusively comprised of letters drawn from a single one of these groups. The desired relations that define $G * H$ as a quotient of $F(UG \sqcup UH)$ identify a word of words of this type with the word obtained by evaluating each subword using the group structure of G or of H.

To encode the desired relations, we define a natural pair of group homomorphisms.

$$F(UFUG \sqcup UFUH) \xrightarrow[\underset{F\kappa}{\qquad} FUF(UG \sqcup UH) \underset{\epsilon_{F(UG \sqcup UH)}}{\qquad}]{F(U\epsilon_G \sqcup U\epsilon_H)} F(UG \sqcup UH)$$

The fact that G and H are groups is encoded by a pair of homomorphisms ϵ_G: $FUG \to G$ and ϵ_H: $FUH \to H$, the components of the counit, that evaluate a word to the group element that it represents. The top map sends an element in the group $F(UFUG \sqcup UFUH)$, a word of words in which these subwords are exclusively in the group G or in the group H, to a word in $UG \sqcup UH$ by evaluating each subword to the corresponding group element.

The bottom composite homomorphism first makes use of the natural map κ of Exercise 3.3.i, which compares the coproduct of the image of two objects under a functor, in this case UF, with the image of the coproduct under that functor. Here this has the effect of regarding a word of words, with subwords exclusively in either G or H, as a word of words whose letters might belong to either G or H, but happen in this case to belong to only one or the other. The map $\epsilon_{F(UG \sqcup UH)}$, again a counit component, then concatenates to produce a single word in letters of G and H. From this point, we see that the coequalizer of these two group homomorphisms imposes exactly the relations desired to define the free product:

$$F(UFUG \sqcup UFUH) \xrightarrow{F(U\epsilon_G \sqcup U\epsilon_H)} F(UG \sqcup UH) \twoheadrightarrow G * H$$
$$\xrightarrow[F\kappa]{} FUF(UG \sqcup UH) \xrightarrow[\epsilon_{F(UG \sqcup UH)}]{}$$

This construction of the coproduct generalizes to any monadic functor $U : A \to C$ over a cocomplete category whenever the coequalizer just described exists in A.

PROPOSITION 5.6.11. *Suppose* C *is cocomplete and* $U : A \to C$ *is monadic. Then the following are equivalent:*

(i) A *is cocomplete.*
(ii) A *has coequalizers.*

PROOF. The implication (i) $\Rightarrow$ (ii) is trivial. For (ii) $\Rightarrow$ (i), we may replace A by the equivalent category of algebras C^T. It suffices by Theorem 3.4.12 to prove that a category of algebras that has coequalizers also has coproducts. The coproduct (A, α) of a family of T-algebras $(A_i, \alpha_i) \in C^T$ is defined to be the coequalizer

$$(T(\textstyle\coprod_i TA_i), \mu_{\coprod_i TA_i}) \xrightarrow{T(\coprod_i \alpha_i)} (T(\textstyle\coprod_i A_i), \mu_{\coprod_i A_i}) \twoheadrightarrow (A, \alpha).$$
$$\xrightarrow[T\kappa]{} (T^2(\textstyle\coprod_i A_i), \mu_{T(\coprod_i A_i)}) \xrightarrow[\mu_{\coprod_i A_i}]{}$$

Note that the objects in the coequalizer diagram are all free algebras. Using the adjunction $F^T \dashv U^T$, it is straightforward to show that (A, α) has the desired universal property of the coproduct of the objects (A_i, α_i). The proof is left as Exercise 5.6.iii. □

It remains to enumerate conditions that suffice to prove that the category of algebras has coequalizers and is therefore cocomplete.

THEOREM 5.6.12. *If* $T : C \to C$ *is a finitary monad on a complete and cocomplete, locally small category* C, *then the category* C^T *of* T-*algebras is also complete and cocomplete.*

The proof presented here can be found in [**Bor94b**].

PROOF. By Theorem 5.6.5, completeness of C implies that C^T is complete. By Proposition 5.6.11, to show that C^T is cocomplete, it suffices to show that C^T has coequalizers. By Proposition 4.5.1, to show that C^T has coequalizers, we must show that the constant diagram functor from the category of T-algebras to the category of parallel pairs of T-algebras

$$C^T \xleftarrow[\Delta]{\overset{\text{colim}}{\underset{\perp}{\longleftarrow}}} (C^T)^{\bullet \rightrightarrows \bullet}$$

has a left adjoint. By Proposition 3.3.9, which says that limits in a functor category are computed objectwise, the constant diagram functor preserves limits. To prove that the desired left adjoint exists, we apply the General Adjoint Functor Theorem 4.6.3; the conclusion follows as soon as we show that the functor Δ satisfies the solution set condition.

To apply Theorem 4.6.3, we must produce a solution set for each object in the category $(\mathsf{C}^T)^{\bullet\rightrightarrows\bullet}$, i.e., for each parallel pair of T-algebra morphisms $f, g \colon (A, \alpha) \rightrightarrows (B, \beta)$. We will produce a solution set consisting of a single element by defining a fork

$$(A, \alpha) \underset{g}{\overset{f}{\rightrightarrows}} (B, \beta) \xrightarrow{\ q\ } (Q, u)$$

so that any other fork under the pair f, g in C^T factors through q. If we knew how to form coequalizers in C^T, this would be easy; instead, the T-algebra (Q, u) and T-algebra morphism q is a reasonable approximation[21] to the coequalizer defined using an inductive construction.

To start, we define the coequalizer $q_0 \colon B \twoheadrightarrow Q_0$ in C of the pair $f, g \colon A \rightrightarrows B$. If Tq_0 were the coequalizer of Tf and Tg in C, then its universal property would enable us to equip Q_0 with the structure of a T-algebra, but this is unlikely as we do not assume that the monad preserves coequalizers. Instead, we form the coequalizer $p_0 \colon TB \twoheadrightarrow P_0$ of $Tf, Tg \colon TA \rightrightarrows TB$ and the induced maps

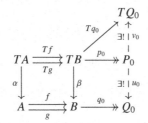

Continuing inductively, for each $n \geq 0$, we will produce a diagram in C

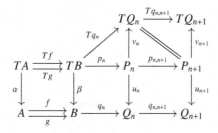

so that

- the p_n and q_n define cones under the parallel pairs (Tf, Tg) and (f, g), respectively,
- $p_{n+1} = p_{n,n+1} \cdot p_n$ and $q_{n+1} = q_{n,n+1} \cdot q_n$, and
- the ω-indexed diagram $(P_n)_{n\in\omega}$ is a shifted version of the ω-indexed diagram $(TQ_n)_{n\in\omega}$.

For the inductive step, define $P_{n+1} = TQ_n$, define Q_{n+1} to be the coequalizer

$$TP_n \underset{Tu_n}{\overset{\mu_{Q_n}\cdot Tv_n}{\rightrightarrows}} TQ_n \xrightarrow{\ u_{n+1}\ } Q_{n+1},$$

define

$$p_{n,n+1} := v_n, \qquad q_{n,n+1} := u_{n+1} \cdot \eta_{Q_n}, \qquad v_{n+1} := Tq_{n,n+1},$$

and define p_{n+1} and q_{n+1} to be the necessary composites.

---

[21]The solution set condition does not demand that factorizations through q are unique.

This construction yields a pair of ω-indexed diagrams $(P_n)_{n\in\omega}$ and $(Q_n)_{n\in\omega}$ in $\mathbf{C}$ with a natural transformation $u\colon P \Rightarrow Q$ between them. We extend this data by forming the colimits:

$$
\begin{array}{ccccccc}
& TQ_0 \xrightarrow{Tq_{0,1}} TQ_1 \xrightarrow{Tq_{1,2}} \cdots \longrightarrow & \mathrm{colim}_n\, TQ_n \cong T\,\mathrm{colim}_n\, Q_n \cong TQ_\omega \\[2pt]
& \| \qquad \| & \|! \\[2pt]
P_0 \xrightarrow{p_{0,1}} P_1 \xrightarrow{p_{1,2}} P_2 \xrightarrow{p_{2,3}} \cdots \longrightarrow & \mathrm{colim}_n\, P_n \\[2pt]
u_0\downarrow\quad u_1\downarrow\quad u_2\downarrow & \exists!\,|\,u_\omega \\[2pt]
Q_0 \xrightarrow{q_{0,1}} Q_1 \xrightarrow{q_{1,2}} Q_2 \xrightarrow{q_{2,3}} \cdots \longrightarrow & \mathrm{colim}_n\, Q_n =: Q_\omega
\end{array}
$$

As the category ω is filtered and the monad T preserves filtered colimits, the functor T carries the colimit cone $(Q_n \xrightarrow{q_{n,\omega}} Q_\omega)_{n\in\omega}$ under the bottom sequence to a colimit cone $(TQ_n \xrightarrow{Tq_{n,\omega}} TQ_\omega)_{n\in\omega}$ under the top one. Adding the initial leg, this data extends to define a colimit cone over the ω-diagram of P_n's. In particular, the natural transformation $u\colon P \Rightarrow Q$ induces a map $u_\omega\colon TQ_\omega \to Q_\omega$ between the colimits so that $u_\omega \cdot Tq_{n,\omega} = q_{n+1,\omega}\cdot u_{n+1}$.

We claim that (Q_ω, u_ω) is a T-algebra. For the unit condition it suffices, by the universal property of Q_ω to show that $u_\omega \cdot \eta_{Q_\omega} \cdot q_{n,\omega} = q_{n,\omega}$ for all sufficiently large n. Indeed,

$$u_\omega \cdot \eta_{Q_\omega} \cdot q_{n,\omega} = u_\omega \cdot Tq_{n,\omega} \cdot \eta_{Q_n} = q_{n+1,\omega} \cdot u_{n+1} \cdot \eta_{Q_n} = q_{n+1,\omega} \cdot q_{n,n+1} = q_{n,\omega}$$

by naturality of η, the definition of u_ω, the definition of $q_{n,n+1}$, and the fact that the $q_{n,\omega}$ form a cone.

For the associativity condition, observe that there is an ω-indexed sequence of coequalizer diagrams defining the maps u_n

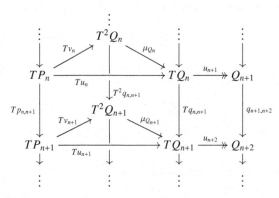

Passing to the sequential colimits, which the finitary monad T preserves, observe that v_ω is the identity. So we conclude that $u_\omega\colon TQ_\omega \to Q_\omega$ is the coequalizer of Tu_ω and μ_{Q_ω}, and so in particular $u_\omega \cdot Tu_\omega = u_\omega \cdot \mu_{Q_\omega}$. Thus, (Q_ω, u_ω) is a T-algebra.

By construction, the map $q_{0,\omega}\colon (B,\beta) \to (Q_\omega, u_\omega)$ is a T-algebra morphism. It remains to show that any other T-algebra morphism $h\colon (B,\beta) \to (C,\gamma)$ that defines a fork under the pair $f, g\colon (A,\alpha) \rightrightarrows (B,\beta)$ factors through $q_{0,\omega}$. By the universal property of the sequential colimit $Q_\omega = \mathrm{colim}_n\, Q_n$, to define the factorization $k\colon Q_\omega \to C$ it suffices to define its components

$$k_n := Q_n \xrightarrow{q_{n,\omega}} Q_\omega \xrightarrow{k} C.$$

For each $n \in \omega$, we will show that the diagram

(5.6.13)

$$
\begin{array}{ccc}
P_n \xrightarrow{\;v_n\;} TQ_n \xrightarrow{\;Tk_n\;} TC \\
\quad \searrow{\scriptstyle u_n} \qquad \qquad \downarrow{\scriptstyle \gamma} \\
\qquad Q_n \xrightarrow[\;k_n\;]{} C
\end{array}
$$

commutes in C. Passing to sequential colimits, v_ω is an identity, so it will follow that $k : (Q_\omega, u_\omega) \to (C, \gamma)$ defines an algebra map.

Define k_0 to be the unique factorization of h through the coequalizer Q_0 of $f, g : A \rightrightarrows B$. To prove (5.6.13) in the case $n = 0$, it suffices to prove commutativity after restricting along the epimorphism $p_0 : TB \twoheadrightarrow P_0$, and indeed

$$
\gamma \cdot Tk_0 \cdot v_0 \cdot p_0 = \gamma \cdot Tk_0 \cdot Tq_0 = \gamma \cdot Th = h \cdot \beta = k_0 \cdot q_0 \cdot \beta = k_0 \cdot u_0 \cdot p_0 .
$$

Inductively, define k_{n+1} to be the unique factorization of $\gamma \cdot Tk_n : TQ_n \to C$ through the coequalizer $u_{n+1} : P_{n+1} \twoheadrightarrow Q_{n+1}$; a straightforward diagram chase shows that $\gamma \cdot Tk_n$ defines a fork under the pair coequalized by u_{n+1}. Another diagram chase shows that the condition (5.6.13) is satisfied by k_{n+1}. A final diagram chase verifies that the maps $k_n : Q_n \to C$ assemble into a cone under the ω-indexed diagram whose colimit is Q_ω. Thus, they induce the desired map $k : Q_\omega \to C$. Because T preserves these filtered colimits, the pentagon conditions (5.6.13) imply that

$$
\begin{array}{ccc}
TQ_\omega & \xrightarrow{\;Tk\;} & TC \\
{\scriptstyle u_\omega}\downarrow & & \downarrow{\scriptstyle \gamma} \\
Q_\omega & \xrightarrow[\;k\;]{} & C
\end{array}
$$

so k defines a T-algebra homomorphism. With the solution set condition so-verified, Theorem 4.6.3 implies that the constant diagram functor $\Delta : \mathsf{C}^T \to (\mathsf{C}^T)^{\bullet \rightrightarrows \bullet}$ has a left adjoint and thus that C^T has coequalizers. $\square$

In particular, by Theorems 3.2.6 and 5.6.9, Set is complete and cocomplete. By Definition 5.5.5, categories of models for an algebraic theory are algebras for a finitary monad. Applying Theorem 5.6.12, we conclude:

COROLLARY 5.6.14. *Any category of models for an algebraic theory is cocomplete.*

While the monad β for compact Hausdorff spaces is not finitary, the category cHaus is nonetheless cocomplete. Example 4.5.14(i) observes that compact Hausdorff spaces form a reflective subcategory of Top. The claim follows from Propositions 3.5.2 and 4.5.15.

Exercises.

EXERCISE 5.6.i. Exercise 3.3.iv suggests an alternate form of the monadicity theorem: a functor $U : \mathsf{D} \to \mathsf{C}$ is monadic if U has a left adjoint, if U reflects isomorphisms, and if D has and U preserves U-split coequalizers. Use this, together with a direct proof that any continuous bijection between compact Hausdorff spaces is a homeomorphism, to re-prove Corollary 5.5.6 by showing that cHaus has U-split coequalizers, constructed as in Top. (Hint: It is easy to see that the coequalizer of maps $f, g : X \rightrightarrows Y$ of compact spaces is compact. To prove that a U-split coequalizer is Hausdorff, use the split coequalizer diagram in Set to prove that the kernel pair $E \hookrightarrow Y \times Y$ of the quotient map $Y \twoheadrightarrow Z$ is closed.)

EXERCISE 5.6.ii. Prove Theorem 5.6.5(ii).

EXERCISE 5.6.iii. Prove that the T-algebra constructed as a coequalizer in the proof of Proposition 5.6.11 defines the coproduct in the category of T-algebras.

CHAPTER 6

All Concepts are Kan Extensions

> The notion of Kan extensions subsumes all the
> other fundamental concepts of category theory.
>
> Saunders Mac Lane, *Categories for the Working
> Mathematician* [**ML98a**]

Extension problems are pervasive in mathematics. For instance, iterated products of a fixed positive real number, such as 2, define its natural number powers. The resulting function $2^-\colon \mathbb{N} \to \mathbb{R}_{>0}$ can be extended to a homomorphism from the additive group of integers to the multiplicative group of non-negative reals by declaring $2^{-n} := (\frac{1}{2})^n$ for $n \in \mathbb{N}$. The resulting function $2^-\colon \mathbb{Z} \to \mathbb{R}_{>0}$ can be extended further to an additive homomorphism defined on the rationals by declaring $2^{\frac{1}{n}} := \sqrt[n]{2}$ for $n \in \mathbb{N}$. Finally, $2^-\colon \mathbb{Q} \to \mathbb{R}_{>0}$ can be extended to an order-preserving function on the reals, not via any explicitly described arithmetic formula, but by using the fact that $\mathbb{R}_{>0}$ has limits of all bounded increasing sequences.

The construction of the function $2^-\colon \mathbb{R} \to \mathbb{R}_{>0}$ as an order-preserving function extending $2^-\colon \mathbb{Q} \to \mathbb{R}_{>0}$ is a special case of a solution to a categorically-defined problem of extending a functor $F\colon \mathsf{C} \to \mathsf{E}$ along another functor $K\colon \mathsf{C} \to \mathsf{D}$, possibly but not necessarily a subcategory inclusion. In general, strict extensions may or may not exist. Obstructions can take several forms: two arrows in C with distinct images in E might be identified in D, or two objects in C might have a morphism between their images in D but not in E. For this reason, it is more reasonable to ask for a "best approximation" to an extension taking the form of a universal natural transformation pointing either from or to F. The resulting categorical notion, quite simple to define, is surprisingly ubiquitous throughout mathematics, as we shall soon discover.

Solutions to this categorical extension problem, the first examples of which were studied in [**Kan58**], are called *Kan extensions* and come in dual left-handed and right-handed forms. The general definition and some basic examples of Kan extensions are introduced in §6.1, which explains how Kan extensions define partial adjoints to the functor $K^*\colon \mathsf{E}^\mathsf{D} \to \mathsf{E}^\mathsf{C}$ defined by restriction along $K\colon \mathsf{C} \to \mathsf{D}$. A formula, defining left Kan extensions as certain colimits and right Kan extensions as certain limits, is presented in §6.2, which in particular provides criteria guaranteeing that certain Kan extensions exist. In §6.3, we discover that the (co)limit formulae imply that the Kan extensions so-defined satisfy a stronger universal property than was originally supposed: namely, Kan extensions defined by the (co)limit formulae are *pointwise* Kan extensions. In §6.4, we study total derived functors, which are defined as certain Kan extensions, and introduce an abstract general framework that constructs "point-set level" lifts of these derived functors. Unexpectedly—since total derived functors take values in *homotopy categories*, which have few limits and colimits— the total derived functors constructed in this manner have the stronger universal property of being pointwise and in fact *absolute* Kan extensions, a fact that has significant consequence

189

for derived adjunctions. We conclude in §6.5 by showing how simple special cases of Kan extensions can be used to redefine adjunctions, limits, colimits, and monads and generalize the Yoneda lemma.

6.1. Kan extensions

The data of a Kan extension consists of a functor and a natural transformation satisfying a universal property relative to other similar pairs. For the reasons discussed in §4.4, this definition can also be stated for 2-categories other than the 2-category of categories, functors, and natural transformations, in which case the eponym "Kan" is frequently dropped. Since we are only interested in this one example of a 2-category at present, we retain the traditional name.

DEFINITION 6.1.1. Given functors $F: C \to E$, $K: C \to D$, a **left Kan extension** of F along K is a functor $\mathrm{Lan}_K F: D \to E$ together with a natural transformation $\eta: F \Rightarrow \mathrm{Lan}_K F \cdot K$ such that for any other such pair $(G: D \to E, \gamma: F \Rightarrow GK)$, γ factors uniquely through η as illustrated.[1]

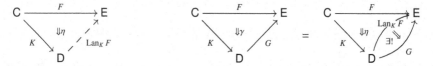

Dually, a **right Kan extension** of F along K is a functor $\mathrm{Ran}_K F: D \to E$ together with a natural transformation $\epsilon: \mathrm{Ran}_K F \cdot K \Rightarrow F$ such that for any $(G: D \to E, \delta: GK \Rightarrow F)$, δ factors uniquely through ϵ as illustrated.

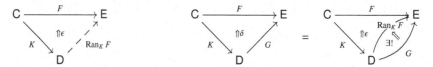

REMARK 6.1.2. When $K: C \to D$ is fully faithful, then assuming certain (co)limits exist, the left and right Kan extensions do in fact extend the functor F along K, at least up to natural isomorphism; see Corollary 6.3.9. But even a genuine on-the-nose extension of F along K might not necessarily define a Kan extension; see Exercises 6.1.i and 6.1.ii.

EXAMPLE 6.1.3. For any object $c \in C$ and any $F: C \to Set$, there is a bijection between elements $x \in Fc$ and natural transformations with boundary as displayed

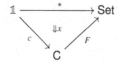

where $*$ denotes the functor that picks out the singleton set. By the Yoneda lemma, the representable functor $C(c, -)$ and the identity $1_c \in C(c, c)$ define the left Kan extension of

---

[1] Writing α for the natural transformation $\mathrm{Lan}_K F \Rightarrow G$, the right-hand *pasting diagrams* express the equality $\gamma = \alpha K \cdot \eta$, i.e., that γ factors as $F \overset{\eta}{\Longrightarrow} \mathrm{Lan}_K F \cdot K \overset{\alpha K}{\Longrightarrow} GK$.

$*: \mathbb{1} \to \mathsf{Set}$ along $c: \mathbb{1} \to \mathsf{C}$. The required unique factorization

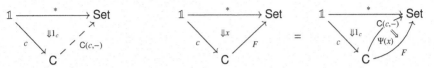

is the natural transformation $\Psi(x): \mathsf{C}(c, -) \Rightarrow F$ defined in the proof of Theorem 2.2.4. The pasting identity asserts that $\Psi(x)_c(1_c) = x$.

Replacing C by C^{op}, this argument shows that the contravariant representable functor $\mathsf{C}(-, c)$ defines the left Kan extension of $*: \mathbb{1} \to \mathsf{Set}$ along $c: \mathbb{1} \to \mathsf{C}^{op}$.

EXAMPLE 6.1.4. For any group G, there is a unique functor $\mathbb{1} \to BG$ picking out the single object in this category. For any category C, object $c \in \mathsf{C}$, and G-object X in C, natural transformations as displayed below-left and below-right

correspond to maps in C between c and the underlying object of the G-object X, with the group action forgotten.

If C has coproducts, then morphisms $c \to X$ correspond to G-equivariant maps $\coprod_G c \to X$, where G acts on the left of the indexing set of the coproduct. This bijection is implemented by restricting the G-equivariant map to the component indexed by the identity element $e \in G$. Thus, we see that

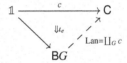

is a left Kan extension diagram.

Dually, if C has products, then morphisms $X \to c$ correspond to G-equivariant maps $X \to \prod_G c$, where G acts on the right[2] of the indexing set of the product. This bijection is implemented by projecting the G-equivariant map to the component indexed by the identity element. Thus, we see that

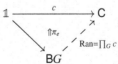

is a right Kan extension diagram.

The universal property that defines Kan extensions posits a bijection between certain classes of natural transformations, implemented by composing with a given natural transformation. In practice, the functor categories E^{C} and E^{D} referred to in Definition 6.1.1 might not be locally small, in which case the bijection may be between a pair of proper classes.

---

[2]By the universal property of the product, a function $f: A \to B$ between indexing sets induces a map $f^*: \prod_B c \to \prod_A c$ defined at the component indexed by $a \in A$ by projecting from the domain to the component indexed by $f(a) \in B$.

Passing to a higher set-theoretical universe, we can nonetheless think of a left Kan extension of $F: C \to E$ along $K: C \to D$ as a representation for the functor

$$E^C(F, - \cdot K): E^D \to SET$$

that sends a functor $D \to E$ to the collection of natural transformations from F to its restriction along K. By the Yoneda lemma, any pair (G, γ) as in Definition 6.1.1 defines a natural transformation

$$E^D(G, -) \overset{\gamma}{\Longrightarrow} E^C(F, - \cdot K).$$

The universal property satisfied by the left Kan extension $(\text{Lan}_K F, \eta)$ is equivalent to the assertion that the corresponding map

$$E^D(\text{Lan}_K F, -) \overset{\eta}{\Longrightarrow} E^C(F, - \cdot K)$$

is a natural isomorphism, i.e., that $(\text{Lan}_K F, \eta)$ represents the functor $E^C(F, - \cdot K)$.

Fixing $K: C \to D$ but letting $F \in E^C$ vary:

PROPOSITION 6.1.5. *If, for fixed $K: C \to D$ and E, the left and right Kan extensions of any functor $F: C \to E$ along K exist, then these define left and right adjoints to the pre-composition functor $K^*: E^D \to E^C$.*
(6.1.6)

$$E^D(\text{Lan}_K F, G) \cong E^C(F, GK) \qquad E^C \underset{\xleftarrow{\;\;\perp\;\;}}{\overset{\text{Lan}_K}{\underset{\xrightarrow{K^*}}{\overset{\perp}{\longleftarrow}}}} E^D \qquad E^C(GK, F) \cong E^D(G, \text{Ran}_K F)$$

PROOF. If the functors $\text{Lan}_K F, \text{Ran}_K F: D \rightrightarrows E$ exist for any $F: C \to E$, then Proposition 4.3.4 applied to the natural isomorphisms of Definition 6.1.1 tells us that these values assemble to define adjoint functors $\text{Lan}_K \dashv K^* \dashv \text{Ran}_K$. The 2-cell $\eta: F \Rightarrow \text{Lan}_K F \cdot K$ is the component of the unit for $\text{Lan}_K \dashv K^*$ indexed by the object $F \in E^C$; dually, $\epsilon: \text{Ran}_K F \cdot K \Rightarrow F$ is a component of the counit for $K^* \dashv \text{Ran}_K$. The universal property of the left Kan extension declares that $\eta: F \Rightarrow \text{Lan}_K F \cdot K$ is initial in the comma category $F \downarrow K^*$; see Lemma 4.6.1.   □

Conversely, by uniqueness of adjoints (Proposition 4.4.1), the objects in the image of any left or right adjoint to a pre-composition functor are Kan extensions. This observation leads to several immediate examples.

EXAMPLE 6.1.7. The objects of the functor category Vect_k^{BG} are G-representations over the field k, and arrows are G-equivariant linear maps. If H is a subgroup of G, restriction $\text{res}_H^G: \text{Vect}_k^{BG} \to \text{Vect}_k^{BH}$ of a G-representation to an H-representation is simply pre-composition by the inclusion functor $BH \hookrightarrow BG$. This functor has a left adjoint, called **induction**, and a right adjoint, called **coinduction**.

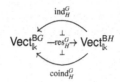

By Proposition 6.1.5, the induction functor is given by left Kan extension along the inclusion $BH \hookrightarrow BG$, while coinduction is given by right Kan extension. The reader unfamiliar with the construction of induced representations need not remain in suspense for very long;

see Theorem 6.2.1 and Example 6.2.8. Similar remarks apply for G-sets, G-spaces, based G-spaces, or indeed G-objects in any category, although if the ambient category has few limits and colimits, these adjoints need not exist.

EXAMPLE 6.1.8. The category Δ of finite non-empty ordinals and order-preserving maps is a full subcategory containing all but the initial object of the category Δ_+, of finite ordinals and order-preserving maps. Presheaves on Δ are **simplicial sets**, while presheaves on Δ_+ are called **augmented simplicial sets**. Left Kan extension defines a left adjoint to restriction

$$\mathsf{Set}^{\Delta_+^{\mathrm{op}}} \underset{\overset{\perp}{\longleftarrow}}{\overset{\pi_0}{\underset{\mathrm{res}}{\overset{\perp}{\longrightarrow}}}} \mathsf{Set}^{\Delta^{\mathrm{op}}}$$

$$\mathrm{triv}$$

that augments a simplicial set X with its set $\pi_0 X$ of path components. Right Kan extension assigns a simplicial set the trivial augmentation built from the one-point set, as can easily be verified by readers familiar with the combinatorics of simplicial sets.

Exercises.

EXERCISE 6.1.i. Construct a toy example to illustrate that if F factors through K along some functor H, it is not necessarily the case that $(H, 1_F)$ is either the left or right Kan extension of F along K.

EXERCISE 6.1.ii. Consider the functor $d^1 \colon 2 \hookrightarrow 3$ whose image omits the middle object. Show that for any category C and any functor $2 \to \mathsf{C}$, there exist both left and right Kan extensions along $d^1 \colon 2 \hookrightarrow 3$ by giving an explicit construction. Observe that both the left and right Kan extensions extend the original diagram

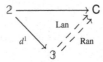

but, nonetheless, these functors are not typically isomorphic.

EXERCISE 6.1.iii. Define the category in which the right Kan extension of $F \colon \mathsf{C} \to \mathsf{E}$ along $K \colon \mathsf{C} \to \mathsf{D}$ is terminal and explain the connection between this category and the functor $\mathsf{E}^{\mathsf{C}}(- \cdot K, F) \colon (\mathsf{E}^{\mathsf{D}})^{\mathrm{op}} \to \mathsf{Set}$, in the case where these functor categories are locally small.

6.2. A formula for Kan extensions

In this section, we will see that right and left Kan extensions along any functor $K \colon \mathsf{C} \to \mathsf{D}$ whose domain is small and whose codomain is locally small exist whenever the target category E has certain limits or colimits, respectively. Moreover, these (co)limits provide an explicit formula for the left and right Kan extension functors.

In the case of left Kan extensions, we are seeking to define a functor $\mathrm{Lan}_K F \colon \mathsf{D} \to \mathsf{E}$ that is the "closest approximation to an extension of F along K from the left," i.e., up to a natural transformation η

that is universal among natural transformations $\gamma\colon F \Rightarrow GK$.

So, to define the value of $\mathrm{Lan}_K F(d)$ at an object $d \in \mathsf{D}$, we should consider all possible approximations to d "from the left" that come from the category C. This leads us to consider the comma category $K \downarrow d$ whose objects are morphisms $Kc \to d$ from the image of a specified $c \in \mathsf{C}$ and whose morphisms are morphisms $c \to c'$ in C that give rise to commutative triangles in D. The objects of the category $K \downarrow d$ parametrize approximations to the object d coming from C from the left.

Recall that $K \downarrow d$ is the category of elements of the functor $\mathsf{D}(K-, d)\colon \mathsf{C}^{\mathrm{op}} \to \mathsf{Set}$, and as such comes with a canonical projection functor $\Pi^d\colon K \downarrow d \to \mathsf{C}$. The composite $K \downarrow d \xrightarrow{\Pi^d} \mathsf{C} \xrightarrow{F} \mathsf{E}$ of the projection functor with F sends each approximation $Kc \to d$ to the corresponding object Fc in E that approximates the yet-to-be-defined $\mathrm{Lan}_K F(d)$. Based on this intuition, it seems reasonable to define $\mathrm{Lan}_K F(d)$ to be the colimit of this diagram, and indeed if these colimits exist, they define the left Kan extension $\mathrm{Lan}_K F\colon \mathsf{D} \to \mathsf{E}$.

THEOREM 6.2.1. *Given functors $F\colon \mathsf{C} \to \mathsf{E}$ and $K\colon \mathsf{C} \to \mathsf{D}$, if for every $d \in \mathsf{D}$ the colimit*

$$(6.2.2) \qquad \mathrm{Lan}_K F(d) := \mathrm{colim}(K \downarrow d \xrightarrow{\Pi^d} \mathsf{C} \xrightarrow{F} \mathsf{E})$$

exists, then they define the left Kan extension $\mathrm{Lan}_K F\colon \mathsf{D} \to \mathsf{E}$, in which case the unit transformation $\eta\colon F \Rightarrow \mathrm{Lan}_K F \cdot K$ can be extracted from the colimit cone. Dually, if for every $d \in \mathsf{D}$ the limit

$$(6.2.3) \qquad \mathrm{Ran}_K F(d) := \mathrm{lim}(d \downarrow K \xrightarrow{\Pi_d} \mathsf{C} \xrightarrow{F} \mathsf{E})$$

exists, then they define the right Kan extension $\mathrm{Ran}_K F\colon \mathsf{D} \to \mathsf{E}$, in which case the counit transformation $\epsilon\colon \mathrm{Ran}_K F \cdot K \Rightarrow F$ can be extracted from the limit cone.

PROOF. The first task is to extend the data (6.2.2) defining the objects $\mathrm{Lan}_K F(d)$, each with a specified colimit cone $\lambda^d\colon F\Pi^d \Rightarrow \mathrm{Lan}_K F(d)$, to a functor $\mathrm{Lan}_K F\colon \mathsf{D} \to \mathsf{E}$. A morphism $g\colon d \to d'$ in D induces a canonical functor, which commutes with the projections to C,

$$K \downarrow d \xrightarrow{g_*} K \downarrow d',$$
$$\Pi^d \searrow \quad \swarrow \Pi^{d'}$$
$$\mathsf{C}$$

and is defined on objects by composing with g. Any cone under a diagram indexed by the category $K \downarrow d'$ also defines a cone under the diagram defined by restricting along $g_*\colon K \downarrow d \to K \downarrow d'$; the legs of the new cone are defined by reindexing the legs of the old cone. In particular, the colimit cone $\lambda^{d'}\colon F\Pi^{d'} \Rightarrow \mathrm{Lan}_K F(d')$ that defines $\mathrm{Lan}_K F(d')$ can be reindexed to define a cone under the diagram $F\Pi^d = F\Pi^{d'} g_*$. By the universal property of the colimit cone $\lambda^d\colon F\Pi^d \Rightarrow \mathrm{Lan}_K F(d)$, there is a unique morphism between their nadirs in E commuting with the legs of the respective colimit cones:

$$\lambda^d_{f\colon Kc\to d} \swarrow \overset{Fc}{\quad} \searrow \lambda^{d'}_{gf\colon Kc\to d'}$$
$$\mathrm{Lan}_K F(d) \dashrightarrow_{\mathrm{Lan}_K F(g)} \mathrm{Lan}_K F(d'),$$

which we take to be $\mathrm{Lan}_K F(g)$. Uniqueness implies that $\mathrm{Lan}_K F$ is functorial. The leg of the unit transformation $\eta_c\colon Fc \to \mathrm{Lan}_K F(Kc)$, indexed by an object $c \in \mathsf{C}$, is defined to be the leg

$$\eta_c := \lambda^{Kc}_{1_{Kc}\colon Kc\to Kc}\colon Fc \to \mathrm{Lan}_K F(Kc)$$

of the colimit cone defining $\mathrm{Lan}_K F(Kc)$. For any $h\colon c \to c'$ in C, there are commuting triangles

$$
\begin{array}{ccc}
Fc & \xrightarrow{\;\;Fh\;\;} & Fc' \\
\lambda^{Kc}_{1_{Kc}} \downarrow & \searrow{\scriptstyle \lambda^{Kc'}_{Kh}} & \downarrow \lambda^{Kc'}_{1_{Kc'}} \\
\mathrm{Lan}_K F(Kc) & \xrightarrow[\mathrm{Lan}_K F(Kh)]{} & \mathrm{Lan}_K F(Kc')
\end{array}
$$

proving that $\eta\colon F \Rightarrow \mathrm{Lan}_K F \cdot K$ is natural.

It remains to show that the pair $(\mathrm{Lan}_K F, \eta)$ has the universal property that defines the left Kan extension of F along K. That is, given any pair (G, γ) as below-left, we wish to construct the unique factorization $\alpha\colon \mathrm{Lan}_K F \Rightarrow G$ displayed below-right:

(6.2.4)

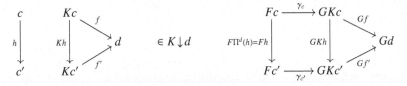

Each leg $\alpha_d\colon \mathrm{Lan}_K F(d) \to Gd$ is defined to be the unique factorization of some cone under $F\Pi^d$ with nadir Gd through the colimit cone $\lambda^d\colon F\Pi^d \Rightarrow \mathrm{Lan}_K F(d)$:

(6.2.5)

$$
\begin{array}{ccc}
Fc & \xrightarrow{\;\lambda^d_{f\colon Kc \to d}\;} & \mathrm{Lan}_K F(d) \\
\gamma_c \downarrow & & \vdots\, \alpha_d \\
GKc & \xrightarrow[\;\;Gf\;\;]{} & Gd
\end{array}
$$

The leg of this cone, indexed by the object $f\colon Kc \to d \in K \downarrow d$, is the left-bottom composite displayed above. For each morphism $h\colon c \to c'$ in $K \downarrow d$, the commutative diagram displayed below-right

$$
\begin{array}{ccc}
c \\
\downarrow h \\
c'
\end{array}
\qquad
\begin{array}{ccc}
Kc & \xrightarrow{\;f\;} & \\
Kh \downarrow & \searrow & d \\
Kc' & \nearrow_{f'} &
\end{array}
\ \in K \downarrow d
\qquad
F\Pi^d(h)=Fh
\qquad
\begin{array}{ccc}
Fc & \xrightarrow{\;\gamma_c\;} & GKc & \\
F\Pi^d(h) \downarrow & GKh \downarrow & \searrow^{Gf} & Gd \\
Fc' & \xrightarrow{\;\gamma_{c'}\;} & GKc' & \nearrow_{Gf'}
\end{array}
$$

implies that these legs indeed define a cone $F\Pi^d \Rightarrow Gd$.

To show that $\alpha\colon \mathrm{Lan}_K F \Rightarrow G$ is natural in morphisms $g\colon d \to d'$ in D, it suffices, by the universal property of $\mathrm{Lan}_K F(d)$, to demonstrate that the outer pentagon

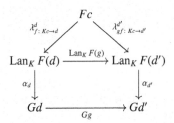

commutes for each $f\colon Kc \to d$. Using the commutative squares (6.2.5) that define the components of α, this follows easily from functoriality of G and naturality of γ.

Finally, to verify the pasting equality (6.2.4), we must show that $\alpha_{Kc} \cdot \eta_c = \gamma_c$ for each $c \in \mathsf{C}$. From our definition of $\eta\colon F \Rightarrow \mathrm{Lan}_K F$, this is immediate, as a special case

$$
\begin{array}{ccc}
Fc & \xrightarrow{\;\;\lambda^{Kc}_{1_{Kc}}\colon Kc \to Kc\;\;} & \mathrm{Lan}_K F(Kc) \\
& {\scriptstyle =:\eta_c} & \\
\gamma_c \downarrow & & \downarrow \alpha_{Kc} \\
GKc & =\!\!=\!\!=\!\!=\!\!= & GKc \\
& {\scriptstyle G1_{Kc}} &
\end{array}
$$

of the commutative squares that define the components of α. To show that $\alpha\colon \mathrm{Lan}_K F \Rightarrow G$ is the unique natural transformation defining a factorization $\alpha_K \cdot \eta = \gamma$, observe that naturality of α in $f\colon Kc \to d$

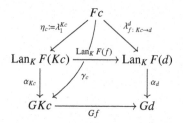

forces us to define α_d as in (6.2.5).   □

COROLLARY 6.2.6. *If $K\colon \mathsf{C} \to \mathsf{D}$ is a functor so that C is small and D is locally small then:*

(i) *If E is cocomplete, then left Kan extensions $\mathsf{E}^{\mathsf{D}} \underset{K^*}{\overset{\mathrm{Lan}_K}{\underset{\perp}{\leftrightarrows}}} \mathsf{E}^{\mathsf{C}}$ exist and are given by the colimit formula (6.2.2).*

(ii) *If E is complete, then right Kan extensions $\mathsf{E}^{\mathsf{D}} \underset{\mathrm{Ran}_K}{\overset{K^*}{\underset{\perp}{\rightleftarrows}}} \mathsf{E}^{\mathsf{C}}$ exist and are given by the limit formula (6.2.3).*

PROOF. If C is small and D is locally small, then for any $d \in \mathsf{D}$ the comma categories $d \downarrow K$ and $K \downarrow d$ are small, so the respective completeness and cocompleteness hypotheses ensure that the limits and colimits of Theorem 6.2.1 exist in E.   □

EXAMPLE 6.2.7. Consider the posets $(\mathbb{Q}, \leq)$ of rationals and $(\mathbb{R}_{>0}, \leq)$ of non-negative reals. We have the functor $2^-\colon \mathbb{Q} \to \mathbb{R}_{>0}$ defined at the introduction to this chapter. The poset $\mathbb{R}_{>0}$ has all limits (infima) and colimits (suprema) of bounded subsets. In particular, we may use Theorem 6.2.1 to define the left Kan extension along $\mathbb{Q} \hookrightarrow \mathbb{R}$. The value of $\mathrm{Lan}\, 2^x$ is the supremum (the colimit) of all 2^q with $q \in \mathbb{Q}$ and $q \leq x$. This is the usual definition of 2^x, so we conclude that the exponential function $2^-\colon \mathbb{R} \to \mathbb{R}_{>0}$ is the left Kan extension of $2^-\colon \mathbb{Q} \to \mathbb{R}_{>0}$ along the inclusion $\mathbb{Q} \hookrightarrow \mathbb{R}$.

In this case, the exponential function $2^-\colon \mathbb{R} \to \mathbb{R}_{>0}$ is also the right Kan extension of $2^-\colon \mathbb{Q} \to \mathbb{R}_{>0}$ along the inclusion $\mathbb{Q} \hookrightarrow \mathbb{R}$; this is because 2^x is also the infimum of 2^q for all $q \in \mathbb{Q}$ with $x \leq q$. Exercise 6.2.ii invites the reader to explore why the left and right Kan extensions coincide in this case.

EXAMPLE 6.2.8. Returning to Example 6.1.7, we can use Theorem 6.2.1 to give explicit formulae for the induction and coinduction functors associated to a subgroup inclusion

$H \hookrightarrow G$. When C is complete and cocomplete, Corollary 6.2.6 supplies adjunctions:

(6.2.9)
$$C^{BG} \xrightarrow[\mathrm{res}_H^G]{} C^{BH}$$
with ind_H^G above and coind_H^G below.

Given $X \in C^{BH}$, Theorem 6.2.1 defines $\mathrm{ind}_H^G X$ as a colimit of a diagram

(6.2.10)
$$BH \downarrow *_G \xrightarrow{\Pi} BH \xrightarrow{X} C$$

indexed by the comma category $BH \downarrow *_G$, where $*_G$ denotes the unique object in BG. Objects in the indexing category are morphisms in BG, i.e., group elements $g \in G$. A morphism $h: g \to g'$ is an element $h \in H$ so that $g'h = g$. The diagram (6.2.10) sends each object to $X \in C$ with morphisms given by the left action of H. Applying Theorem 3.4.12, this colimit is defined by the coequalizer diagram

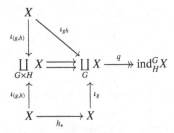

where the top map includes the component indexed by (g,h) to the component indexed by gh, while the bottom map acts by h and then includes into the component indexed by g.

By inspection, the object $\mathrm{ind}_H^G X$ is isomorphic in C to $\coprod_{G/H} X$ via an isomorphism that allows us to explicitly describe the epimorphism q. Fix a complete set of left coset representatives $G/H = \{g_i H \mid g_i \in G\}$. Then each $g \in G$ factors uniquely as $g_j h'$ with g_j among the chosen representatives and $h' \in H$. The g-component of the quotient map q is defined as

$$
\begin{array}{ccc}
X & \xrightarrow{(h')_*} & X \\
{\scriptstyle \iota_g}\downarrow & & \downarrow{\scriptstyle \iota_{g_j}} \\
\coprod_G X & \xrightarrow{q} & \coprod_{G/H} X \cong \mathrm{ind}_H^G X
\end{array}
$$

The action of an element $g' \in G$ on the object $\mathrm{ind}_H^G X \cong \coprod_{G/H} X \in C$ is the unique quotient, with respect to this map q, of the action of g' on $\coprod_G X$ by left multiplication on the indexing set:

$$
\begin{array}{ccccc}
\coprod_{G\times H} X & \rightrightarrows & \coprod_G X & \xrightarrow{q} & \coprod_{G/H} X \cong \mathrm{ind}_H^G X \\
{\scriptstyle (g,h)\mapsto(g'g,h)}\downarrow & & {\scriptstyle g\mapsto g'g}\downarrow & & \vdots\,{\scriptstyle (g')_*} \\
\coprod_{G\times H} X & \rightrightarrows & \coprod_G X & \xrightarrow{q} & \coprod_{G/H} X \cong \mathrm{ind}_H^G X
\end{array}
$$

Dually, Theorem 6.2.1 defines $\text{coin}_H^G X$ as an equalizer

$$\text{coin}_H^G X \cong \coprod_{H\backslash G} X \overset{m}{\rightarrowtail} \coprod_G X \rightrightarrows \coprod_{H\times G} X$$

There is an isomorphism $\text{coin}_H^G X \cong \coprod_{H\backslash G} X$ of underlying objects of C, under which the monomorphism $m\colon \coprod_{H\backslash G} X \rightarrowtail \coprod_G X$ admits a dual description, given by choosing right coset representatives. The action of G on $\text{coin}_H^G X \cong \coprod_{H\backslash G} X$ is defined by restricting the right action of G on the indexing set G of $\coprod_G X$ along this map m.

In the category $\mathsf{Vect}_\Bbbk$, finite products and finite coproducts coincide: these are just direct sums of vector spaces. If $V \in \mathsf{Vect}_\Bbbk^{BH}$ is an H-representation and H is a finite index subgroup of G, then the colimit and limit formulae of (6.2.2) and (6.2.3) both yield a direct sum of copies of V indexed by a set whose size is the index of H in G. Moreover, the G-actions are compatible with this isomorphism $\text{ind}_H^G V \cong \text{coin}_H^G V$. Thus, for finite index subgroups, the left and right adjoints of (6.2.9) are the same; i.e., induction from a finite index subgroup is both left and right adjoint to restriction.

EXAMPLE 6.2.11. Recall from Example 1.3.15 that any group G has an orbit category O_G whose objects are subgroups $H \subset G$, which we identify with the left G-sets G/H of left cosets of H. Morphisms $G/H \to G/K$ are G-equivariant maps, each of which is represented by a mapping $gH \mapsto g\gamma K$, where $\gamma \in G$ is an element so that $\gamma^{-1}H\gamma \subset K$. From this description, we see that the monoid of endomorphisms of the object $G/e \in O_G$ is the group G^{op}. So there is an embedding $BG^{\text{op}} \hookrightarrow O_G$ sending the single object to G/e.

For any left G-set $X \in \mathsf{Set}^{BG}$, Theorem 6.2.1 can be used to compute the right Kan extension

$$
\begin{array}{ccc}
BG & \overset{X}{\longrightarrow} & \mathsf{Set} \\
& \Downarrow & \nearrow_{\text{Ran }X} \\
& O_G^{\text{op}} &
\end{array}
$$

The comma category that indexes the diagram whose limit defines the value of Ran X at a subgroup $H \subset G$ turns out to be isomorphic to the translation groupoid $T_G(G/H)$ for the left G-set G/H of left cosets; see Example 1.5.18. As a category, the translation groupoid is equivalent to the group of automorphisms of any of its objects; in this case, we choose the coset H whose automorphism group is H. The equivalence of categories $BH \simeq T_G(G/H)$ allows us to reduce the limit that defines Ran $X(G/H)$ to the limit of the diagram $\text{res}_H^G X\colon BH \to \mathsf{Set}$ restricting the G-set X to the subgroup H. By Example 3.2.12, this limit is X^H, the subset of H-fixed points.

In summary, the right Kan extension Ran $X\colon O_G^{\text{op}} \to \mathsf{Set}$ is the fixed point functor, mapping G/H to X^H and carrying a morphism $\gamma\colon G/H \to G/K$ to the direct-image function $\gamma_*\colon X^K \to X^H$; if $x \in X^K$, then $\gamma x \in X^H$ because $\gamma^{-1}h\gamma \in K$ for all $h \in H$. In the case where G is the Galois group for a finite Galois extension E/F, the right Kan extension of $E\colon BG \to \mathsf{Field}_F^E$ defines the isomorphism of categories $O_G^{\mathrm{op}} \cong \mathsf{Field}_F^E$ that categorifies the fundamental theorem of Galois theory.[3]

EXAMPLE 6.2.12. Write $\mathbb{\Delta}_{\leq n}$ for the full subcategory spanned by the first $n + 1$ ordinals $[0], \ldots, [n]$ in $\mathbb{\Delta}$. Restriction along the inclusion functor $i_n\colon \mathbb{\Delta}_{\leq n} \hookrightarrow \mathbb{\Delta}$ is called n-**truncation**. As Set is complete and cocomplete, Corollary 6.2.6 implies that n-truncation admits both left and right adjoints:

$$\mathsf{Set}^{\mathbb{\Delta}^{\mathrm{op}}} \underset{\mathrm{Ran}_{i_n}}{\overset{\mathrm{Lan}_{i_n}}{\underset{\perp}{\overset{\perp}{\xrightarrow{\;i_n^*\;}}}}} \mathsf{Set}^{\mathbb{\Delta}_{\leq n}^{\mathrm{op}}}$$

The composite comonad on $\mathsf{Set}^{\mathbb{\Delta}^{\mathrm{op}}}$ is sk_n, the functor that maps a simplicial set to its n-**skeleton**. The composite monad on $\mathsf{Set}^{\mathbb{\Delta}^{\mathrm{op}}}$ is cosk_n, the functor that maps a simplicial set to its n-**coskeleton**. Furthermore, sk_n is left adjoint to cosk_n, as is the case for any comonad and monad arising in this way (see Exercise 4.1.iii).

Again, Theorem 6.2.1 provides an explicit description of the skeleton and coskeleton functors. We leave the details to those readers familiar with simplicial sets.

Exercises.

EXERCISE 6.2.i. Directed graphs are functors from the category with two objects E, V and a parallel pair of maps $s, t\colon E \rightrightarrows V$ to Set. A natural transformation between two such functors is a graph homomorphism. The forgetful functor $\mathsf{DirGraph} \to \mathsf{Set}$ that maps a graph to its set of vertices is given by restricting along the functor from the terminal category $\mathbb{1}$ that picks out the object V. Use Theorem 6.2.1 to compute left and right adjoints to this forgetful functor.

EXERCISE 6.2.ii. Give conditions on a monotone function $f\colon \mathbb{Q} \to \mathbb{R}$, meaning a functor $f\colon (\mathbb{Q}, \leq) \to (\mathbb{R}, \leq)$, so that the left and right Kan extensions along $\mathbb{Q} \hookrightarrow \mathbb{R}$ exist and coincide.

6.3. Pointwise Kan extensions

Unusually for a mathematical object defined by a universal property, generic Kan extensions are rather poorly behaved, particularly in the case of derived functors, which are introduced in the next section. On account of this, in his much cited book on enriched category theory, Kelly reserves the name "Kan extension" for pairs satisfying the condition to be introduced shortly, calling those of our Definition 6.1.1 "weak" and writing:

> Our present choice of nomenclature is based on our failure to find a
> single instance where a weak Kan extension plays any mathematical
> role whatsoever. [**Kel82**, §4]

By the categorical community's consensus, the important Kan extensions are *pointwise* Kan extensions. To explain this, a preliminary notion is needed:

---

[3]The category Field_F^E has very few limits but it does have the limits of Theorem 6.2.1 in this instance; see Theorem 6.3.7.

DEFINITION 6.3.1. A functor $L\colon \mathsf{E} \to \mathsf{F}$ **preserves** a left Kan extension $(\operatorname{Lan}_K F, \eta)$ if the whiskered composite $(L \operatorname{Lan}_K F, L\eta)$ is the left Kan extension of LF along K.

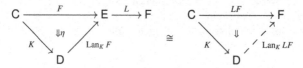

Theorem 6.2.1 constructs certain left Kan extensions using colimits. As left adjoints preserve colimits (Theorem 4.5.3), it follows that left adjoints preserve such left Kan extensions. Indeed, left adjoints preserve all left Kan extensions, regardless of how they are constructed:

LEMMA 6.3.2. *Left adjoints preserve left Kan extensions.*

PROOF 1. Suppose given a left Kan extension $(\operatorname{Lan}_K F, \eta)$ with codomain E and suppose further that $L\colon \mathsf{E} \to \mathsf{F}$ has a right adjoint R with unit $\iota\colon 1_\mathsf{E} \Rightarrow RL$ and counit $\nu\colon LR \Rightarrow 1_\mathsf{F}$. Then given $H\colon \mathsf{D} \to \mathsf{F}$ there are natural isomorphisms

$$\mathsf{F}^\mathsf{D}(L \operatorname{Lan}_K F, H) \cong \mathsf{E}^\mathsf{D}(\operatorname{Lan}_K F, RH) \cong \mathsf{E}^\mathsf{C}(F, RHK) \cong \mathsf{F}^\mathsf{C}(LF, HK).$$

Taking $H = L \operatorname{Lan}_K F$, these isomorphisms act on the identity natural transformation as follows:

$$1_{L \operatorname{Lan}_K F} \mapsto \iota_{\operatorname{Lan}_K F} \mapsto \iota_{\operatorname{Lan}_K F \cdot K} \cdot \eta \mapsto \nu_{L \operatorname{Lan}_K F \cdot K} \cdot L\iota_{\operatorname{Lan}_K F \cdot K} \cdot L\eta = L\eta.$$

Hence, by the Yoneda lemma, $(L \operatorname{Lan}_K F, L\eta)$ is a left Kan extension of LF along K. □

Unpacking the natural isomorphisms in Proof 1 yields an explicit construction of the unique factorization of a natural transformation $\alpha\colon LF \Rightarrow HK$ through $L\eta$, which is more easily visualized in a second proof that displays the composite natural transformations as pasting diagrams.

PROOF 2. Given a natural transformation $\alpha\colon LF \Rightarrow HK$, form the pasted composite with the unit $\iota\colon 1_E \Rightarrow RL$ displayed below-left.

By the universal property of the left Kan extension $(\operatorname{Lan}_K F, \eta)$, there is a unique factorization through η as displayed above-right. Composing each of these natural transformations with the counit $\nu\colon LR \Rightarrow 1_F$ and applying the triangle identity $\nu L \cdot L\iota = 1_L$,

reveals that α factors through $L\eta$ via the whiskered composite of $\nu H \cdot L\zeta \colon L \operatorname{Lan}_K F \Rightarrow H$ with K.

To demonstrate uniqueness, note that given any such factorization

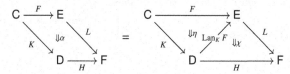

composing each of these natural transformations with the unit $\iota\colon 1_E \Rightarrow RL$

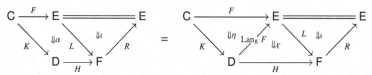

yields a factorization of $R\alpha \cdot \iota_F\colon F \Rightarrow RHK$ through η, whence $\zeta = R\chi \cdot \iota\, \mathrm{Lan}_K\, F$. Composing with ν and applying a triangle identity

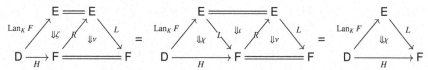

we see that the factorization $\chi\colon L\, \mathrm{Lan}_K\, F \Rightarrow H$ must agree with the factorization constructed above. □

EXAMPLE 6.3.3. The forgetful functor $U\colon \mathsf{Top} \to \mathsf{Set}$ has both left and right adjoints. It follows from Lemma 6.3.2 that U preserves the left and right Kan extensions of Example 6.1.7, as constructed in Example 6.2.8.

EXAMPLE 6.3.4. The forgetful functor $U\colon \mathsf{Vect}_{\Bbbk} \to \mathsf{Set}$ preserves limits, because it is a right adjoint, but not colimits because the underlying set of a direct sum is not simply the coproduct of the underlying sets of vectors. Indeed, it is clear from the construction given in Example 6.2.8 that the underlying set of a G-representation induced from an H-representation is not equal to the G-set induced from the underlying H-set. Lemma 6.3.2 implies, however, that the underlying set of the coinduced representation does agree with the coinduced G-set from the underlying H-set.

The notion of preservation of Kan extensions is used to define what it means for a Kan extension to be "pointwise."

DEFINITION 6.3.5. When E is locally small, a right Kan extension, as displayed below-right, is a **pointwise right Kan extension** if it is preserved by all representable functors $\mathsf{E}(e, -)$.

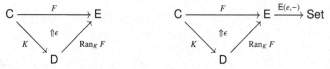

Left Kan extensions are dual to right Kan extensions, not under reversal of the direction of functors (the morphisms in CAT), but under reversal of the direction of natural transformations (the 2-morphisms in the 2-category CAT). A left Kan extension may be converted to a right Kan extension by "replacing every category by its opposite": a functor $K\colon \mathsf{C} \to \mathsf{D}$ is equally a functor $K\colon \mathsf{C}^{\mathrm{op}} \to \mathsf{D}^{\mathrm{op}}$ but the process of replacing each category by its opposite reverses the direction of natural transformations, because their components are morphisms in the opposites of the target categories.[4] A left Kan extension is **pointwise**, if the corresponding right Kan extension is pointwise:

DEFINITION 6.3.6. When E is locally small, a left Kan extension, as displayed below-left, is a **pointwise left Kan extension** if for all $e \in$ E the data displayed below-right is a right Kan extension:

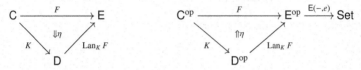

Because covariant representables preserve all limits (Proposition 3.4.5), it is clear that if a right Kan extension is given by the formula of Theorem 6.2.1, then that Kan extension is pointwise; dually, the left Kan extensions defined by Theorem 6.2.1 are pointwise. The surprise is that the converse also holds. This characterization justifies the terminology: a pointwise Kan extension can be computed pointwise as a limit in E.

THEOREM 6.3.7. *If* D *and* E *are locally small, a right Kan extension of* $F \colon$ C $\to$ E *along* $K \colon$ C $\to$ D *is pointwise if and only if it can be computed by*

$$\mathrm{Ran}_K F(d) \cong \lim\left(d \downarrow K \xrightarrow{\Pi_d} C \xrightarrow{F} E\right),$$

in which case, in particular, this limit exists. Dually, a left Kan extension of $F \colon$ C $\to$ E *along* $K \colon$ C $\to$ D *is pointwise if and only if it can be computed by*

$$\mathrm{Lan}_K F(d) \cong \mathrm{colim}\left(K \downarrow d \xrightarrow{\Pi^d} C \xrightarrow{F} E\right),$$

in which case, in particular, this colimit exists.

PROOF. If $\mathrm{Ran}_K F$ is pointwise, then for any $e \in$ E, $E(e, \mathrm{Ran}_K F)$ is the right Kan extension of $E(e, F-) \colon$ C $\to$ Set along K. By the Yoneda lemma[5] and the defining universal property of this right Kan extension

$$E(e, \mathrm{Ran}_K F(d)) \cong \mathrm{Set}^D(D(d, -), E(e, \mathrm{Ran}_K F)) \cong \mathrm{Set}^C(D(d, K-), E(e, F-))$$
$$\cong \mathrm{Cone}(e, F\Pi_d),$$

where the final natural isomorphism is the content of Lemma 6.3.8 below. This isomorphism exhibits $\mathrm{Ran}_K F(d)$ as the limit of $F\Pi_d \colon d \downarrow K \to$ E.   □

LEMMA 6.3.8. *Given functors* $F \colon$ C $\to$ E *and* $K \colon$ C $\to$ D *with* D *and* E *locally small and an object* $d \in$ D*, from which we define* $\Pi_d \colon d \downarrow K \to$ C*, there is a natural isomorphism*

$$\mathrm{Cone}(e, F\Pi_d) \cong \mathrm{Set}^C(D(d, K-), E(e, F-)).$$

Note that if C is not small, then $d \downarrow K$ is not small, so even if D and E are locally small there may be more than a set's worth of cones with summit e over $F\Pi_d$ and more than a set's worth of natural transformations $D(d, K-) \Rightarrow E(e, F-)$. However, the isomorphisms exhibited in the proof of Theorem 6.3.7 imply that the collections being considered are in fact sets.

---

[4]Succinctly, "op" is a 2-functor $(-)^{\mathrm{op}} \colon$ CAT$^{\mathrm{co}} \to$ CAT, where "co" is used to denote the dual of a 2-category obtained by reversing the direction of the 2-morphisms but not the 1-morphisms.

[5]The hypothesis that D is locally small is included so that the Yoneda lemma can be used here, but in a sufficient robust set-theoretical framework in which the proof of Theorem 2.2.4 can be extended to classes, this hypothesis can be dropped.

PROOF. To define the legs of a cone with summit e over $F\Pi_d$ is to choose, for every $f \in D(d, Kc)$, a morphism $\lambda_f^c \in E(e, Fc)$ so that for every $h: c \to c'$ in C, commutativity of the left-hand triangle in D implies commutativity of the right-hand triangle in E:

$$
\begin{array}{ccc}
& Kc & \\
f \nearrow & \downarrow Kh & \\
d & & \\
f' \searrow & & \\
& Kc' &
\end{array}
\qquad \rightsquigarrow \qquad
\begin{array}{ccc}
& Fc & \\
\lambda_f^c \nearrow & \downarrow Fh & \\
e & & \\
\lambda_{f'}^{c'} \searrow & & \\
& Fc' &
\end{array}
$$

This defines precisely the data of a natural family of functions $\lambda^c: D(d, Kc) \to E(e, Fc)$ so that for all $h: c \to c'$ in C, the square

$$
\begin{array}{ccc}
D(d, Kc) & \xrightarrow{\lambda^c} & E(e, Fc) \\
{\scriptstyle (Kh)_*} \downarrow & & \downarrow {\scriptstyle (Fh)_*} \\
D(d, Kc') & \xrightarrow{\lambda^{c'}} & E(e, Fc')
\end{array}
$$

commutes. $\qquad\qquad\qquad\qquad\qquad\qquad\qquad\qquad\qquad\qquad\qquad\qquad\qquad\qquad\quad\square$

Importantly, all pointwise Kan extensions along fully faithful functors, by which we mean Kan extensions defined by the (co)limit formulae of Theorem 6.2.1, define genuine extensions, up to natural isomorphism:

COROLLARY 6.3.9. *If K is fully faithful, then the counit of any pointwise right Kan extension defines a natural isomorphism* $\text{Ran}_K F \cdot K \cong F$, *and dually the unit of any pointwise left Kan extension defines a natural isomorphism* $F \cong \text{Lan}_K F \cdot K$.

PROOF. If K is fully faithful, then for any $c \in C$ the canonical functor $C \downarrow c \to K \downarrow Kc$ defines an isomorphism of categories, and this isomorphism commutes with the projection functors to C. By Theorem 6.3.7, we then have

$$\text{Lan}_K F(Kc) := \text{colim}(C \downarrow c \xrightarrow{\Pi} C \xrightarrow{F} E),$$

but in this case the indexing category $C \downarrow c$ has a terminal object 1_c. Therefore, by Exercise 3.1.ix, the leg of the colimit cone indexed by the terminal object is an isomorphism; in particular, the colimit can be computed by evaluating at the terminal object. From the definition of the unit given in Theorem 6.2.1, this tells us that the component of the unit defines an isomorphism $\eta_c: Fc \cong \text{Lan}_K F(Kc)$. $\qquad\qquad\qquad\qquad\qquad\qquad\quad\square$

The **cone** under a small category J is the category $J^\triangleright$ defined as the pushout of the inclusion $i_1: J \hookrightarrow J \times 2$ of J as the "right end" of the cylinder $J \times 2$ along the unique functor $J \to 1$.

$$
\begin{array}{ccc}
J & \xrightarrow{\;!\;} & 1 \\
{\scriptstyle i_1} \downarrow & & \downarrow \\
J \times 2 & \xrightarrow{\;r\;} & J^\triangleright
\end{array}
$$

The effect of this pushout is to collapse the right end of the cylinder to a single object. The category $J^\triangleright$ has one new object, a freely-adjoined terminal object t that serves as the nadir of a new cone under the inclusion $J \hookrightarrow J^\triangleright$. There are no additional new objects or morphisms. A dual pushout along the other inclusion $i_0: J \hookrightarrow J \times 2$ defines a category $J^\triangleleft$ with a freely-adjoined initial object. See Exercise 3.5.iv.

As mentioned in Remark 3.1.8, the category $J^{\triangleright}$ indexes cones under diagrams of shape J. The informal intuition that a colimit of a diagram should be the "closest" extension of the diagram to a cone under it is formalized by the following result.

PROPOSITION 6.3.10. *A category* C *admits all colimits of diagrams indexed by a small category* J *if and only if the restriction functor* $C^{J^{\triangleright}} \to C^J$ *admits a left adjoint, defined by pointwise left Kan extension:*

$$C^{J^{\triangleright}} \xleftarrow[\text{res}]{\overset{\text{colim}}{\underset{\perp}{\longleftarrow}}} C^J$$

Dually, C *admits all* J*-indexed limits if and only if the restriction functor* $C^{J^{\triangleleft}} \to C^J$ *admits a right adjoint, defined by pointwise right Kan extension:*

$$C^{J^{\triangleleft}} \xleftarrow[\text{lim}]{\overset{\text{res}}{\underset{\perp}{\longleftarrow}}} C^J$$

PROOF. By construction, there is a fully faithful inclusion $J \hookrightarrow J^{\triangleright}$. Consider a diagram $F: J \to C$. For $j \in J \subset J^{\triangleright}$, Corollary 6.3.9 implies that the colimit (6.2.2) defining the left Kan extension always exists: it is simply Fj. For the cone point $t \in J^{\triangleright}$, the comma category $J \downarrow t$ is isomorphic to J, by construction. So if $\text{colim}(J \xrightarrow{F} C)$ exists, then Theorem 6.2.1 tells us that these values define the left Kan extension functor, which we call colim: $C^J \to C^{J^{\triangleright}}$ in this context because it maps a J-indexed diagram to its $J^{\triangleright}$-indexed colimit diagram. □

Most commonly, pointwise Kan extensions are found whenever the codomain category is cocomplete (for left Kan extensions) or complete (for right), but this is not the only case as we shall now discover.

Exercises.

EXERCISE 6.3.i. Prove that if E is locally small and has all small copowers, then all right Kan extensions with codomain E are pointwise right Kan extensions. Dually, if E is locally small and has all small powers, then all left Kan extensions with codomain E are pointwise.

EXERCISE 6.3.ii. Proposition 6.3.10 expresses a stronger universal property of the colimit cone than is usually stated. What is it?

6.4. Derived functors as Kan extensions

In this section, we explain how derived functors in homological algebra or algebraic topology are defined using Kan extensions. A *derived functor* is derived from a functor whose domain is a category equipped with a specified class of morphisms called "weak equivalences" and whose codomain is either another category with weak equivalences or an ordinary category, in which case the weak equivalences are taken to be the class of isomorphisms.

We introduce two varieties of derived functors. The first of these are the *total derived functors*, which are defined to be Kan extensions along "localization" functors that formally invert the weak equivalences. In practice, total derived functors are frequently accompanied by *point-set derived functors*, which are lifts of total derived functors to weak-equivalence-preserving functors between the original categories.

After stating these definitions, we give an abstract account of a commonly applied construction that can be used to construct point-set derived functors. Unexpectedly, the total derived functors that are constructed in this way are pointwise Kan extensions, despite the fact that homotopy categories have notoriously few limits and colimits. In fact, the universal property enjoyed by the derived functors that are defined via this construction is even stronger and has important implications to the theory of derived adjunctions, which we discuss.

Throughout, we reserve the term "weak equivalences" for a specified class of morphisms in a category that contains all the identities and satisfies one further condition. The idea, that the weak equivalences are like isomorphisms but weaker, is encoded by a condition that the class of weak equivalences must satisfy:

DEFINITION 6.4.1. A class of **weak equivalences** $\mathcal{W}$ is a class of morphisms in a category that contains the identities and satisfies the **2-of-6 property**: if for any composable triple of arrows

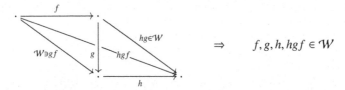

$$\Rightarrow \quad f, g, h, hgf \in \mathcal{W}$$

if hg and gf are in $\mathcal{W}$ so are f, g, h, and hgf.

LEMMA 6.4.2. *Any class of morphisms in a category that contains the identities and satisfies the 2-of-6 property necessarily also contains the isomorphisms.*

PROOF. Exercise 6.4.i.   □

The modern viewpoint, that one can "do homotopy theory" in any category with a class of weak equivalences, explains the following terminology.

DEFINITION 6.4.3. A **homotopical category** is a category equipped with a class of weak equivalences. A functor between such categories is **homotopical** if it preserves the weak equivalences.

REMARK 6.4.4 (the universal property of the homotopy category). Any homotopical category C has an associated **homotopy category** HoC characterized by the following universal property: the homotopy category HoC is initial among categories E equipped with a homotopical functor C → E, that is, a functor that sends weak equivalences in C to isomorphisms.[6] In analogy with ring theory, the universal functor C → HoC is referred to as a **localization functor**.

The universal property of the localization functor—pre-composition with C → HoC induces a bijective correspondence between functors HoC → E and homotopical functors C → E—also has a 2-dimensional aspect. Pre-composition with the localization functor also induces a bijection between natural transformations between functors HoC → E and natural transformations between the corresponding homotopical functors C → E; see Exercise 6.4.ii.

The following explicit construction of the homotopy category is due to Gabriel and Zisman [**GZ67**, 1.1]. The category HoC has the same objects as C. Its morphisms are equivalence classes of finite zig-zags of morphisms in C, with only weak equivalences permitted to go backward, modulo the following relations:

- Adjacent arrows pointing in the same direction may be composed.

---

[6]Set-theoretic concerns make it somewhat tricky to specify what functor HoC represents. In the examples of greatest interest, C is not small, in which case HoC is a priori not even locally small, though in practice it frequently is.

- Adjacent pairs $\xleftarrow{w}\xrightarrow{w}$ or $\xrightarrow{w}\xleftarrow{w}$ labelled by a weak equivalence w may be removed.
- Identities pointing either forward or backward may be removed.

There is a canonical identity-on-objects localization functor

$$\mathsf{C} \longrightarrow \mathsf{HoC}$$

in which the morphisms in C are sent to unary zig-zags pointing forward.

EXAMPLE 6.4.5. When C has no specified subcategory of weak equivalences, as a default we assign it the **minimal homotopical structure**, where the weak equivalences are precisely the isomorphisms and the localization functor is the identity.

EXAMPLE 6.4.6. There are two homotopical structures of interest on the category Ch_R of chain complexes of R-modules. In the first, the class of weak equivalences is taken to be the chain homotopy equivalences. In the second, the class of weak equivalences is enlarged to contain all **quasi-isomorphisms**, that is, chain maps $f \colon A_{\bullet} \to B_{\bullet}$ that induce an isomorphism on graded abelian groups after applying the homology functor introduced in Example 1.3.2(viii).

EXAMPLE 6.4.7. The category of topological spaces also has two homotopical structures of interest. In the first, the class of weak equivalences is taken to be the class of homotopy equivalences. In the second, this class is enlarged to contain all **weak homotopy equivalences**, which are maps inducing isomorphisms on all homotopy groups. The **homotopy category of spaces** is the homotopy category associated to the second homotopical structure. It is equivalent to the category $\mathsf{Htpy}_{\mathsf{CW}}$ of CW complexes and homotopy classes of maps that appears in Examples 2.1.6(vi) and (vii).

It is in this context that we may define total derived functors.

DEFINITION 6.4.8 (total derived functors). Consider a functor $F \colon \mathsf{C} \to \mathsf{D}$ between homotopical categories.

- When the *right* Kan extension

$$
\begin{array}{ccc}
\mathsf{C} & \xrightarrow{\;F\;} & \mathsf{D} \\
\downarrow & \Uparrow & \downarrow \\
\mathsf{HoC} & \underset{\mathbf{L}F}{\dashrightarrow} & \mathsf{HoD}
\end{array}
$$

  exists, it defines the **total left derived functor** $\mathbf{L}F$ of F.

- When the *left* Kan extension

$$
\begin{array}{ccc}
\mathsf{C} & \xrightarrow{\;F\;} & \mathsf{D} \\
\downarrow & \Downarrow & \downarrow \\
\mathsf{HoC} & \underset{\mathbf{R}F}{\dashrightarrow} & \mathsf{HoD}
\end{array}
$$

  exists, it defines the **total right derived functor** $\mathbf{R}F$ of F.

Note the conflict in handedness in Definition 6.4.8. A justification for calling the right Kan extension a total *left* derived functor is that left derived functors approximate the original functor from the left.

By the universal property of the localization functor $\mathsf{C} \to \mathsf{HoC}$, $\mathbf{L}F$ and $\mathbf{R}F$ are equivalently expressible as homotopical functors $\mathbf{L}F, \mathbf{R}F \colon \mathsf{C} \rightrightarrows \mathsf{HoD}$; these functors are often called the **left** or **right derived functors** of F. Sometimes, though by no means always, there exists a lift of a left or right derived functor along $\mathsf{D} \to \mathsf{HoD}$. For emphasis, we will call these lifts **point-set derived functors**, though other authors choose to call them simply "derived functors" [**Rie14**].

DEFINITION 6.4.9 (point-set derived functors). Consider a pair of homotopical categories and localization functors $\gamma\colon \mathsf{C} \to \mathsf{HoC}$ and $\delta\colon \mathsf{D} \to \mathsf{HoD}$.

- A **point-set left derived functor** of $F\colon \mathsf{C} \to \mathsf{D}$ is a homotopical functor $\mathbb{L}F\colon \mathsf{C} \to \mathsf{D}$ equipped with a natural transformation $\lambda\colon \mathbb{L}F \Rightarrow F$ so that $\delta\mathbb{L}F$ (which by Remark 6.4.4 is equivalently encoded by a functor $\delta\mathbb{L}F\colon \mathsf{HoC} \to \mathsf{HoD}$) and $\delta\lambda\colon \delta\mathbb{L}F \Rightarrow \delta F$ define a total left derived functor of F.

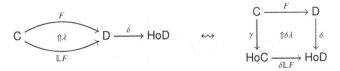

- A **point-set right derived functor** of $F\colon \mathsf{C} \to \mathsf{D}$ is a homotopical functor $\mathbb{R}F\colon \mathsf{C} \to \mathsf{D}$ equipped with a natural transformation $\rho\colon F \Rightarrow \mathbb{R}F$ so that $\delta\mathbb{R}F$ (which by Remark 6.4.4 is equivalently encoded by a functor $\delta\mathbb{R}F\colon \mathsf{HoC} \to \mathsf{HoD}$) and $\delta\rho\colon \delta F \Rightarrow \delta\mathbb{R}F$ define a total right derived functor of F.

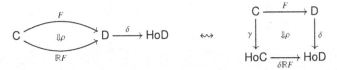

There is a common setting in which derived functors exist and admit a simple construction. Such categories have a collection of "fat" objects on which the functor of interest becomes homotopical and a functorial reflection into this full subcategory. The details are encoded in the following axiomatization due to [**DHKS04**].

DEFINITION 6.4.10.

- A **left deformation** on a homotopical category C consists of an endofunctor Q together with a natural weak equivalence $q\colon Q \overset{\sim}{\Rightarrow} 1_\mathsf{C}$.
- A **left deformation** for a functor $F\colon \mathsf{C} \to \mathsf{D}$ between homotopical categories consists of a left deformation (Q, q) for C such that F preserves all weak equivalences between objects in a full subcategory containing the image of Q.

There is an obvious dual notion of **right deformation** $r\colon 1_\mathsf{C} \overset{\sim}{\Rightarrow} R$.

Interestingly, left deformations can be used to construct point-set left derived functors, a homotopical universal property that is independent of the deformation used.

PROPOSITION 6.4.11. *If $F\colon \mathsf{C} \to \mathsf{D}$ has a left deformation $q\colon Q \overset{\sim}{\Rightarrow} 1_\mathsf{C}$, then*

$$(\mathbb{L}F = FQ, Fq\colon \mathbb{L}F \Rightarrow F)$$

is a point-set left derived functor of F.

PROOF. Write $\delta\colon \mathsf{D} \to \mathsf{HoD}$ for the localization functor. To show that (FQ, Fq) is a point-set left derived functor, we must show that the functor δFQ and natural transformation $\delta Fq\colon \delta FQ \Rightarrow \delta F$ define a right Kan extension. The verification makes use of Remark 6.4.4, which identifies the functor category $\mathsf{HoD}^{\mathsf{HoC}}$ with the full subcategory of HoD^C spanned by the homotopical functors. Suppose $G\colon \mathsf{C} \to \mathsf{HoD}$ is homotopical and consider $\alpha\colon G \Rightarrow \delta F$.

Because G is homotopical and $q: Q \Rightarrow 1_C$ is a natural weak equivalence, $Gq: GQ \Rightarrow G$ is a natural isomorphism. Using naturality of α, it follows that α factors through δFq as

$$\alpha: G \xRightarrow{(Gq)^{-1}} GQ \xRightarrow{\alpha_Q} \delta FQ \xRightarrow{\delta Fq} \delta F$$

To prove uniqueness, suppose α factors as

$$\alpha: G \xRightarrow{\beta} \delta FQ \xRightarrow{\delta Fq} \delta F$$

Naturality of β provides a commutative square of natural transformations:

$$
\begin{array}{ccc}
GQ & \xRightarrow{\beta_Q} & \delta FQ^2 \\
{\scriptstyle Gq} \big\Downarrow & & \big\Downarrow {\scriptstyle \delta FQq} \\
G & \xRightarrow[\beta]{} & \delta FQ
\end{array}
$$

Because q is a natural weak equivalence and the functors G and δFQ are homotopical, the vertical arrows are natural isomorphisms, so β is determined by β_Q. This restricted natural transformation is uniquely determined: q_Q is a natural weak equivalence between objects in the image of Q. Since F is homotopical on this subcategory, this means that Fq_Q is a natural weak equivalence and thus δFq_Q is an isomorphism, so β_Q must equal the composite of the inverse of this natural isomorphism with α_Q. □

Unusually for constructions characterized by a universal property, generic total derived functors are poorly behaved: for instance, the composite of the total left derived functors of a pair of composable functors is not necessarily a total left derived functor for the composite. The defining universal property is also insufficient to prove that the total left derived functor of a left adjoint is left adjoint to the total right derived functor of its right adjoint. The problem with the standard definition is that total derived functors are not typically required to be pointwise Kan extensions. In light of Theorem 6.3.7, this seems reasonable because homotopy categories have few limits and colimits.[7]

Somewhat surprisingly, the particular construction of total derived functors via deformations given in Proposition 6.4.11 implies that the total derived functors produced in this manner are pointwise and indeed **absolute Kan extensions**: not only are these Kan extensions preserved by representables but they are preserved by all functors.

PROPOSITION 6.4.12. *The total left derived functor of a left deformable functor is an absolute right Kan extension, and in particular pointwise.*

PROOF. If $F: C \to D$ is left deformable, it has a total left derived functor $(\delta FQ, \delta Fq)$ constructed using a left deformation (Q, q) for F and the localization functor $\delta: D \to \mathrm{HoD}$. Because Kan extensions are characterized by a universal property, any total derived functor of F is isomorphic to this one. To show that this right Kan extension is absolute, our task is to show that for any functor $H: \mathrm{HoD} \to E$, the pair $(H\delta FQ, H\delta Fq)$ again defines a right Kan extension.

Note that (Q, q) also defines a left deformation for $H\delta F$, simply because the functor $H: \mathrm{HoD} \to E$ preserves isomorphisms. The proof of Proposition 6.4.11 now demonstrates that $(H\delta FQ, H\delta Fq)$ is a right Kan extension, as claimed. □

---

[7]With the exception of products and coproducts, the so-called "homotopy limits" and "homotopy colimits" do not define limits and colimits in the homotopy category.

In homological algebra or homotopy theory, deformable functors often come in adjoint pairs, with the left adjoint admitting a left deformation and the right adjoint admitting a right deformation. In such cases, the total derived functors are again adjoints, but interestingly the proof requires Proposition 6.4.12: the mere universal property of being a left or right Kan extension is not enough. This is the main theorem from a paper of Maltsiniotis [**Mal07**]. As with many non-trivial discoveries in category theory, once the statement is known, the proof is elementary enough to leave to the reader:

PROPOSITION 6.4.13. *Suppose $F \dashv G$ is an adjunction between homotopical categories and suppose also that F has a total left derived functor $\mathbf{L}F$, G has a total right derived functor $\mathbf{R}G$, and both derived functors are absolute Kan extensions. Then the total derived functors form an adjunction $\mathbf{L}F \dashv \mathbf{R}G$ between the homotopy categories.*

PROOF. Exercise 6.4.iv.   □

REMARK 6.4.14. If $F \dashv G$ is an adjunction satisfying the hypotheses of Proposition 6.4.13, then the adjunction

$$
\begin{array}{ccc}
\mathsf{C} & \xrightarrow[\underset{G}{\perp}]{F} & \mathsf{D} \\
\gamma \downarrow & & \downarrow \delta \\
\mathsf{HoC} & \xrightarrow[\underset{\mathbf{R}G}{\perp}]{\mathbf{L}F} & \mathsf{HoD}
\end{array}
$$

between the total derived functors is the unique adjunction compatible with the localization functors $\gamma\colon \mathsf{C} \to \mathsf{HoC}$ and $\delta\colon \mathsf{D} \to \mathsf{HoD}$ in the sense that the diagram

$$
\begin{array}{ccc}
\mathsf{D}(Fc,d) & \cong & \mathsf{C}(c,Gd) \\
\delta \downarrow & & \downarrow \gamma \\
\mathsf{HoD}(Fc,d) & & \mathsf{HoC}(c,Gd) \\
Fq^* \downarrow & & \cdot \downarrow Gr_* \\
\mathsf{HoD}(\mathbf{L}Fc,d) & \cong & \mathsf{HoC}(c,\mathbf{R}Gd)
\end{array}
$$

commutes for each pair $c \in \mathsf{C}$, $d \in \mathsf{D}$.

Exercises.

EXERCISE 6.4.i. Prove Lemma 6.4.2.

EXERCISE 6.4.ii. Prove the 2-dimensional aspect of the universal property of the localization functor described in Remark 6.4.4. That is, consider a homotopical category C and a homotopical functor $\gamma\colon \mathsf{C} \to \mathsf{HoC}$ satisfying the universal property that restriction along γ induces a bijection between functors $\mathsf{HoC} \to \mathsf{E}$ and homotopical functors $\mathsf{C} \to \mathsf{E}$. Prove that restriction along γ also induces a bijection between natural transformations between functors $\mathsf{HoC} \to \mathsf{E}$ and natural transformations between the associated homotopical functors $\mathsf{C} \to \mathsf{E}$.

EXERCISE 6.4.iii. Prove that any endofunctor $Q\colon \mathsf{C} \to \mathsf{C}$ on a homotopical category that is equipped with a natural weak equivalence $q\colon Q \xRightarrow{\sim} 1_{\mathsf{C}}$ is a homotopical functor.

EXERCISE 6.4.iv. Prove Proposition 6.4.13.

6.5. All concepts

To wrap up this chapter, and with it this book, we see that simple special cases of Kan extensions can be used to define the other basic categorical concepts introduced in Chapters 2–5. This justifies Mac Lane's famous assertion that "The notion of Kan extensions subsumes all the other fundamental concepts of category theory" [**ML98a**,

§X.7]. Because much of basic category theory can be encoded in this way, generalizations of ordinary category theory to more exotic contexts often commence with the development of a suitable calculus of Kan extensions.

PROPOSITION 6.5.1 ((co)limits as Kan extensions). *The left Kan extension of* $F\colon \mathsf{C} \to \mathsf{D}$ *along the unique functor* $!\colon \mathsf{C} \to \mathbb{1}$ *defines the colimit of* F *in* D, *each existing if and only if the other does.*

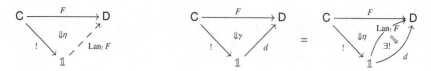

Dually $\mathrm{Ran}_! F$ *defines the limit of* F.

PROOF. The composite of a functor $d\colon \mathbb{1} \to \mathsf{D}$ with the functor $!\colon \mathsf{C} \to \mathbb{1}$ yields the constant functor at the object $d \in \mathsf{D}$. Hence, the universal property of the left Kan extension of F along $!\colon \mathsf{C} \to \mathbb{1}$ specifies that cones $\gamma\colon F \Rightarrow d$ under F with nadir d correspond to morphisms $\mathrm{Lan}_! F \to d$ in D, and that this correspondence is implemented by composing with the displayed cone $\eta\colon F \Rightarrow \mathrm{Lan}_! F$. Thus, we recognize $\mathrm{Lan}_! F$ as the colimit of F with colimit cone η. $\square$

PROPOSITION 6.5.2 (adjunctions as Kan extensions).

(i) *If* $F \dashv G$ *is an adjunction with unit* $\eta\colon 1 \Rightarrow GF$ *and counit* $\epsilon\colon FG \Rightarrow 1$, *then* (G, η) *is a left Kan extension of the identity functor along* F *and* (F, ϵ) *is a right Kan extension of the identity functor along* G.

Moreover, both Kan extensions are absolute, and in particular pointwise.

(ii) *Conversely, if* $(G, \eta\colon 1_\mathsf{C} \Rightarrow GF)$ *is a left Kan extension of the identity along* F *and if* F *preserves this Kan extension, then* $F \dashv G$ *with unit* η.

PROOF. For (i), Proposition 4.4.6 tells us that an adjunction $F \dashv G$ gives rise to an adjunction

$$\mathsf{C}^\mathsf{D} \underset{F^*}{\overset{G^*}{\underset{\perp}{\longleftrightarrow}}} \mathsf{C}^\mathsf{C}$$

By uniqueness of adjoints (Proposition 4.4.1), left Kan extension along F is given by pre-composition with G, with the components of the unit of $G^* \dashv F^*$ defining the unit for the left Kan extension. This tells us that (G, η) defines a left Kan extension of 1_C along F. This Kan extension is absolute because pre-composition commutes with post-composition with any functor $L\colon \mathsf{C} \to \mathsf{E}$. Put differently, there is a similar adjunction $G^* \dashv F^*$ between functor categories E^D and E^C for any E.

For (ii), recall from Definition 4.2.5 that the data of an adjunction is comprised of a pair of functors and a pair of natural transformations so that the "triangle identity" pasting identities hold:

Suppose that (G, η) is a left Kan extension of 1_C preserved by F, meaning that $(FG, F\eta)$ is a left Kan extension of F along F. Using the universal property of this latter Kan extension, there is a unique factorization

of 1_F through $F\eta$ defining a natural transformation $\epsilon\colon FG \Rightarrow 1_D$. Immediately from this definition, we see that $\epsilon F \cdot F\eta = 1_F$, proving one of the two triangle identities. For the other, note that by the first triangle identity, the left-hand pasted composite

reduces to the right-hand ones. From this, we see that $1_G\colon G \Rightarrow G$ and $G\epsilon \cdot \eta G\colon G \Rightarrow G$ both define factorizations of η through itself. By the uniqueness in the universal property of the left Kan extension (G, η), we conclude that $G\epsilon \cdot \eta G = 1_G$, proving that $F \dashv G$ with unit η. □

The next aim is to use a simple pointwise right Kan extension to reprove and then generalize the Yoneda lemma. From the defining universal property, the right Kan extension of a functor $F\colon C \to D$ along the identity $1_C\colon C \to C$ is (isomorphic to) F. It follows immediately from Definition 6.3.5 that this Kan extension is also pointwise, so we can apply the limit formula of Theorem 6.3.7 to conclude that

$$(6.5.3) \qquad Fc \cong \lim(c/C \xrightarrow{\Pi} C \xrightarrow{F} D).$$

In the case of a functor $F\colon C \to \mathsf{Set}$, this limit formula proves the Yoneda lemma. By the defining universal property of the limit (6.5.3),

$$Fc \cong \mathsf{Set}(*, Fc) \cong \mathrm{Cone}(*, F\Pi) \cong \mathsf{Set}^C(C(c, -), \mathsf{Set}(*, F-)) \cong \mathsf{Set}^C(C(c, -), F),$$

where the third isomorphism is a special case of Lemma 6.3.8. This is the Yoneda lemma.

In the case of a small diagram $F\colon C \to D$ valued in a category that has products and equalizers, we can apply Theorem 3.4.12 to re-express this limit formula as an equalizer diagram. On account of Exercise 3.2.v, which expresses the set of natural transformations between small diagrams as a similar equalizer, this limit formula is regarded as a generalization of the Yoneda lemma.

PROPOSITION 6.5.4 (Yoneda lemma). *For any small diagram* $F\colon C \to D$ *valued in a category with products and equalizers, there is an equalizer diagram*

$$Fc \rightarrowtail \prod_{c \to x} Fx \rightrightarrows \prod_{c \to x \to y} Fy$$

in which one of the parallel maps projects to the component indexed by the composite $c \to x \to y$ *in* C *while the other projects to the component indexed by the map* $c \to x$ *and then acts by* F *on the second map* $x \to y$.

PROOF. Apply Theorem 3.4.12 to (6.5.3).   □

Like any abstract nonsense, this argument can be dualized. Dually, any $F: C \to D$ is a pointwise left Kan extension of itself along 1_C. The colimit formula of Theorem 6.3.7 applies, so

$$(6.5.5) \qquad Fc \cong \operatorname{colim}(C/c \xrightarrow{\Pi} C \xrightarrow{F} D),$$

which is appropriately called the **coYoneda lemma**:

PROPOSITION 6.5.6 (coYoneda lemma). *For any small diagram* $F: C \to D$ *valued in a category with coproducts and coequalizers, there is a coequalizer diagram*

$$\coprod_{y \to x \to c} Fy \rightrightarrows \coprod_{x \to c} Fx \twoheadrightarrow Fc.$$

in which one of the parallel maps includes the component indexed by $y \to x \to c$ *as the component indexed by the composite morphism* $y \to c$, *while the other acts by* F *on the map* $y \to x$ *and then includes as the component indexed by the map* $x \to c$.

PROOF. Apply Theorem 3.4.12 to (6.5.5).   □

The coYoneda lemma has an important consequence in the special case of a set-valued functor called the **density theorem**.

THEOREM 6.5.7 (density I). *For any locally small category* C, *any functor* $F: C \to \mathsf{Set}$ *is the colimit of the diagram*

$$\textstyle\int F^{\mathrm{op}} \xrightarrow{\Pi} C^{\mathrm{op}} \xhookrightarrow{y} \mathsf{Set}^{C},$$

which is indexed by the opposite of the category $\int F$ *of elements of* F *and takes values in the subcategory of representable functors. Dually, any* $F: C^{\mathrm{op}} \to \mathsf{Set}$ *is the colimit of*

$$\textstyle\int F \xrightarrow{\Pi} C \xhookrightarrow{y} \mathsf{Set}^{C^{\mathrm{op}}}.$$

PROOF. For sets S and T, there are isomorphisms $\coprod_S T \cong S \times T \cong \coprod_T S$, so the coequalizer given by the coYoneda lemma is isomorphic to the coequalizer[8]

$$\coprod_{Fy \times C(y,x)} C(x,c) \rightrightarrows \coprod_{Fx} C(x,c) \twoheadrightarrow Fc,$$

where the indexing sets have been swapped. Letting c vary, we conclude that F is canonically a colimit of a diagram of representable functors. Reversing Theorem 3.4.12, the indexing category for this diagram of representables is one whose objects are elements of the set Fx for some $x \in C$ and whose morphisms are maps $y \to x$ in C that send the chosen element of Fy to the corresponding element of Fx. Thus, F is the colimit of the canonical diagram

$$\textstyle\int F^{\mathrm{op}} \xrightarrow{\Pi} C^{\mathrm{op}} \xhookrightarrow{y} \mathsf{Set}^{C}.$$ □

---

[8]If C is not small, then the displayed coproducts, which are tacitly indexed by the objects of C, are not small. Nonetheless, Theorem 6.3.7, which expresses Fc as a colimit of a large diagram (6.5.5), tells us that F is a colimit of a diagram indexed by the large category $\int F$.

The name "density theorem" refers to a second formulation of this result, which asserts that the representable functors form a **dense subcategory** of the presheaf category $\mathsf{Set}^{\mathsf{C}^{\mathrm{op}}}$. Informally speaking, a subcategory is **dense** if any object in the larger category is a colimit of a diagram valued in the subcategory. The precise meaning is given by the statement of the following theorem:

THEOREM 6.5.8 (density II). *For any small category* C, *the identity functor defines the left Kan extension of the Yoneda embedding* $\mathsf{C} \hookrightarrow \mathsf{Set}^{\mathsf{C}^{\mathrm{op}}}$ *along itself.*

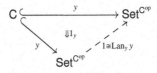

PROOF. By Theorem 6.2.1, for any $F \in \mathsf{Set}^{\mathsf{C}^{\mathrm{op}}}$,

$$\mathrm{Lan}_y\, y(F) \cong \mathrm{colim}\left(y \downarrow F \overset{\Pi}{\to} \mathsf{C} \overset{y}{\to} \mathsf{Set}^{\mathsf{C}^{\mathrm{op}}}\right).$$

By Lemma 2.4.7, the comma category $y \downarrow F$ is isomorphic to the category of elements $\int F$, so Theorem 6.5.7 proves that $\mathrm{Lan}_y\, y(F) \cong F$. □

REMARK 6.5.9. By Corollary 6.2.6, any functor $F: \mathsf{C} \to \mathsf{E}$ whose domain is small and whose codomain is locally small and cocomplete admits a left Kan extension along the Yoneda embedding $y: \mathsf{C} \to \mathsf{Set}^{\mathsf{C}^{\mathrm{op}}}$

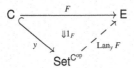

Essentially by Corollary 6.3.9, because the Yoneda embedding is full and faithful, the functor $\mathrm{Lan}_y F: \mathsf{Set}^{\mathsf{C}^{\mathrm{op}}} \to \mathsf{E}$ can be defined to be a genuine extension of F along y, so that the unit of the pointwise left Kan extension is an identity natural transformation. Moreover, this left Kan extension admits a right adjoint $R: \mathsf{E} \to \mathsf{Set}^{\mathsf{C}^{\mathrm{op}}}$. To define $Re \in \mathsf{Set}^{\mathsf{C}^{\mathrm{op}}}$, note that the desired adjunction and the Yoneda lemma demand natural isomorphisms

$$Re(c) \cong \mathsf{Set}^{\mathsf{C}^{\mathrm{op}}}(\mathsf{C}(-,c), Re) \cong \mathsf{E}(\mathrm{Lan}_y F(\mathsf{C}(-,c)), e) \cong \mathsf{E}(Fc, e),$$

which necessitates the definition

$$Re := \mathsf{E}(F-, e): \mathsf{C}^{\mathrm{op}} \to \mathsf{Set}.$$

For each $e \to e' \in \mathsf{E}$, post-composition defines a map $Re \to Re'$, so our construction of $R: \mathsf{E} \to \mathsf{Set}^{\mathsf{C}^{\mathrm{op}}}$ is functorial.

To prove that the functors $\mathrm{Lan}_y\, F \dashv R$ are indeed adjoints, we use the density theorem. For any $X \in \mathsf{Set}^{C^{op}}$ and $e \in \mathsf{E}$,

$$\mathsf{Set}^{C^{op}}(X, Re) \cong \mathsf{Set}^{C^{op}}\Big(\operatorname*{colim}_{(c,x)\in \int X} C(-,c), Re\Big) \qquad\qquad \text{by Theorem 6.5.7}$$

$$\cong \operatorname*{lim}_{(c,x)\in \int X^{op}} \mathsf{Set}^{C^{op}}(C(-,c), Re) \qquad\qquad \text{by Theorem 3.4.7}$$

$$\cong \operatorname*{lim}_{(c,x)\in \int X^{op}} \mathsf{E}(Fc, e) \qquad\qquad\qquad \text{by the definition of } R$$

$$\cong \mathsf{E}\Big(\operatorname*{colim}_{(c,x)\in \int X} Fc, e\Big) \qquad\qquad\qquad \text{by Theorem 3.4.7}$$

$$\cong \mathsf{E}(\mathrm{Lan}_y F(X), e) \qquad\qquad \text{by Theorem 6.2.1 and } y\!\downarrow\! F \cong \int F .$$

Examples of adjunctions constructed in this manner are described in Exercises 6.5.iii-6.5.v.

Finally, we turn to monads. We have seen that a right adjoint functor $G\colon \mathsf{D} \to \mathsf{C}$ induces a monad on C whose endofunctor is the composite of G with its left adjoint. More generally, assuming C has sufficient limits, any functor $G\colon \mathsf{D} \to \mathsf{C}$ induces a monad on C. If G has a left adjoint F, this monad is the monad GF but this construction applies more generally.

DEFINITION 6.5.10 (monads as Kan extensions). The **codensity**[9] **monad** of $G\colon \mathsf{D} \to \mathsf{C}$ is given by the right Kan extension of G along itself, whenever this exists.[10]

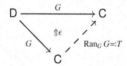

The unit and multiplication natural transformations are defined using the universal property of $\epsilon\colon TG \Rightarrow G$ as follows:

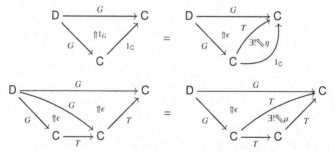

EXAMPLE 6.5.11. Consider the inclusion $\mathsf{Field} \hookrightarrow \mathsf{CRing}$. It has no left adjoint because such a functor would carry $\mathbb{Z}$ to an initial field, which does not exist. But it does have a density monad T defined by

$$TA := \prod_{\mathfrak{p}\in \mathrm{Spec}(A)} \mathrm{Frac}(A/\mathfrak{p})$$

[9]A functor $G\colon \mathsf{D} \to \mathsf{C}$ is **codense** if its right Kan extension along itself is the identity 1_{C}; compare with the statement of Theorem 6.5.8.

[10]In particular, "sufficient limits" in C means those necessary to define $\mathrm{Ran}_G\, G$ as a pointwise right Kan extension.

where $\mathrm{Spec}(A)$ is the set of prime ideals $\mathfrak{p} \subset A$ and $\mathrm{Frac}(A/\mathfrak{p})$ is the field of fractions of the quotient ring $A/\mathfrak{p}$, which is an integral domain because the ideal $\mathfrak{p}$ is prime.

EXAMPLE 6.5.12. Familiar monads also arise as codensity monads. For instance, the codensity monad of the inclusion $\mathsf{Fin} \hookrightarrow \mathsf{Set}$ is the ultrafilter monad β of Example 5.1.4(v). Theorem 6.2.1 characterizes the set of ultrafilters on a set X as a limit

$$\beta(X) \cong \lim\left(X \downarrow \mathsf{Fin} \xrightarrow{\Pi} \mathsf{Fin} \hookrightarrow \mathsf{Set}\right).$$

By Theorem 3.2.6, this limit is isomorphic to the set of cones over the diagram $X \downarrow \mathsf{Fin} \to \mathsf{Set}$ with summit the singleton set. By Lemma 6.3.8, this is isomorphic to the set of natural transformations from $\mathsf{Fin}(X, -)$ to the inclusion $\mathsf{Fin} \hookrightarrow \mathsf{Set}$. The components of such a transformation take a function $f\colon X \to \{1, \ldots, n\}$ and return an element of the set $\{1, \ldots, n\}$. An ultrafilter on X gives rise to such a natural transformation: the element is the unique $i \in \{1, \ldots, n\}$ whose fiber $f^{-1}(i)$ is in the ultrafilter. See [KG71].

Proofs of the claims made in Examples 6.5.11 and 6.5.12 and considerably more can be found in [Lei13].

EXAMPLE 6.5.13. The Giry monad of Example 5.1.5(iv) is also a codensity monad for the inclusion of a subcategory of $\mathsf{C} \hookrightarrow \mathsf{Meas}$. The objects of C are the finite powers of the unit interval $I = [0, 1] \subset \mathbb{R}$, equipped with the Borel σ-algebra, plus the measurable space of sequences in I that converge to 0. The objects can be regarded as convex subspaces of Euclidean space, from which perspective the morphisms are defined to be all affine maps between these spaces. Omitting the space of sequences, the subcategory of finite powers of the interval and affine maps generates a variant of the Giry monad for finitely additive probability measures as its codensity monad. See [Ave16].

Exercises.

EXERCISE 6.5.i. Give a second proof of Proposition 6.5.2(i) using pasting diagrams: show directly that the counit of an adjunction defines an absolute right Kan extension.

EXERCISE 6.5.ii. Use the Yoneda lemma to define the colimit cone for the canonical colimit described in Theorem 6.5.7.

EXERCISE 6.5.iii. For a fixed topological space X, there is a natural inclusion $O(X) \to \mathsf{Top}/X$ that sends an open subset $U \subset X$ to the continuous function $U \hookrightarrow X$. Apply the construction of Remark 6.5.9 to this functor to define an adjunction

$$\mathsf{Top}/X \xrightleftharpoons[\longrightarrow]{\longleftarrow}{\perp} \mathsf{Set}^{O(X)^{\mathrm{op}}}$$

between the category of presheaves on $O(X)$ and the category of spaces over X. By Exercise 4.2.i, this adjunction, like all adjunctions, restricts to define an adjoint equivalence of categories, in this case between the category of **sheaves** on X and the category of **étale spaces** over X.

EXERCISE 6.5.iv. Apply the construction of Remark 6.5.9 to the embedding $\Delta \hookrightarrow \mathsf{Cat}$ of the category of finite non-empty ordinals and order-preserving maps to define an adjunction

$$\mathsf{Cat} \xrightleftharpoons[\xrightarrow{N}]{\xleftarrow{h}}{\perp} \mathsf{Set}^{\Delta^{\mathrm{op}}}.$$

The right adjoint defines the **nerve** of a category, while the left adjoint constructs the **homotopy category** of a simplicial set. The counit of this adjunction is an isomorphism, so Lemma 4.5.13 implies that the nerve is fully faithful. Hence, Cat defines a reflective subcategory of the category of simplicial sets, proving the claim made in Example 4.5.14(vi).

EXERCISE 6.5.v. Apply the construction of Remark 6.5.9 to the functor $\Delta\colon \mathbb{A} \to \mathsf{Top}$ that sends the ordinal $[n] = 0 \to 1 \to \cdots \to n$ to the topological n-simplex

$$\Delta^n := \left\{ (x_0, \ldots, x_n) \in \mathbb{R}^{n+1} \,\middle|\, \sum_i x_i = 1, x_i \geq 0 \right\}.$$

The left adjoint, defined by left Kan extension, forms the **geometric realization** of a simplicial set. The right adjoint is the **total singular complex functor,** which is used to define singular homology:

$$\mathsf{Top} \underset{\mathrm{Sing}}{\overset{|-|}{\underset{\perp}{\rightleftarrows}}} \mathsf{Set}^{\mathbb{A}^{op}}.$$

EXERCISE 6.5.vi. Show that the triple $(T, \eta\colon 1_{\mathsf{C}} \Rightarrow T, \mu\colon T^2 \Rightarrow T)$ constructed in Definition 6.5.10 defines a monad on C.

EXERCISE 6.5.vii. If $F \dashv G$, show that the codensity monad $\mathrm{Ran}_G\, G$ of G exists and is equal to the monad induced by the adjunction $F \dashv G$.

Epilogue: Theorems in Category Theory

> We did not then regard [category theory] as a field for further research efforts, but just as a language and an orientation—a limitation which we followed for a dozen years or so, till the advent of adjoint functors.

Saunders Mac Lane, "Concepts and categories in perspective" [**ML88**]

The author is told with distressing regularity that "there are no theorems in category theory," which typically means that the speaker does not know any theorems in category theory. This attitude is certainly forgivable, as it was held for the first dozen years of the subject's existence by its two founders and might reasonably persist among those with only a casual acquaintance with this area of mathematics. But since Kan's discovery of adjoint functors [**Kan58**] and Grothendieck's contemporaneous work on abelian categories [**Gro57**]—innovations that led to a burst of research activity in the 1960s—without question category theory has not lacked for significant theorems.

What follows is a modest collection of counterexamples to the claim of "no theorems": §E.1 extracts some of the main theorems from the preceding text, §E.2 discusses coherence for symmetric monoidal categories, §E.3 introduces a new universal property of the unit interval, §E.4 characterizes Grothendieck toposes, and §E.5 explores the embedding theorem for abelian categories. Of course, this list is by no means exhaustive: indeed only one of the results appearing here was proven after the early 1970s. As with any mature mathematical discipline, this introductory textbook is rather far from the current frontiers of category theory (see, e.g., [**AR94**], [**Joh02a, Joh02b**], [**Lac10**], [**Lei04**], [**Lur09, Lur16**], [**Sim11b**]).

E.1. Theorems in basic category theory

Significant theorems in basic category theory include, first and foremost, the Yoneda lemma, which characterizes natural transformations whose domain is a represented functor.

THEOREM (2.2.4, 2.2.8). *For any functor $F \colon \mathsf{C}^{\mathrm{op}} \to \mathsf{Set}$, whose domain C is locally small, there is a bijection*

$$\mathrm{Hom}(\mathsf{C}(-, c), F) \cong Fc$$

that identifies a natural transformation $\alpha \colon \mathsf{C}(-, c) \Rightarrow F$ with the element $\alpha_c(1_c) \in Fc$ that is natural in both c and F. In particular, any natural transformation $\mathsf{C}(-, c) \Rightarrow \mathsf{C}(-, c')$ between represented functors is given by composing with a unique morphism $c \to c'$ in C.

A set-valued functor is *representable* just when it is naturally isomorphic to a represented functor, covariant or contravariant as the case dictates. The idea that representable functors encode universal properties of their representing objects is made precise by the following result:

THEOREM (2.4.8). *A representation for a functor $F: \mathsf{C} \to \mathsf{Set}$ is an initial object in its category of elements. Dually, a representation for a functor $F: \mathsf{C}^{\mathrm{op}} \to \mathsf{Set}$ is a terminal object in its category of elements.*

To calculate particular limits and colimits, it is very useful to know that they can be built out of simpler limit and colimit constructions, when these exist.

THEOREM (3.4.12). *A category with small coproducts and coequalizers has small colimits of any shape, and dually a category with small products and equalizers has all small limits.*

In most cases, the cardinality of a category (meaning the cardinality of its set of morphisms) places a sharp bound on the sizes of limits and colimits it may possess.

THEOREM (3.7.3). *Any κ-small category that admits limits or colimits of all diagrams indexed by κ-small categories is equivalent to a poset. In particular, the only small categories that are complete or cocomplete are equivalent to posets.*

The following theorem, which is quite easy to prove, has consequences throughout mathematics.

THEOREM (4.5.2, 4.5.3). *Right adjoints preserve limits, and left adjoints preserve colimits.*

This theorem implies that *continuity* or *cocontinuity* are necessary conditions for a functor to admit a left or right adjoint. The adjoint functor theorems supply additional conditions that are sufficient for an adjoint to exist.

THEOREM (4.6.1, 4.6.3, 4.6.10). *Let $U: \mathsf{A} \to \mathsf{S}$ be a continuous functor whose domain is complete and whose domain and codomain are locally small. If either*

  (i) *for each $s \in \mathsf{S}$ the comma category $s \downarrow U$ has an initial object,*
 (ii) *for every $s \in \mathsf{S}$ there exists a set of morphisms $\Phi_s = \{ f_i : s \to U a_i \}$ so that any $f: s \to U a$ factors through some $f_i \in \Phi_s$ along a morphism $a_i \to a$ in A, or*
(iii) A *has a small cogenerating set and every collection of subobjects of a fixed object in A admits an intersection,*

then U admits a left adjoint.

As a corollary:

THEOREM (4.6.17). *A functor $F: \mathsf{C} \to \mathsf{D}$ between locally presentable categories*

  (i) *admits a right adjoint if and only if it is cocontinuous.*
 (ii) *admits a left adjoint if and only if it is continuous and accessible.*

The proofs of the adjoint functor theorems also lead to proofs of related representability criteria.

THEOREM (4.6.14, 4.6.15). *If $F: \mathsf{C} \to \mathsf{Set}$ preserves limits and C is locally small and complete, then if either*

  (i) C *has a small cogenerating set and has the property that every collection of subobjects of a fixed object has an intersection, or if*
 (ii) *there exists a set Φ of objects of C so that for any $c \in \mathsf{C}$ and any element $x \in Fc$ there exists an $s \in \Phi$, an element $y \in Fs$, and a morphism $f: s \to c$ so that $Ff(y) = x$,*

then F is representable.

The following theorem collects together several properties of monadic functors.

THEOREM (5.5.1, 5.6.1, 5.6.5, 5.6.12). *If $U: \mathsf{A} \to \mathsf{S}$ admits a left adjoint and creates coequalizers of U-split pairs, then:*

(i) *A morphism f in* A *is an isomorphism if and only if Uf is an isomorphism in* S.

(ii) *U creates all limits that exist in* S.

(iii) *U creates all colimits that exist in* S *and are preserved by the induced monad on* S *and its square.*

(iv) *If* S *is locally small, complete, and cocomplete, and if U preserves filtered colimits, then* A *admits all colimits.*

The following theorem establishes the existence of and provides a formula for adjoints to restriction functors, given by *pointwise left* and *right Kan extensions*.

THEOREM (6.2.1, 6.2.6). *Consider a functor K* : C → D *where* C *is small and* D *is locally small. Whenever* E *is cocomplete, the restriction functor K** : E^D → E^C *admits a left adjoint, defined by a particular colimit formula, and whenever* E *is complete, the restriction functor along K admits a right adjoint, defined by a dual limit formula.*

Other candidates for inclusion on this list include theorems, such as 4.2.6 or 6.3.7, that prove that two a priori distinct categorical definitions are equivalent, or the results describing the universal properties of the derived functors constructed via deformations established by 6.4.11 and 6.4.12.

E.2. Coherence for symmetric monoidal categories

There are a number of coherence theorems in the categorical literature, the first and most famous being the following theorem of Mac Lane [**ML63**].

The data of a **symmetric monoidal category** (V, ⊗, ∗) consists of a category V, a bifunctor − ⊗ − : V × V → V called the **monoidal product**, and a **unit object** ∗ ∈ V together with specified natural isomorphisms

(E.2.1) $v \otimes w \cong_{\gamma} w \otimes v$ $u \otimes (v \otimes w) \cong_{\alpha} (u \otimes v) \otimes w$ $* \otimes v \cong_{\lambda} v \cong_{\rho} v \otimes *$

witnessing symmetry, associativity, and unit conditions on the monoidal product. The natural transformations (E.2.1) must satisfy certain "coherence conditions" that will be discussed momentarily. A **monoidal category** is defined similarly, except that the first symmetry natural isomorphism is omitted.

Examples of symmetric monoidal categories include all categories with finite products, such as (Set, ×, ∗), (Top, ×, ∗), or (Cat, ×, 𝟙), where the monoidal unit in each case is the terminal object. Dually, any category with finite coproducts defines a symmetric monoidal category, with the initial object as the monoid unit, e.g., (Man, ⊔, ∅). The category of modules over a ring admits multiple monoidal structures: (Mod_R, ⊕, 0) and (Mod_R, $\otimes_R$, R), the last only in the case where R is commutative. The category of unbounded chain complexes of R-modules, with R commutative, is also a monoidal category (Ch_R, $\otimes_R$, R), where "$\otimes_R$" denotes the usual graded tensor product of chain complexes and R is the chain complex with the R-module R in degree zero and zeros elsewhere. In this case, the symmetry natural isomorphism can be defined in two distinct ways depending on whether or not a sign is introduced in odd degrees.

A commutative monoid, regarded as the objects of a discrete category, defines a trivial special case of a symmetric monoidal category in which the natural isomorphisms (E.2.1) are identities. The natural isomorphisms (E.2.1) defining a symmetric monoidal category are categorifications of the commutativity, associativity, and unit laws. As Mac Lane writes in the introduction to [**ML63**]:

> The usual associative law $a(bc) = (ab)c$ is known to imply the "general associative law," which states that any two iterated products of the same factors in the same order are equal, irrespective of the arrangement of parentheses.

Similarly, in a symmetric monoidal category, composites of the coherence natural isomorphisms define higher associativity and higher symmetry isomorphisms. A final axiom requires these natural isomorphisms to satisfy the following coherence diagrams, expressed in terms of components at objects $u, v, w, x \in \mathsf{V}$, which demand for instance that there is a unique natural isomorphism built from this data that witnesses the symmetry of a three-fold product:

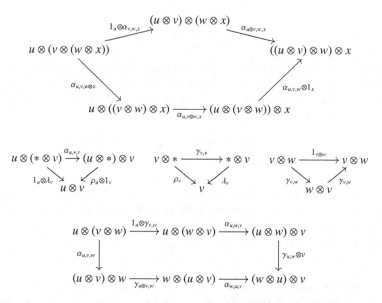

The coherence theorem for monoidal categories, first proven by Mac Lane and then improved by Kelly, tells us that once these conditions are satisfied, then all diagrams whose edges are comprised of composites of these natural isomorphisms, parenthesized in any order, commute; see [**ML98a**, §XI.1] for a precise statement and a proof and [**ML76**] for a general discussion.

Theorem E.2.2 (Mac Lane, Kelly). *Any diagram in a symmetric monoidal category that is comprised of associators α, unitors γ or ρ, or symmetors γ that is "formal" commutes.*

The upshot of Theorem E.2.2 is that any two "n-ary tensor product" functors $\mathsf{V}^{\times n} \to \mathsf{V}$ that are built iteratively from the bifunctor $\otimes$ and the unit object $*$, each permuting and parenthesizing the inputs in some manner, are connected by a unique natural isomorphism that is built out of the given natural isomorphisms α, λ, ρ, and γ. In practice, this naturality means that we need not concern ourselves with particular parenthesizations or orderings when defining the n-ary tensor product. Hence, the structural isomorphisms (E.2.1) are seldom emphasized.

E.3. The universal property of the unit interval

The **unit interval** is the topological space $I = [0, 1] \subset \mathbb{R}$ regarded as a subspace of the real line, with the standard Euclidean metric topology. It is used to define the fundamental groupoid $\Pi_1(X)$ of paths in a topological space X. A **path** in X is simply a continuous function $p: I \to X$. The path has two endpoints $p(0), p(1) \in X$ defined by evaluating at the endpoints $0, 1 \in I$. If $q: I \to X$ is a second path with the property that $p(1) = q(0)$, then there exists a composite path $p * q: I \to X$ defined by the composite continuous function

$$I \xrightarrow[\cong]{\delta} I \vee I \xrightarrow{p \vee q} X.$$

Here $I \vee I$ is the space formed by gluing two copies of I together by identifying the point 1 in the left-hand copy with the point 0 in the right-hand copy:

$$
\begin{array}{ccc}
* & \xrightarrow{\ 0\ } & I \\
{\scriptstyle 1}\downarrow & & \downarrow \\
I & \longrightarrow & I \vee I
\end{array}
$$

The space $I \vee I$ is homeomorphic to the space $[0, 2] \subset \mathbb{R}$. The map $\delta: I \to I \vee I$ is the homeomorphism $t \mapsto 2t$. Note that this map sends the endpoints of the domain I to the endpoints of the fattened interval $I \vee I$. Thus, $(p * q)(0) = p(0)$ and $(p * q)(1) = q(1)$; that is, $p * q$ is a path in X from the starting point of the path p to the ending point of the path q.

The **fundamental groupoid** $\Pi_1(X)$ is the category whose objects are points of X and whose morphisms are endpoint-preserving homotopy classes of paths in X, with composition defined via the gluing of the unit interval just described. The reason that morphisms are endpoint-preserving homotopy classes, rather than simply paths, is that path composition is not strictly associative or unital. Endpoint-preserving homotopies allow in particular for re-parameterizations of the unit interval.

By a theorem of Freyd [**Fre99**], refined by Leinster [**Lei11**], the structure on the unit interval that is used to define composition of paths also describes the universal property of this topological space. A **bipointed space** is a topological space I with two distinct closed points $0 \neq 1 \in I$, one designated as the "left" and the other as the "right." A map of bipointed spaces is a continuous function that sends the left point to the left point and the right point to the right point. For any bipointed space (X, x_0, x_1), we can define $X \vee X$ to be the pushout

$$
\begin{array}{ccc}
* & \xrightarrow{\ x_0\ } & X \\
{\scriptstyle x_1}\downarrow & & \downarrow \\
X & \longrightarrow & X \vee X
\end{array}
$$

THEOREM E.3.1 (Freyd–Leinster). *The unit interval is the terminal bipointed space equipped with a map of bipointed spaces $X \to X \vee X$. That is, for any other bipointed space (X, x_0, x_1), there is a unique bipointed continuous map $r: X \to I$ so that*

$$
\begin{array}{ccc}
X & \longrightarrow & X \vee X \\
{\scriptstyle r}\downarrow & & \downarrow{\scriptstyle r \vee r} \\
I & \xrightarrow[\cong]{\delta} & I \vee I
\end{array}
$$

commutes.

Put more concisely, Theorem E.3.1 says that $(I, 0, 1)$ is the *terminal coalgebra* for the endofunctor $X \mapsto X \vee X$ on the category of bipointed spaces. See [**Lei11**] for a discussion and proof. The fact that the map $I \to I \vee I$ is a homeomorphism is no coincidence: a theorem of Lambek [**Lam68**], whose elementary proof is left as Exercise 1.6.vi, proves that the structure map of any terminal coalgebra is always an isomorphism.

E.4. A characterization of Grothendieck toposes

Given a small category C, a **presheaf** is another name for a contravariant set-valued functor on C. A **Grothendieck topos** is a reflective full subcategory E of a presheaf category

$$\mathsf{E} \underset{\longrightarrow}{\overset{L}{\longleftarrow}} \mathsf{Set}^{\mathsf{C}^{\mathrm{op}}}$$

with the property that the left adjoint preserves finite limits.[11] Objects in E can be characterized as sheaves on a small *site*, which specifies a "covering family" of morphisms $(f_i \colon U_i \to U)_i$ for each object $U \in \mathsf{C}$ satisfying a weak pullback condition. A typical example might take $\mathsf{C} = O(X)$ to be the poset of open sets for a topological space X. A presheaf $P \colon O(X)^{\mathrm{op}} \to \mathsf{Set}$ assigns a set $P(U)$ to each open set $U \subset X$ so that this assignment is functorial with respect to restrictions along inclusion $V \subset U \subset X$ of open subsets. A presheaf P is a **sheaf** if and only if the diagram of restriction maps

$$P(U) \longrightarrow \prod_\alpha P(U_\alpha) \rightrightarrows \prod_{\alpha,\beta} P(U_\alpha \cap U_\beta)$$

is an equalizer for every open cover $U = \cup_\alpha U_\alpha$ of a $U \in O(X)$; see Definition 3.3.4.

Giraud's theorem, which first appeared in print in [**GV72**, IV.1.2], states that Grothendieck toposes can be completely characterized by a combination of "exactness" and size conditions.

THEOREM E.4.1 (Giraud). *A category E is a Grothendieck topos if and only if it satisfies the following conditions:*

- *(i) E is locally small.*
- *(ii) E has finite limits.*
- *(iii) E has all small coproducts and they are disjoint and universal.*
- *(iv) Equivalence relations in E have universal coequalizers.*
- *(v) Every equivalence relation in E is effective, and every epimorphism in E is a coequalizer.*
- *(vi) E has a set of generators.*

Some explanation is in order:

- A coproduct $\coprod_\alpha A_\alpha$ is **disjoint** if each inclusion $A_\alpha \to \coprod_\alpha A_\alpha$ is a monomorphism and if the pullback of any two distinct inclusions is an initial object.
- A colimit cone $(A_\alpha \to A)_\alpha$ is **universal** if the pullbacks of these maps along any $B \to A$ define a colimit cone $(B \times_A A_\alpha \to B)_\alpha$.
- An **effective** equivalence relation is one that arises, as in Example 3.5.10, as a kernel pair of some morphism.
- Finally, recall from Definition 4.6.7 that a set of **generators** for E is a set of objects $\mathcal{G} \subset \mathrm{ob}\,\mathsf{E}$ that is jointly separating: for any $f, g \colon B \rightrightarrows A$ with $f \neq g$ there is some $h \colon G \to B$, with $G \in \mathcal{G}$, so that $fh \neq gh$.

---

[11]Propositions 4.5.15 and 3.3.9 imply that E is cocomplete, so this adjunction can be defined as in Remark 6.5.9 from the functor $Ly \colon \mathsf{C} \to \mathsf{E}$ that is the restriction of the left adjoint along the Yoneda embedding.

A proof of Theorem E.4.1 can be found in [**Joh14**, §0.4], which also presents a very clear exposition of the ideas involved.

E.5. Embeddings of abelian categories

Properties of the categories Ab of abelian groups, or Mod_R of modules over a unital ring, or Ch_R of chain complexes of R-modules are abstracted in the notion of an **abelian category**. Abelian categories were introduced by Buchsbaum [**Buc55**] and developed by Grothendieck [**Gro57**] with the aim of unifying the "coefficients" of the various cohomology theories then under development.

DEFINITION E.5.1. A category is **abelian** if

- it has a **zero object** 0, that is both initial and terminal,
- it has all binary products and binary coproducts,
- it has all **kernels** and **cokernels**, defined respectively to be the equalizer and coequalizer of a map $f\colon A \to B$ with the zero map $A \to 0 \to B$, and
- all monomorphisms and epimorphisms arise as kernels or cokernels, respectively.

In an abelian category A, finite products and finite coproducts coincide (in the sense that the canonical map from the latter to the former described in Remark 3.1.27 is an isomorphism) and are called **biproducts** or **direct sums**. These axioms imply that the hom-sets in A canonically inherit the structure of an abelian group, with the zero map in each hom-set serving as the additive identity, in such a way that composition is bilinear. This gives A the structure of a category *enriched over*[12] the monoidal category $(\mathsf{Ab}, \otimes_{\mathbb{Z}}, \mathbb{Z})$.

The **image** of a morphism is defined to be the kernel of its cokernel, or equivalently, the cokernel of its kernel; these objects are always isomorphic in an abelian category. This permits the definition of an **exact sequence** in an abelian category, a sequence of composable morphisms

$$\cdots \xrightarrow{f_{n+2}} A_{n+1} \xrightarrow{f_{n+1}} A_n \xrightarrow{f_n} A_{n-1} \xrightarrow{f_{n-1}} \cdots$$

so that $\ker f_n = \operatorname{im} f_{n+1}$. Classical results such as the *five lemma*, which is used to detect isomorphisms between a pair of exact sequences, can be proven by general abstract nonsense in any abelian category using the universal properties of kernels and cokernels.

A functor is **exact** if it preserves exact sequences or, equivalently, if it is both left and right exact in the sense introduced in Definition 4.5.9.

THEOREM E.5.2 (Freyd–Mitchell). *If* A *is a small abelian category, then there is a ring*[13] R *and an exact, fully faithful functor* A $\hookrightarrow \mathsf{Mod}_R$, *which embeds* A *as a full subcategory.*

A proof appears as the very last result of the book [**Fre03**]. Unfortunately, most of the material presented in the preceding 150 pages appears to be necessary.

To conclude, we invite the reader to reflect upon a lovely quote from the introduction to Johnstone's *Topos Theory* [**Joh14**, pp. xii–xiii] which describes the significance of the Freyd–Mitchell embedding theorem and, by analogy, the broader project of category theory:

> Incidentally, the Freyd–Mitchell embedding theorem is frequently
> regarded as a culmination rather than a starting point; this is because
> of what seems to me a misinterpretation (or at least an inversion) of
> its true significance. It is commonly thought of as saying "If you

---

[12]Ordinary categorical notions can be extended to categories in which each $\operatorname{Hom}(X, Y)$ is an object of a general monoidal category $(\mathsf{V}, \otimes, *)$, in place of $(\mathsf{Set}, \times, *)$. See [**Rie14**, Chapter 3] for an introduction to enriched category theory.

[13]As per our conventions elsewhere, R is unital and associative but not necessarily commutative.

want to prove something about an abelian category, you might as well assume it is a category of modules"; whereas I believe its true import is "If you want to prove something about categories of modules, you might as well work in a general abelian category"—for the embedding theorem ensures that your result will be true in this generality, and by forgetting the explicit structure of module categories you will be forced to concentrate on the essential aspects of this problem. As an example, compare the module-theoretic proof of the Snake Lemma in [**HS97**] with the abelian-category proof in [**ML98a**].

Bibliography

[AR94]   Jiří Adámek and Jiří Rosický. *Locally presentable and accessible categories*. London Mathematical Society lecture note series. Cambridge University Press, Cambridge, England, New York, NY, 1994.

[Ati74]   M.F. Atiyah. How research is carried out. *Bull. IMA.*, 10:232–234, 1974.

[Ave16]   Tom Avery. Codensity and the Giry monad. *Journal of Pure and Applied Algebra*, 220(3):1229–1251, 2016.

[Awo96]   Steve Awodey. Structure in mathematics and logic: a categorical perspective. *Philos. Math. (3)*, 4(3):209–237, 1996.

[Awo10]   Steve Awodey. *Category theory*, volume 52 of *Oxford Logic Guides*. Oxford University Press, Oxford, second edition, 2010.

[Bae06]   John Baez. Quantum quandaries: a category-theoretic perspective. In *The structural foundations of quantum gravity*, pages 240–265. Oxford Univ. Press, Oxford, 2006.

[Bec67]   Jonathan Mock Beck. *Triples, Algebras and Cohomology*. PhD thesis, Columbia University, 1967.

[BJLS15]   Marie Bjerrum, Peter Johnstone, Tom Leinster, and William F. Sawin. Notes on commutation of limits and colimits. *Theory Appl. Categ.*, 30(15):527–532, 2015.

[Bor94a]   Francis Borceux. *Handbook of categorical algebra. 1*, volume 50 of *Encyclopedia of Mathematics and its Applications*. Cambridge University Press, Cambridge, 1994. Categories and structures.

[Bor94b]   Francis Borceux. *Handbook of categorical algebra. 2*, volume 51 of *Encyclopedia of Mathematics and its Applications*. Cambridge University Press, Cambridge, 1994.

[Buc55]   D. A. Buchsbaum. Exact categories and duality. *Trans. Amer. Math. Soc.*, 80:1–34, 1955.

[CE56]   Henri Cartan and Samuel Eilenberg. *Homological Algebra*. Princeton University Press, Princeton, N. J., 1956.

[CGR14]   Eugenia Cheng, Nick Gurski, and Emily Riehl. Cyclic multicategories, multivariable adjunctions and mates. *J. K-theory*, 13(2):337–396, 2014.

[CM13]   Gunnar Carlsson and Facundo Mémoli. Classifying clustering schemes. *Foundations of Computational Mathematics*, 13(2):221–252, 2013.

[CMW09]   David Clark, Scott Morrison, and Kevin Walker. Fixing the functoriality of Khovanov homology. *Geometry and Topology*, 13:1499–1582, 2009.

[DHKS04]   W. G. Dwyer, P. S. Hirschhorn, D. M. Kan, and J. H. Smith. *Homotopy limit functors on model categories and homotopical categories*, volume 113 of *Mathematical Surveys and Monographs*. American Mathematical Society, Providence, RI, 2004.

[Ehr65]   Charles Ehresmann. *Catégories et structures*. Dunod, Paris, 1965.

[EM42a]   Samuel Eilenberg and Saunders MacLane. Group extensions and homology. *Ann. of Math. (2)*, 43(757–831), 1942.

[EM42b]   Samuel Eilenberg and Saunders MacLane. Natural isomorphisms in group theory. *Proc. Nat. Acad. Sci. U. S. A.*, 28:537–543, 1942.

[EM45]   Samuel Eilenberg and Saunders MacLane. General theory of natural equivalences. *Trans. Amer. Math. Soc.*, 58:231–294, 1945.

[ES52]   Samuel Eilenberg and Norman Steenrod. *Foundations of algebraic topology*. Princeton University Press, Princeton, New Jersey, 1952.

[Faw86]   Barry Fawcett. A categorical characterization of the four colour theorem. *Canad. Math. Bull.*, 29(4):426–431, 1986.

[Fre99]   Peter J. Freyd. Real coalgebra. post on the categories mailing list, 22 December 1999, available via www.math.ca/~cat-dist, 1999.

[Fre03]   Peter J. Freyd. Abelian categories. *Repr. Theory Appl. Categ.*, (3):1–190, 2003.

[Fre04]   Peter J. Freyd. Homotopy is not concrete. *Reprints in Theory and Applications of Categories*, (6):1–10, 2004.

[FS90]   Peter J. Freyd and Andre Scedrov. *Categories, Allegories*. North-Holland Mathematical Library (Book 39). North Holland, 1990.

[Ghr14]   Robert Ghrist. *Elementary Applied Topology*. CreateSpace Independent Publishing Platform, 2014.

[Gir82]   M. Giry. A categorical approach to probability theory. In *Categorical Aspects of Topology and Analysis*, volume 915 of *Lecture Notes in Mathematics*, pages 68–85. Springer, 1982.

[Gro57]   Alexander Grothendieck. Sur quelques points d'algèbre homologique. *Tôhoku Math. J. (2)*, 9:119–221, 1957.

[Gro58]   Alexander Grothendieck. A general theory of fibre spaces with structure sheaf. National Science Foundation Research Project on Geometry of Function Space, Report No. 4, Second Edition, May 1958.

[Gro60]   Alexander Grothendieck. Technique de descente et théorèmes d'existence en géométrie algébriques. ii. le théorème d'existence en théorie formelle des modules. *Séminaire Bourbaki*, 5(195):195, 1958-1960.

[GV72]   A. Grothendieck and J. L. Verdier. *Théorie des topos et cohomologie étale des schémas. Tome 1: Théorie des topos*, volume Séminaire de Géométrie Algébrique du Bois-Marie 1963–1964 (SGA 4) of *Lecture Notes in Mathematics, Vol. 269*, chapter Exposé IV: Topos. Springer-Verlag, Berlin-New York, 1972.

[GZ67]   P. Gabriel and M. Zisman. *Calculus of fractions and homotopy theory*. Ergebnisse der Mathematik und ihrer Grenzgebiete, Band 35. Springer-Verlag New York, Inc., New York, 1967.

[Har83]   Donald G. Hartig. The Riesz representation theorem revisited. *The American Mathematical Monthly*, 90(4), 1983.

[HS97]   P. J. Hilton and U. Stammbach. *A course in homological algebra*, volume 4 of *Graduate Texts in Mathematics*. Springer-Verlag, New York, second edition, 1997.

[Iry09]   James Iry. A brief, incomplete, and mostly wrong history of programming languages, May 2009.

[Joh02a]   P.T. Johnstone. *Sketches of an Elephant: A Topos Theory Compendium. Vol. 1*, volume 43 of *Oxford Logic Guides*. The Clarendon Press Oxford University Press, New York, 2002.

[Joh02b]   P.T. Johnstone. *Sketches of an Elephant: A Topos Theory Compendium. Vol. 2*, volume 44 of *Oxford Logic Guides*. The Clarendon Press Oxford University Press, New York, 2002.

[Joh14]   P.T. Johnstone. *Topos theory*. reprint of London Mathematical Society Monographs, Vol. 10. Dover Publications, 2014.

[Joy81]   André Joyal. Une théorie combinatoire des séries formelles. *Adv. in Math.*, 42(1):1–82, 1981.

[Kan58]   Daniel M. Kan. Adjoint functors. *Trans. Amer. Math. Soc.*, 87:294–329, 1958.

[Kel82]   G. M. Kelly. *Basic concepts of enriched category theory*. Reprints in Theory and Applications of Categories. 1982. Reprint of the 1982 original [Cambridge Univ. Press, Cambridge; MR0651714].

[KG71]   J. F. Kennison and Dion Gildenhuys. Equational completion, model induced triples and pro-objects. *J. Pure Appl. Algebra*, 1(4):317–346, 1971.

[Kle03]   Jon M. Kleinberg. An impossibility theorem for clustering. In S. Becker, S. Thrun, and K. Obermayer, editors, *Advances in Neural Information Processing Systems 15*, pages 463–470. MIT Press, 2003.

[KS74]   G. M. Kelly and R. Street. Review of the elements of 2-categories. In *Category Seminar (Proc. Sem., Sydney, 1972/1973)*, pages 75–103. Lecture Notes in Math., Vol. 420. Springer, Berlin, 1974.

[Lac10]   Stephen Lack. A 2-categories companion. In J.C. Baez and J.P. May, editors, *Towards Higher Categories*, volume 152 of *The IMA Volumes in Mathematics and its Applications*, pages 105–192, New York, 2010. Springer.

[Lam68]   Joachim Lambek. A fixpoint theorem for complete categories. *Mathematische Zeitschrift*, 103:151–161, 1968.

[Lan84]   Serge Lang. *Algebra*. Addison-Wesley Publishing Company, Advanced Book Program, Reading, MA, second edition, 1984.

[Lan02]   Serge Lang. *Algebra*, volume 211 of *Graduate Texts in Mathematics*. Springer-Verlag, New York, third edition, 2002.

[Law05]   F. William Lawvere. An elementary theory of the category of sets (long version) with commentary. *Repr. Theory Appl. Categ.*, 11:1–35, 2005.

[Lei04]   Tom Leinster. *Higher Operads, Higher Categories*. Number 298 in London Mathematical Society Lecture Note Series. Cambridge University Press, Cambridge, 2004.

[Lei08]   Tom Leinster. Doing without diagrams. available from http://www.maths.ed.ac.uk/~tl/elements.pdf, 2008.

[Lei11]   Tom Leinster. A general theory of self-similarity. *Adv. Math.*, 226(4):2935–3017, 2011.

[Lei13]   Tom Leinster. Codensity and the ultrafilter monad. *Theory Appl. Categ.*, 28:No. 13, 332–370, 2013.

[Lei14]   Tom Leinster. *Basic category theory*. Cambridge Studies in Advanced Mathematics. Cambridge University Press, 2014.

[Lin66]   Fred Linton. Some aspects of equational theories. In *Proceedings of the Conference on Categorical Algebra, La Jolla*, pages 84–95, 1966.

[Lin70]   C. E. Linderholm. A group epimorphism is surjective. *Amer. Math. Monthly*, 77(176–177), 1970.

[Lur09]   Jacob Lurie. *Higher topos theory*, volume 170 of *Annals of Mathematics Studies*. Princeton University Press, Princeton, NJ, 2009.

[Lur16]   Jacob Lurie. Higher Algebra. http://www.math.harvard.edu/~lurie/papers/HigherAlgebra.pdf, March 2016.

[Mal07]   G. Maltsiniotis. Le théorème de Quillen, d'adjonction des foncteurs dérivés, revisité. *C. R. Math. Acad. Sci. Paris*, 344(9):549–552, 2007.

[Mar09]   Jean-Pierre Marquis. *From a Geometrical Point of View, A Study of the History and Philosophy of Category Theory*, volume 14 of *Logic, Epistemology, and the Unity of Science*. Springer, 2009.

[May01]   J. P. May. Picard groups, Grothendieck rings, and Burnside rings of categories. *Adv. Math.*, 163(1):1–16, 2001.

[Maz08]   Barry Mazur. When is one thing equal to some other thing? In B. Gold and Simons R, editors, *Proof and other dilemmas*, MAA Spectrum, pages 221–241. Math. Assoc. America, Washington, DC, 2008.

[Maz16]   Barry Mazur. Thinking about Grothendieck. *Notices of the AMS*, 63(4):404–405, 2016.

[ML50]   Saunders Mac Lane. Duality for groups. *Bull. Amer. Math. Soc.*, 56:485–516, 1950.

[ML63]   Saunders Mac Lane. Natural associativity and commutativity. *Rice Univ. Studies*, 49(4):28–46, 1963.

[ML76]   Saunders Mac Lane. Topology and logic as a source of algebra. *Bull. Amer. Math. Soc.*, 82(1):1–40, 1976.

[ML88]   Saunders Mac Lane. Concepts and categories in perspective. In *A century of mathematics in America, Part I*, volume 1 of *Hist. Math.*, pages 323–365. Amer. Math. Soc., Providence, RI, 1988.

[ML98a]   Saunders Mac Lane. *Categories for the working mathematician*, volume 5 of *Graduate Texts in Mathematics*. Springer-Verlag, New York, second edition, 1998.

[ML98b]   Saunders Mac Lane. The Yoneda lemma. *Mathematica Japonica*, 47:156, 1998.

[ML05]   Saunders Mac Lane. *Saunders Mac Lane—a mathematical autobiography*. A K Peters, Ltd., Wellesley, MA, 2005.

[Par71]   Robert Paré. On absolute colimits. *Journal of Algebra*, 19:80–95, 1971.

[Par74]   Robert Paré. Colimits in topoi. *Bulletin of the American Mathematical Society*, 80(3):556–561, 1974.

[Poi08]   Henri Poincaré. L'avenir des mathématiques. *Rendiconti del Circolo Matematico di Palermo (1884–1940)*, 26(1):152–168, December 1908.

[Poo14]   Bjorn Poonen. Why all rings should have a 1. arXiv:1404.0135 [math.RA], 2014.

[Rei88]   Miles Reid. *Undergraduate algebraic geometry*, volume 12 of *London Mathematical Society Student Texts*. Cambridge University Press, Cambridge, 1988.

[Rie14]   Emily Riehl. *Categorical homotopy theory*, volume 24 of *New Mathematical Monographs*. Cambridge University Press, Cambridge, 2014.

[Seg74]   Graeme Segal. Categories and cohomology theories. *Topology*, 13:293–312, 1974.

[Shu08]   Michael Shulman. Set theory for category theory. arXiv:0810.1279, 2008.

[Sim11a] Harold Simmons. *An introduction to category theory*. Cambridge University Press, Cambridge, 2011.

[Sim11b] Carlos Simpson. *Homotopy Theory of Higher Categories: From Segal Categories to n-Categories and Beyond*. New Mathematical Monographs. Cambridge University Press, Cambridge, 2011.

[Smi]   Peter Smith. The galois connection between syntax and semantics. http://www.logicmatters.net/resources/pdfs/Galois.pdf.

[Ste67]   N. E. Steenrod. A convenient category of topological spaces. *Michigan Math. J.*, 14:133–152, 1967.

[TW02]   R. Theiele and L. Wos. Hilbert's twenty-fourth problem. *J. Automated Reasoning*, 29:67–89, 2002.

Catalog of Categories

229

Glossary of Notation

$=$, xvi
$\Rightarrow$, xvi
$\cdot$, 3
$:=$, xvi
$\cong$, xvi, 7
$\dashrightarrow$, xvi
$\dashv$, 117
$\leftrightarrows$, xvi
$\mapsto$, xvi
$\rightarrowtail$, 11
$\rightrightarrows$, xvi
$\rightsquigarrow$, xvi, 162
$\simeq$, 30
$\rightarrow$, xvi
$\twoheadrightarrow$, 11

$\mathbb{0}$, 5
$\mathbb{1}$, 5
$\mathbb{2}$, 5

$\mathbb{A}^n$, 63
$\alpha: F \Rightarrow G$, 23
$\alpha: F \cong G$, 24

$\mathsf{B}G$, 5

$\mathsf{C}(X, Y)$, 7
$\mathsf{C}(c, -)$, 19
C/c, 8
c/C, 8
$\mathsf{C}(-, c)$, 19
C^J, 44
cod, 6
colim F, 75
$\text{colim}_\mathsf{J} F$, 76
C^{op}, 9
$\coprod_{j \in J}$, 81
cosk_n, 199
C^T, 160
C_T, 162

Δ, 74
Δ_+, 193
Δ, 18
D^n, 73
dom, 6

$\int F$, 66

f^*, 10
f_*, 10
$F \downarrow G$, 22
$(-)^\flat$, 116
$f: X \to Y$, 3

$\text{Hom}(X, Y)$, 7
$\text{Hom}(A, B)$, 130
$\text{Hom}(V, W)$, 63

id, 6
1_X, 3
$\mathbb{I}$, 29

$\mathsf{J}^\triangleleft$, 76
$\mathsf{J}^\triangleright$, 76

K_n, 49

$\text{Lan}_K F$, 190
$\mathbb{L}F$, 207
$\mathbf{L}F$, 206
$\lim F$, 75
$\lim_\mathsf{J} F$, 76

$\text{Map}(X, Y)$, 130

$[n]$, 18
$\mathbb{n}$, 5

O_G, 21
ω, 5
$\oplus$, 83
$\otimes$, 63
$O(X)$, 18

$\prod_{j \in J}$, 77
$\lrcorner$, 79
$\ulcorner$, 81

$\text{Ran}_K F$, 190
$\mathbb{R}F$, 207
$\mathbf{R}F$, 206

$(-)^\sharp$, 116
sk_n, 199
$\mathsf{sk}\mathsf{C}$, 34

231

Index

A CATALOG OF SELECTED
DOVER BOOKS
IN SCIENCE AND MATHEMATICS

Engineering

FUNDAMENTALS OF ASTRODYNAMICS, Roger R. Bate, Donald D. Mueller, and Jerry E. White. Teaching text developed by U.S. Air Force Academy develops the basic two-body and n-body equations of motion; orbit determination; classical orbital elements, coordinate transformations; differential correction; more. 1971 edition. 455pp. 5 3/8 x 8 1/2. 0-486-60061-0

INTRODUCTION TO CONTINUUM MECHANICS FOR ENGINEERS: Revised Edition, Ray M. Bowen. This self-contained text introduces classical continuum models within a modern framework. Its numerous exercises illustrate the governing principles, linearizations, and other approximations that constitute classical continuum models. 2007 edition. 320pp. 6 1/8 x 9 1/4. 0-486-47460-7

ENGINEERING MECHANICS FOR STRUCTURES, Louis L. Bucciarelli. This text explores the mechanics of solids and statics as well as the strength of materials and elasticity theory. Its many design exercises encourage creative initiative and systems thinking. 2009 edition. 320pp. 6 1/8 x 9 1/4. 0-486-46855-0

FEEDBACK CONTROL THEORY, John C. Doyle, Bruce A. Francis and Allen R. Tannenbaum. This excellent introduction to feedback control system design offers a theoretical approach that captures the essential issues and can be applied to a wide range of practical problems. 1992 edition. 224pp. 6 1/2 x 9 1/4. 0-486-46933-6

THE FORCES OF MATTER, Michael Faraday. These lectures by a famous inventor offer an easy-to-understand introduction to the interactions of the universe's physical forces. Six essays explore gravitation, cohesion, chemical affinity, heat, magnetism, and electricity. 1993 edition. 96pp. 5 3/8 x 8 1/2. 0-486-47482-8

DYNAMICS, Lawrence E. Goodman and William H. Warner. Beginning engineering text introduces calculus of vectors, particle motion, dynamics of particle systems and plane rigid bodies, technical applications in plane motions, and more. Exercises and answers in every chapter. 619pp. 5 3/8 x 8 1/2. 0-486-42006-X

ADAPTIVE FILTERING PREDICTION AND CONTROL, Graham C. Goodwin and Kwai Sang Sin. This unified survey focuses on linear discrete-time systems and explores natural extensions to nonlinear systems. It emphasizes discrete-time systems, summarizing theoretical and practical aspects of a large class of adaptive algorithms. 1984 edition. 560pp. 6 1/2 x 9 1/4. 0-486-46932-8

INDUCTANCE CALCULATIONS, Frederick W. Grover. This authoritative reference enables the design of virtually every type of inductor. It features a single simple formula for each type of inductor, together with tables containing essential numerical factors. 1946 edition. 304pp. 5 3/8 x 8 1/2. 0-486-47440-2

THERMODYNAMICS: Foundations and Applications, Elias P. Gyftopoulos and Gian Paolo Beretta. Designed by two MIT professors, this authoritative text discusses basic concepts and applications in detail, emphasizing generality, definitions, and logical consistency. More than 300 solved problems cover realistic energy systems and processes. 800pp. 6 1/8 x 9 1/4. 0-486-43932-1

THE FINITE ELEMENT METHOD: Linear Static and Dynamic Finite Element Analysis, Thomas J. R. Hughes. Text for students without in-depth mathematical training, this text includes a comprehensive presentation and analysis of algorithms of time-dependent phenomena plus beam, plate, and shell theories. Solution guide available upon request. 672pp. 6 1/2 x 9 1/4. 0-486-41181-8

HELICOPTER THEORY, Wayne Johnson. Monumental engineering text covers vertical flight, forward flight, performance, mathematics of rotating systems, rotary wing dynamics and aerodynamics, aeroelasticity, stability and control, stall, noise, and more. 189 illustrations. 1980 edition. 1089pp. 5 5/8 x 8 1/4.　　0-486-68230-7

MATHEMATICAL HANDBOOK FOR SCIENTISTS AND ENGINEERS: Definitions, Theorems, and Formulas for Reference and Review, Granino A. Korn and Theresa M. Korn. Convenient access to information from every area of mathematics: Fourier transforms, Z transforms, linear and nonlinear programming, calculus of variations, random-process theory, special functions, combinatorial analysis, game theory, much more. 1152pp. 5 3/8 x 8 1/2.　　0-486-41147-8

A HEAT TRANSFER TEXTBOOK: Fourth Edition, John H. Lienhard V and John H. Lienhard IV. This introduction to heat and mass transfer for engineering students features worked examples and end-of-chapter exercises. Worked examples and end-of-chapter exercises appear throughout the book, along with well-drawn, illuminating figures. 768pp. 7 x 9 1/4.　　0-486-47931-5

BASIC ELECTRICITY, U.S. Bureau of Naval Personnel. Originally a training course; best nontechnical coverage. Topics include batteries, circuits, conductors, AC and DC, inductance and capacitance, generators, motors, transformers, amplifiers, etc. Many questions with answers. 349 illustrations. 1969 edition. 448pp. 6 1/2 x 9 1/4.
0-486-20973-3

BASIC ELECTRONICS, U.S. Bureau of Naval Personnel. Clear, well-illustrated introduction to electronic equipment covers numerous essential topics: electron tubes, semiconductors, electronic power supplies, tuned circuits, amplifiers, receivers, ranging and navigation systems, computers, antennas, more. 560 illustrations. 567pp. 6 1/2 x 9 1/4.　　0-486-21076-6

BASIC WING AND AIRFOIL THEORY, Alan Pope. This self-contained treatment by a pioneer in the study of wind effects covers flow functions, airfoil construction and pressure distribution, finite and monoplane wings, and many other subjects. 1951 edition. 320pp. 5 3/8 x 8 1/2.　　0-486-47188-8

SYNTHETIC FUELS, Ronald F. Probstein and R. Edwin Hicks. This unified presentation examines the methods and processes for converting coal, oil, shale, tar sands, and various forms of biomass into liquid, gaseous, and clean solid fuels. 1982 edition. 512pp. 6 1/8 x 9 1/4.　　0-486-44977-7

THEORY OF ELASTIC STABILITY, Stephen P. Timoshenko and James M. Gere. Written by world-renowned authorities on mechanics, this classic ranges from theoretical explanations of 2- and 3-D stress and strain to practical applications such as torsion, bending, and thermal stress. 1961 edition. 560pp. 5 3/8 x 8 1/2.　　0-486-47207-8

PRINCIPLES OF DIGITAL COMMUNICATION AND CODING, Andrew J. Viterbi and Jim K. Omura. This classic by two digital communications experts is geared toward students of communications theory and to designers of channels, links, terminals, modems, or networks used to transmit and receive digital messages. 1979 edition. 576pp. 6 1/8 x 9 1/4.　　0-486-46901-8

LINEAR SYSTEM THEORY: The State Space Approach, Lotfi A. Zadeh and Charles A. Desoer. Written by two pioneers in the field, this exploration of the state space approach focuses on problems of stability and control, plus connections between this approach and classical techniques. 1963 edition. 656pp. 6 1/8 x 9 1/4.
0-486-46663-9

Browse over 9,000 books at www.doverpublications.com

Mathematics-Bestsellers

HANDBOOK OF MATHEMATICAL FUNCTIONS: with Formulas, Graphs, and Mathematical Tables, Edited by Milton Abramowitz and Irene A. Stegun. A classic resource for working with special functions, standard trig, and exponential logarithmic definitions and extensions, it features 29 sets of tables, some to as high as 20 places. 1046pp. 8 x 10 1/2. 0-486-61272-4

ABSTRACT AND CONCRETE CATEGORIES: The Joy of Cats, Jiri Adamek, Horst Herrlich, and George E. Strecker. This up-to-date introductory treatment employs category theory to explore the theory of structures. Its unique approach stresses concrete categories and presents a systematic view of factorization structures. Numerous examples. 1990 edition, updated 2004. 528pp. 6 1/8 x 9 1/4. 0-486-46934-4

MATHEMATICS: Its Content, Methods and Meaning, A. D. Aleksandrov, A. N. Kolmogorov, and M. A. Lavrent'ev. Major survey offers comprehensive, coherent discussions of analytic geometry, algebra, differential equations, calculus of variations, functions of a complex variable, prime numbers, linear and non-Euclidean geometry, topology, functional analysis, more. 1963 edition. 1120pp. 5 3/8 x 8 1/2. 0-486-40916-3

INTRODUCTION TO VECTORS AND TENSORS: Second Edition–Two Volumes Bound as One, Ray M. Bowen and C.-C. Wang. Convenient single-volume compilation of two texts offers both introduction and in-depth survey. Geared toward engineering and science students rather than mathematicians, it focuses on physics and engineering applications. 1976 edition. 560pp. 6 1/2 x 9 1/4. 0-486-46914-X

AN INTRODUCTION TO ORTHOGONAL POLYNOMIALS, Theodore S. Chihara. Concise introduction covers general elementary theory, including the representation theorem and distribution functions, continued fractions and chain sequences, the recurrence formula, special functions, and some specific systems. 1978 edition. 272pp. 5 3/8 x 8 1/2. 0-486-47929-3

ADVANCED MATHEMATICS FOR ENGINEERS AND SCIENTISTS, Paul DuChateau. This primary text and supplemental reference focuses on linear algebra, calculus, and ordinary differential equations. Additional topics include partial differential equations and approximation methods. Includes solved problems. 1992 edition. 400pp. 7 1/2 x 9 1/4. 0-486-47930-7

PARTIAL DIFFERENTIAL EQUATIONS FOR SCIENTISTS AND ENGINEERS, Stanley J. Farlow. Practical text shows how to formulate and solve partial differential equations. Coverage of diffusion-type problems, hyperbolic-type problems, elliptic-type problems, numerical and approximate methods. Solution guide available upon request. 1982 edition. 414pp. 6 1/8 x 9 1/4. 0-486-67620-X

VARIATIONAL PRINCIPLES AND FREE-BOUNDARY PROBLEMS, Avner Friedman. Advanced graduate-level text examines variational methods in partial differential equations and illustrates their applications to free-boundary problems. Features detailed statements of standard theory of elliptic and parabolic operators. 1982 edition. 720pp. 6 1/8 x 9 1/4. 0-486-47853-X

LINEAR ANALYSIS AND REPRESENTATION THEORY, Steven A. Gaal. Unified treatment covers topics from the theory of operators and operator algebras on Hilbert spaces; integration and representation theory for topological groups; and the theory of Lie algebras, Lie groups, and transform groups. 1973 edition. 704pp. 6 1/8 x 9 1/4. 0-486-47851-3

Browse over 9,000 books at www.doverpublications.com

A SURVEY OF INDUSTRIAL MATHEMATICS, Charles R. MacCluer. Students learn how to solve problems they'll encounter in their professional lives with this concise single-volume treatment. It employs MATLAB and other strategies to explore typical industrial problems. 2000 edition. 384pp. 5 3/8 x 8 1/2. 0-486-47702-9

NUMBER SYSTEMS AND THE FOUNDATIONS OF ANALYSIS, Elliott Mendelson. Geared toward undergraduate and beginning graduate students, this study explores natural numbers, integers, rational numbers, real numbers, and complex numbers. Numerous exercises and appendixes supplement the text. 1973 edition. 368pp. 5 3/8 x 8 1/2. 0-486-45792-3

A FIRST LOOK AT NUMERICAL FUNCTIONAL ANALYSIS, W. W. Sawyer. Text by renowned educator shows how problems in numerical analysis lead to concepts of functional analysis. Topics include Banach and Hilbert spaces, contraction mappings, convergence, differentiation and integration, and Euclidean space. 1978 edition. 208pp. 5 3/8 x 8 1/2. 0-486-47882-3

FRACTALS, CHAOS, POWER LAWS: Minutes from an Infinite Paradise, Manfred Schroeder. A fascinating exploration of the connections between chaos theory, physics, biology, and mathematics, this book abounds in award-winning computer graphics, optical illusions, and games that clarify memorable insights into self-similarity. 1992 edition. 448pp. 6 1/8 x 9 1/4. 0-486-47204-3

SET THEORY AND THE CONTINUUM PROBLEM, Raymond M. Smullyan and Melvin Fitting. A lucid, elegant, and complete survey of set theory, this three-part treatment explores axiomatic set theory, the consistency of the continuum hypothesis, and forcing and independence results. 1996 edition. 336pp. 6 x 9. 0-486-47484-4

DYNAMICAL SYSTEMS, Shlomo Sternberg. A pioneer in the field of dynamical systems discusses one-dimensional dynamics, differential equations, random walks, iterated function systems, symbolic dynamics, and Markov chains. Supplementary materials include PowerPoint slides and MATLAB exercises. 2010 edition. 272pp. 6 1/8 x 9 1/4. 0-486-47705-3

ORDINARY DIFFERENTIAL EQUATIONS, Morris Tenenbaum and Harry Pollard. Skillfully organized introductory text examines origin of differential equations, then defines basic terms and outlines general solution of a differential equation. Explores integrating factors; dilution and accretion problems; Laplace Transforms; Newton's Interpolation Formulas, more. 818pp. 5 3/8 x 8 1/2. 0-486-64940-7

MATROID THEORY, D. J. A. Welsh. Text by a noted expert describes standard examples and investigation results, using elementary proofs to develop basic matroid properties before advancing to a more sophisticated treatment. Includes numerous exercises. 1976 edition. 448pp. 5 3/8 x 8 1/2. 0-486-47439-9

THE CONCEPT OF A RIEMANN SURFACE, Hermann Weyl. This classic on the general history of functions combines function theory and geometry, forming the basis of the modern approach to analysis, geometry, and topology. 1955 edition. 208pp. 5 3/8 x 8 1/2. 0-486-47004-0

THE LAPLACE TRANSFORM, David Vernon Widder. This volume focuses on the Laplace and Stieltjes transforms, offering a highly theoretical treatment. Topics include fundamental formulas, the moment problem, monotonic functions, and Tauberian theorems. 1941 edition. 416pp. 5 3/8 x 8 1/2. 0-486-47755-X

Browse over 9,000 books at www.doverpublications.com

Mathematics–Logic and Problem Solving

PERPLEXING PUZZLES AND TANTALIZING TEASERS, Martin Gardner. Ninety-three riddles, mazes, illusions, tricky questions, word and picture puzzles, and other challenges offer hours of entertainment for youngsters. Filled with rib-tickling drawings. Solutions. 224pp. 5 3/8 x 8 1/2.　　　　　　　　　　0-486-25637-5

MY BEST MATHEMATICAL AND LOGIC PUZZLES, Martin Gardner. The noted expert selects 70 of his favorite "short" puzzles. Includes The Returning Explorer, The Mutilated Chessboard, Scrambled Box Tops, and dozens more. Complete solutions included. 96pp. 5 3/8 x 8 1/2.　　　　　　　　　　0-486-28152-3

THE LADY OR THE TIGER?: and Other Logic Puzzles, Raymond M. Smullyan. Created by a renowned puzzle master, these whimsically themed challenges involve paradoxes about probability, time, and change; metapuzzles; and self-referentiality. Nineteen chapters advance in difficulty from relatively simple to highly complex. 1982 edition. 240pp. 5 3/8 x 8 1/2.　　　　　　　　　　0-486-47027-X

SATAN, CANTOR AND INFINITY: Mind-Boggling Puzzles, Raymond M. Smullyan. A renowned mathematician tells stories of knights and knaves in an entertaining look at the logical precepts behind infinity, probability, time, and change. Requires a strong background in mathematics. Complete solutions. 288pp. 5 3/8 x 8 1/2.

0-486-47036-9

THE RED BOOK OF MATHEMATICAL PROBLEMS, Kenneth S. Williams and Kenneth Hardy. Handy compilation of 100 practice problems, hints and solutions indispensable for students preparing for the William Lowell Putnam and other mathematical competitions. Preface to the First Edition. Sources. 1988 edition. 192pp. 5 3/8 x 8 1/2.　　　　　　　　　　0-486-69415-1

KING ARTHUR IN SEARCH OF HIS DOG AND OTHER CURIOUS PUZZLES, Raymond M. Smullyan. This fanciful, original collection for readers of all ages features arithmetic puzzles, logic problems related to crime detection, and logic and arithmetic puzzles involving King Arthur and his Dogs of the Round Table. 160pp. 5 3/8 x 8 1/2.

0-486-47435-6

UNDECIDABLE THEORIES: Studies in Logic and the Foundation of Mathematics, Alfred Tarski in collaboration with Andrzej Mostowski and Raphael M. Robinson. This well-known book by the famed logician consists of three treatises: "A General Method in Proofs of Undecidability," "Undecidability and Essential Undecidability in Mathematics," and "Undecidability of the Elementary Theory of Groups." 1953 edition. 112pp. 5 3/8 x 8 1/2.　　　　　　　　　　0-486-47703-7

LOGIC FOR MATHEMATICIANS, J. Barkley Rosser. Examination of essential topics and theorems assumes no background in logic. "Undoubtedly a major addition to the literature of mathematical logic." – *Bulletin of the American Mathematical Society.* 1978 edition. 592pp. 6 1/8 x 9 1/4.　　　　　　　　　　0-486-46898-4

INTRODUCTION TO PROOF IN ABSTRACT MATHEMATICS, Andrew Wohlgemuth. This undergraduate text teaches students what constitutes an acceptable proof, and it develops their ability to do proofs of routine problems as well as those requiring creative insights. 1990 edition. 384pp. 6 1/2 x 9 1/4.　　0-486-47854-8

FIRST COURSE IN MATHEMATICAL LOGIC, Patrick Suppes and Shirley Hill. Rigorous introduction is simple enough in presentation and context for wide range of students. Symbolizing sentences; logical inference; truth and validity; truth tables; terms, predicates, universal quantifiers; universal specification and laws of identity; more. 288pp. 5 3/8 x 8 1/2.　　　　　　　　　　0-486-42259-3

Browse over 9,000 books at www.doverpublications.com

Mathematics–History

THE WORKS OF ARCHIMEDES, Archimedes. Translated by Sir Thomas Heath. Complete works of ancient geometer feature such topics as the famous problems of the ratio of the areas of a cylinder and an inscribed sphere; the properties of conoids, spheroids, and spirals; more. 326pp. 5 3/8 x 8 1/2. 0-486-42084-1

THE HISTORICAL ROOTS OF ELEMENTARY MATHEMATICS, Lucas N. H. Bunt, Phillip S. Jones, and Jack D. Bedient. Exciting, hands-on approach to understanding fundamental underpinnings of modern arithmetic, algebra, geometry and number systems examines their origins in early Egyptian, Babylonian, and Greek sources. 336pp. 5 3/8 x 8 1/2. 0-486-25563-8

THE THIRTEEN BOOKS OF EUCLID'S ELEMENTS, Euclid. Contains complete English text of all 13 books of the Elements plus critical apparatus analyzing each definition, postulate, and proposition in great detail. Covers textual and linguistic matters; mathematical analyses of Euclid's ideas; classical, medieval, Renaissance and modern commentators; refutations, supports, extrapolations, reinterpretations and historical notes. 995 figures. Total of 1,425pp. All books 5 3/8 x 8 1/2.

Vol. I: 443pp. 0-486-60088-2
Vol. II: 464pp. 0-486-60089-0
Vol. III: 546pp. 0-486-60090-4

A HISTORY OF GREEK MATHEMATICS, Sir Thomas Heath. This authoritative two-volume set that covers the essentials of mathematics and features every landmark innovation and every important figure, including Euclid, Apollonius, and others. 5 3/8 x 8 1/2.

Vol. I: 461pp. 0-486-24073-8
Vol. II: 597pp. 0-486-24074-6

A MANUAL OF GREEK MATHEMATICS, Sir Thomas L. Heath. This concise but thorough history encompasses the enduring contributions of the ancient Greek mathematicians whose works form the basis of most modern mathematics. Discusses Pythagorean arithmetic, Plato, Euclid, more. 1931 edition. 576pp. 5 3/8 x 8 1/2.

0-486-43231-9

CHINESE MATHEMATICS IN THE THIRTEENTH CENTURY, Ulrich Libbrecht. An exploration of the 13th-century mathematician Ch'in, this fascinating book combines what is known of the mathematician's life with a history of his only extant work, the Shu-shu chiu-chang. 1973 edition. 592pp. 5 3/8 x 8 1/2.

0-486-44619-0

PHILOSOPHY OF MATHEMATICS AND DEDUCTIVE STRUCTURE IN EUCLID'S ELEMENTS, Ian Mueller. This text provides an understanding of the classical Greek conception of mathematics as expressed in Euclid's Elements. It focuses on philosophical, foundational, and logical questions and features helpful appendixes. 400pp. 6 1/2 x 9 1/4. 0-486-45300-6

BEYOND GEOMETRY: Classic Papers from Riemann to Einstein, Edited with an Introduction and Notes by Peter Pesic. This is the only English-language collection of these 8 accessible essays. They trace seminal ideas about the foundations of geometry that led to Einstein's general theory of relativity. 224pp. 6 1/8 x 9 1/4. 0-486-45350-2

HISTORY OF MATHEMATICS, David E. Smith. Two-volume history – from Egyptian papyri and medieval maps to modern graphs and diagrams. Non-technical chronological survey with thousands of biographical notes, critical evaluations, and contemporary opinions on over 1,100 mathematicians. 5 3/8 x 8 1/2.

Vol. I: 618pp. 0-486-20429-4
Vol. II: 736pp. 0-486-20430-8

Physics

THEORETICAL NUCLEAR PHYSICS, John M. Blatt and Victor F. Weisskopf. An uncommonly clear and cogent investigation and correlation of key aspects of theoretical nuclear physics by leading experts: the nucleus, nuclear forces, nuclear spectroscopy, two-, three- and four-body problems, nuclear reactions, beta-decay and nuclear shell structure. 896pp. 5 3/8 x 8 1/2.   0-486-66827-4

QUANTUM THEORY, David Bohm. This advanced undergraduate-level text presents the quantum theory in terms of qualitative and imaginative concepts, followed by specific applications worked out in mathematical detail. 655pp. 5 3/8 x 8 1/2.
0-486-65969-0

ATOMIC PHYSICS AND HUMAN KNOWLEDGE, Niels Bohr. Articles and speeches by the Nobel Prize–winning physicist, dating from 1934 to 1958, offer philosophical explorations of the relevance of atomic physics to many areas of human endeavor. 1961 edition. 112pp. 5 3/8 x 8 1/2.   0-486-47928-5

COSMOLOGY, Hermann Bondi. A co-developer of the steady-state theory explores his conception of the expanding universe. This historic book was among the first to present cosmology as a separate branch of physics. 1961 edition. 192pp. 5 3/8 x 8 1/2.
0-486-47483-6

LECTURES ON QUANTUM MECHANICS, Paul A. M. Dirac. Four concise, brilliant lectures on mathematical methods in quantum mechanics from Nobel Prize–winning quantum pioneer build on idea of visualizing quantum theory through the use of classical mechanics. 96pp. 5 3/8 x 8 1/2.   0-486-41713-1

THE PRINCIPLE OF RELATIVITY, Albert Einstein and Frances A. Davis. Eleven papers that forged the general and special theories of relativity include seven papers by Einstein, two by Lorentz, and one each by Minkowski and Weyl. 1923 edition. 240pp. 5 3/8 x 8 1/2.   0-486-60081-5

PHYSICS OF WAVES, William C. Elmore and Mark A. Heald. Ideal as a classroom text or for individual study, this unique one-volume overview of classical wave theory covers wave phenomena of acoustics, optics, electromagnetic radiations, and more. 477pp. 5 3/8 x 8 1/2.   0-486-64926-1

THERMODYNAMICS, Enrico Fermi. In this classic of modern science, the Nobel Laureate presents a clear treatment of systems, the First and Second Laws of Thermodynamics, entropy, thermodynamic potentials, and much more. Calculus required. 160pp. 5 3/8 x 8 1/2.   0-486-60361-X

QUANTUM THEORY OF MANY-PARTICLE SYSTEMS, Alexander L. Fetter and John Dirk Walecka. Self-contained treatment of nonrelativistic many-particle systems discusses both formalism and applications in terms of ground-state (zero-temperature) formalism, finite-temperature formalism, canonical transformations, and applications to physical systems. 1971 edition. 640pp. 5 3/8 x 8 1/2.   0-486-42827-3

QUANTUM MECHANICS AND PATH INTEGRALS: Emended Edition, Richard P. Feynman and Albert R. Hibbs. Emended by Daniel F. Styer. The Nobel Prize–winning physicist presents unique insights into his theory and its applications. Feynman starts with fundamentals and advances to the perturbation method, quantum electrodynamics, and statistical mechanics. 1965 edition, emended in 2005. 384pp. 6 1/8 x 9 1/4.   0-486-47722-3

Browse over 9,000 books at www.doverpublications.com

Physics

INTRODUCTION TO MODERN OPTICS, Grant R. Fowles. A complete basic undergraduate course in modern optics for students in physics, technology, and engineering. The first half deals with classical physical optics; the second, quantum nature of light. Solutions. 336pp. 5 3/8 x 8 1/2. 0-486-65957-7

THE QUANTUM THEORY OF RADIATION: Third Edition, W. Heitler. The first comprehensive treatment of quantum physics in any language, this classic introduction to basic theory remains highly recommended and widely used, both as a text and as a reference. 1954 edition. 464pp. 5 3/8 x 8 1/2. 0-486-64558-4

QUANTUM FIELD THEORY, Claude Itzykson and Jean-Bernard Zuber. This comprehensive text begins with the standard quantization of electrodynamics and perturbative renormalization, advancing to functional methods, relativistic bound states, broken symmetries, nonabelian gauge fields, and asymptotic behavior. 1980 edition. 752pp. 6 1/2 x 9 1/4. 0-486-44568-2

FOUNDATIONS OF POTENTIAL THERY, Oliver D. Kellogg. Introduction to fundamentals of potential functions covers the force of gravity, fields of force, potentials, harmonic functions, electric images and Green's function, sequences of harmonic functions, fundamental existence theorems, and much more. 400pp. 5 3/8 x 8 1/2.
0-486-60144-7

FUNDAMENTALS OF MATHEMATICAL PHYSICS, Edgar A. Kraut. Indispensable for students of modern physics, this text provides the necessary background in mathematics to study the concepts of electromagnetic theory and quantum mechanics. 1967 edition. 480pp. 6 1/2 x 9 1/4. 0-486-45809-1

GEOMETRY AND LIGHT: The Science of Invisibility, Ulf Leonhardt and Thomas Philbin. Suitable for advanced undergraduate and graduate students of engineering, physics, and mathematics and scientific researchers of all types, this is the first authoritative text on invisibility and the science behind it. More than 100 full-color illustrations, plus exercises with solutions. 2010 edition. 288pp. 7 x 9 1/4. 0-486-47693-6

QUANTUM MECHANICS: New Approaches to Selected Topics, Harry J. Lipkin. Acclaimed as "excellent" (*Nature*) and "very original and refreshing" (*Physics Today*), these studies examine the Mössbauer effect, many-body quantum mechanics, scattering theory, Feynman diagrams, and relativistic quantum mechanics. 1973 edition. 480pp. 5 3/8 x 8 1/2. 0-486-45893-8

THEORY OF HEAT, James Clerk Maxwell. This classic sets forth the fundamentals of thermodynamics and kinetic theory simply enough to be understood by beginners, yet with enough subtlety to appeal to more advanced readers, too. 352pp. 5 3/8 x 8 1/2. 0-486-41735-2

QUANTUM MECHANICS, Albert Messiah. Subjects include formalism and its interpretation, analysis of simple systems, symmetries and invariance, methods of approximation, elements of relativistic quantum mechanics, much more. "Strongly recommended." – *American Journal of Physics.* 1152pp. 5 3/8 x 8 1/2. 0-486-40924-4

RELATIVISTIC QUANTUM FIELDS, Charles Nash. This graduate-level text contains techniques for performing calculations in quantum field theory. It focuses chiefly on the dimensional method and the renormalization group methods. Additional topics include functional integration and differentiation. 1978 edition. 240pp. 5 3/8 x 8 1/2.
0-486-47752-5

Physics

MATHEMATICAL TOOLS FOR PHYSICS, James Nearing. Encouraging students' development of intuition, this original work begins with a review of basic mathematics and advances to infinite series, complex algebra, differential equations, Fourier series, and more. 2010 edition. 496pp. 6 1/8 x 9 1/4. 0-486-48212-X

TREATISE ON THERMODYNAMICS, Max Planck. Great classic, still one of the best introductions to thermodynamics. Fundamentals, first and second principles of thermodynamics, applications to special states of equilibrium, more. Numerous worked examples. 1917 edition. 297pp. 5 3/8 x 8. 0-486-66371-X

AN INTRODUCTION TO RELATIVISTIC QUANTUM FIELD THEORY, Silvan S. Schweber. Complete, systematic, and self-contained, this text introduces modern quantum field theory. "Combines thorough knowledge with a high degree of didactic ability and a delightful style." – *Mathematical Reviews*. 1961 edition. 928pp. 5 3/8 x 8 1/2. 0-486-44228-4

THE ELECTROMAGNETIC FIELD, Albert Shadowitz. Comprehensive under-graduate text covers basics of electric and magnetic fields, building up to electromagnetic theory. Related topics include relativity theory. Over 900 problems, some with solutions. 1975 edition. 768pp. 5 5/8 x 8 1/4. 0-486-65660-8

THE PRINCIPLES OF STATISTICAL MECHANICS, Richard C. Tolman. Definitive treatise offers a concise exposition of classical statistical mechanics and a thorough elucidation of quantum statistical mechanics, plus applications of statistical mechanics to thermodynamic behavior. 1930 edition. 704pp. 5 5/8 x 8 1/4. 0-486-63896-0

INTRODUCTION TO THE PHYSICS OF FLUIDS AND SOLIDS, James S. Trefil. This interesting, informative survey by a well-known science author ranges from classical physics and geophysical topics, from the rings of Saturn and the rotation of the galaxy to underground nuclear tests. 1975 edition. 320pp. 5 3/8 x 8 1/2. 0-486-47437-2

STATISTICAL PHYSICS, Gregory H. Wannier. Classic text combines thermodynamics, statistical mechanics, and kinetic theory in one unified presentation. Topics include equilibrium statistics of special systems, kinetic theory, transport coefficients, and fluctuations. Problems with solutions. 1966 edition. 532pp. 5 3/8 x 8 1/2. 0-486-65401-X

SPACE, TIME, MATTER, Hermann Weyl. Excellent introduction probes deeply into Euclidean space, Riemann's space, Einstein's general relativity, gravitational waves and energy, and laws of conservation. "A classic of physics." – *British Journal for Philosophy and Science*. 330pp. 5 3/8 x 8 1/2. 0-486-60267-2

RANDOM VIBRATIONS: Theory and Practice, Paul H. Wirsching, Thomas L. Paez and Keith Ortiz. Comprehensive text and reference covers topics in probability, statistics, and random processes, plus methods for analyzing and controlling random vibrations. Suitable for graduate students and mechanical, structural, and aerospace engineers. 1995 edition. 464pp. 5 3/8 x 8 1/2. 0-486-45015-5

PHYSICS OF SHOCK WAVES AND HIGH-TEMPERATURE HYDRO DYNAMIC PHENOMENA, Ya B. Zel'dovich and Yu P. Raizer. Physical, chemical processes in gases at high temperatures are focus of outstanding text, which combines material from gas dynamics, shock-wave theory, thermodynamics and statistical physics, other fields. 284 illustrations. 1966–1967 edition. 944pp. 6 1/8 x 9 1/4. 0-486-42002-7

Browse over 9,000 books at www.doverpublications.com